The Urban Book Series

The Urban Book Series is a resource for urban studies and geography research worldwide. It provides a unique and innovative resource for the latest developments in the field, nurturing a comprehensive and encompassing publication venue for urban studies, urban geography, planning and regional development.

The series publishes peer-reviewed volumes related to urbanization, sustainability, urban environments, sustainable urbanism, governance, globalization, urban and sustainable development, spatial and area studies, urban management, transport systems, urban infrastructure, urban dynamics, green cities and urban landscapes. It also invites research which documents urbanization processes and urban dynamics on a national, regional and local level, welcoming case studies, as well as comparative and applied research.

The series will appeal to urbanists, geographers, planners, engineers, architects, policy makers, and to all of those interested in a wide-ranging overview of contemporary urban studies and innovations in the field. It accepts monographs, edited volumes and textbooks.

Now Indexed by Scopus!

More information about this series at http://www.springer.com/series/14773

Magdalena Matysek-Imielińska

Warsaw Housing Cooperative

City in Action

Springer

Magdalena Matysek-Imielińska
Faculty of Historical and Pedagogical Science
Institute of Cultural Studies
University of Wrocław
Wrocław, Poland

Translated by Monika Fryszkowska
Introduction and Conclusions translated by Marcin Starnawski
The translation into English was financed by the Faculty of Historical and Pedagogical Science University of Wrocław

The English version of the work is an extended and revised version of "Miasto w działaniu. Warszawska Spółdzielnia Mieszkaniowa – dobro wspólne w epoce nowoczesnej" (City in action. The Warsaw Housing Cooperative – a common good in modern times) published by Fundacja Nowej Kultury Bęc Zmiana (Bęc Zmiana Foundation for New Culture).

ISSN 2365-757X ISSN 2365-7588 (electronic)
The Urban Book Series
ISBN 978-3-030-23076-0 ISBN 978-3-030-23077-7 (eBook)
https://doi.org/10.1007/978-3-030-23077-7

This Springer imprint is published by the registered company Springer Nature Switzerland AG
The registered company address is: Gewerbestrasse 11, 6330 Cham, Switzerland

For Zygmunt Bauman, in memoriam

Contents

List of Figures

List of Tables

Chapter 1
Introduction

Abstract The Warsaw Housing Cooperative ('Żoliborz republic') is a historic example of an attempt to replace the extensive system of representative democracy and majority rule with the principles of direct democracy and the self-governance of various entities/subjects, an example of creating a commons based on biopolitical production of customs, norms, tastes, principles of cooperation and, finally, subjectivity. A common good produced and distributed contrary to the logic of capital in a modern city, at a time it was desperately lacking its resources. The cooperative's ambition was to show people how to live and to turn (passive) residents into (active) citizens. Soon, the cooperative's founders have transpired to be—as we would now say—activists, or critical spatial practitioners (Markus Miessen), and the development became a laboratory for modernist urban planning practices. What began as an estate for working-class residents, thanks to the involvement of intellectuals keen on the concept of 'sociology in action', saw the introduction of numerous experimental forms of urban communal living. An analysis of this historical case relates to questions often posed today by both the urban grass-roots movements and researchers such as David Harvey and Andy Merrifield: Who owns the city? Who does the city belong to? Who should manage the city and how? The performative perspective allows me to see 'architecture in action' and dwelling as a process, rather than a form. The resident-turned citizen can be seen as engaged in subversive and freedom-oriented undertakings (social, political, economic and educational). In turn, the critical perspective (inseparable from the performative one) revealed the extraordinary power of rebellion and the desire for change resulting from the combination of thought and action, thanks to culture understood as *praxis*. I assume that the WHC residents' strategies, described in this book, may prove valuable particularly today, when modern cities are implementing a model of governance based on urban entrepreneurship, while local governments are eager to dismantle the municipal social welfare system, privatise public goods and collective consumption and cooperate with private investors more often than with residents.

Keywords Performative perspective · Criticism of modernity · Preposterous history · Urban studies · Utopian studies

M. Matysek-Imielińska, *Warsaw Housing Cooperative*, The Urban Book Series,
https://doi.org/10.1007/978-3-030-23077-7_1

1.1 How to Create Cities?

A new model of governance based on urban entrepreneurship, which has been implemented in modern cities, gives local governments in Western Europe freedom to dismantle urban systems of welfare, to privatise public goods and collective consumption, or to undertake far-reaching collaboration with private investors. Parallel to this process, there are multiple attempts at constructing 'exceptional sites' or 'creative cities' typical for the neoliberal-capitalist mode of production, which consistently commodifies various aspects of human lives: knowledge, information, lifestyles, or even forms of resistance.

It seems, however, that the only chance so far to keep the city for its residents lies in numerous efforts by people themselves to 'reclaim' official institutions in order to expand access to basic services and goods. Sometimes such efforts take the form of 'joyful encounters', a kind of urban celebration, and they often fail to bring expected results or long-term effects. Such was the case with Teatro Valle in Rome. Facing the perspective of privatisation of the theatre building, an object of historical heritage status, a group of play directors, actors and dancers occupied the site and transformed it into the Teatro Valle Occupato, a centre of activism far beyond theatrical art, a shared space of urban movements claiming the right to the common. They foresaw a commons managed not so much by public institutions but rather by the very users of the city facilities through direct democracy. Eventually, after long negotiations the activists were evicted.

This does not mean, however, that such urban struggles are not necessary, and that communities emerging through resistance against authorities are doomed to fail. To the contrary, such efforts prove that urbanity should be seen as networked and condensed urban culture, with its material fabric and immaterial labour of residents, a meeting space. It can also be seen as a set of tools for expansion of the power of capital, but also as an instrument to contest this power. Such struggles are also important because they contribute significantly to our understanding of the role of common goods, that is, non-commercial ways of satisfying the needs of urban dwellers.

When we talk about these efforts, we point to the fact that urban communities of citizens are stakeholders of those goods, while the interests of those communities cannot be subject to economic discourse (De Angelis 2003; Bollier 2007). They require research because it is in them that one may seek an answer to the question of how to create a governance model for public property overseen by civic organisations and cooperation-based communities. The question is all the more sound in the light of warning expressed by David Harvey in his *Rebel Cities* against a naive faith in polycentrism and decentralisation of common goods governed entirely bottom-up, with no higher-rank authority (above the city district level). Thus, we need to explore not only the changing strategies of the existing urban rivalries but also the ways of thinking about new urban subjects. We need to address the following questions: Who are the major players in the urban power struggle? Who does the city belong to? Are new communities now in the process of taking over the urban space?

In his article *The New Urbanism and the Communitarian Trap* (Harvey 1997), published in 1997, Harvey still cast doubt on the question concerning the organisational efficiency of struggling communities in transforming the city or implementing new solutions of the urban question. Yet in *Rebel Cities* he moved towards the analysis of urban struggles in Latin America. And asking 'how to organise the city?', he recognises the indispensability of urban communities. Similarly, Michael Hardt and Antonio Negri emphasise the urban productivities, both material and immaterial, which reconfigure the urban fabric and turn the city into a kind of social factory.[1] This is why many scholars, including Lefebvre (2014), speak up for the recognition of urban communities of day-to-day resistance, which organise their lives beyond capital. Andy Merrifield, who belongs to a younger generation in the field of critical urban studies,[2] offers, in the spirit of 'magical Marxism', his analysis of lived urban practice beyond capitalism and explores the processes of making urban solidarity among diverse people, regardless of their working-class background (Merrifield 2011).

Merrifield suggests a new approach to the urban question and challenges the very reasonableness of the struggle over what Manuel Castells called the right to 'collective consumption', that is, the struggle over public goods, because these eventually become subject to municipal policies of urban entrepreneurship, or, when already privatised, serve the investors in creative capitalism. Merrifield's *The New Urban Question* reveals his abandonment of faith in the possibility of exercising the 'right to the city' and demonstrates his new focus on the logic of encounters that no longer fit the logic of rights. If there is no return to the state's welfare policy protecting equal access to public goods, then in the conflict between capital and citizens there is only one winner: capital. This results in the situation where growing privatisation, 'parasite urbanisation' and neo-Haussmannisation (Merrifield 2014, pp. ix–x) force residents to take entirely different actions than those typical for the struggle over the right to 'collective consumption'. How then to organise the city? Can it be nowadays grounded in residents' engagement, direct democracy, grass-roots action or relative local autonomy? What form should residents' organising take? Do strategies of indignation communities, city occupations, urban guerrillas, 'joyful encounters' and even 'jacqueries' proposed by the authors of *Commons*, constitute the only alternative to the dismantling of the 'old urban question'?

Following the financial crisis of 2008, Barcelona faced the exigency of the bottom-up and day-to-day autarkic administration of the 'housing issue'. The city started its experiments with space-sharing, which thus promoted social economy and non-profit housing. The La Borda cooperative, along with the LaCol association of architects, initiated the construction of flats based on residents' participation. They created flexible infrastructure, adaptable to the changing social and economic conditions. In order to facilitate better living conditions in the community and to achieve higher quality of

[1] See Hardt and Negri (2009), especially the chapter '*De Corpore* 2: Metropolis'.

[2] See, for example, Brenner (2009). 'Critical urban theory' as a subdiscipline of urban studies is often linked with concepts of 'radical geography' (David Harvey, Neil Smith, Erik Swyngedouw, Noel Castree, Andy Merrifield). These scholars represent different methodological approaches, but what they have in common is Marxian inspiration.

services, including common spaces, several decisions on architecture were made, such as designing courtyards and common spaces on middle floors. The Sostre Civic cooperative launched the 'Princesa49' project, aimed at families in need. Several participatory design studios have joined this project, which allowed each family to take an active part in design and cooperate with technicians in developing their own model of co-residency, including community spaces such as urban gardens or civic centres (Civic 2013).[3] What many initiatives in Barcelona have in common is, first and foremost, the need to create neighbourhood relations, sharing space and being together.

This also proves that cities replace the nation state in terms of social space. Being urban residents bears much more significance for our experiences, day-to-day activities and identities than does our living as citizens of a state. 'National' models of citizenship do not prove effective in times of globalisation. As Krzysztof Nawratek argues, we need to return to the city as a political idea. And this means that apart from 'joyful encounters', the city must provide a kind of 'empty oppression', an institutionalised framework, which resident-citizens can rely on. City as a radical space, as an area of agency, a political idea, the common, as well as a contemporary mode of co-dwelling in urban collectives, urban subversions and autarkies, is the most popular topic of analysis within today's engaged humanities and critical urban studies.

1.2 Żoliborz—A Preposterous History and Utopian Turn

That is why I argue that urban strategies, almost hundred years old, from the Żoliborz housing estate can prove appealing to various groups, from engaged activist-researchers[4] and art curators to urban reformers, cultural managers and representatives of urban movements; in other words, anyone to whom we could apply Markus Miessen's term: critical spatial practitioners (Miessen 2017, p. 75). Analysis of the Warsaw Housing Cooperative (*Warszawska Spółdzielnia Mieszkaniowa*) provides a range of useful information on the most significant points of continuity and rupture between historical and contemporary struggles that reclaim urban space. The Żoliborz story helps one notice an enormous package of conceptual tools that make it possible to rethink creatively and democratically the political potential of organising citizens. Although the interwar experience of the WHC now belongs to the realm of history, questions addressed by this initiative inspire today's repertoire of possible urban actions, for example in terms of cooperative and anti-capitalist strategies of resistance, forms of organisation and ways of communication, immaterial work for the neighbourhood community, urban education and even emancipation. All this enables us to make use of historical resources and potential of urban dissent (contrary to a common-sense view, the WHC constituted a form of resistance not

[3] From June to October 2015, Centro de Cultura Contemporanea hosted the excellent 'Piso Piloto' exhibition under supervision of Oriolo Bohigas, presenting housing experiments, cooperatives, and co-housing, organised in Barcelona and Medellin (Colombia).

[4] An activist research in cultural studies has been proposed by Skórzyńska (2016).

merely of the working class but, indeed, of the urban type, a 'housing' or cooperative residents' movement) based on the category of biopolitical labour.

The stake is, therefore, to restore (remember) a kind of awareness about possibilities to exercise the right to the city. It is about sharing knowledge and experiences, which had been created from the grass roots and which were constantly changing in the process of the formation of the city's living fabric. It is true that Żoliborz developed thanks to the professional involvement of avant-garde architects and modern sociologists guided by members of the cooperative whose position parallels that of Gramsci's 'organic intellectuals'. Their knowledge, however, had not been born in university lecture halls, but rather had been created collectively and tested through daily practice of making the housing estate and through day-to-day struggles to keep it alive. One may say that it was 'amateurs'—in Andy Merrifield's sense of the term (Merrifield 2017)—who passionately created the Warsaw Housing Cooperative.

On the other hand, analysing the life of Żoliborz through contemporary urban studies (a field that raises questions about the problems of modern cities) helps us to consider ideological and broader cultural entanglements of emerging norms and practices related to the process of becoming the citizen. Indeed, such an analysis allows exploration of the Żoliborz estate as an example of practised cooperation, as rituals, social actions and scenarios, focused primarily on new institutional constructions of productive, responsible, publicly oriented and controllable subjectivity. One can call this type of research preposterous history (for 'preposterous history' see Bal 2002). The advancement of current urban studies and developments within engaged humanities help us not only to appreciate this particular historical endeavour, but also to reveal its previously unnoticeable aspects. This means that one can examine it somehow retrospectively by uncovering new interpretative possibilities in the Żoliborz history.

My purpose is therefore to present the comprehensive nature of the Żoliborz project, but not in order to expose its totalising profile or debunk some oppressively modern social order. I rather address the question whether its events, institutions, daily rhythm of practices or qualities of dwelling in the estate, put together, go in line with research perspectives based on analysis of a community of performative subjects, multitude, biopolitical labour and the common; whether they fit the new conceptualisations of urbanity. My narrative will thus focus not so much on individuals (strong subjects) nor on ideologies, but on a community (or rather multitude) making an urban art of living thanks to well-designed and sophisticated actions effective in a disciplinary order of labour, education, hygiene, consumption and recreation.

The period under study stretches essentially from the establishment of early cooperative houses in 1926 up to the outset of World War II. The wartime period is the state of emergency, when institutionalised order falls, existence becomes provisional, and new forms of social action emerge, which are typical for this kind of emergency state and far from normal everyday life. What follows is the totalitarian Stalinist regime, when the cooperative 'loses its island nature', as Stanisław Ossowski noted. The project is taken over by the institutions of the state power and ceases to function as a grass-roots and self-governing initiative. Indeed, cooperatives cannot be established top-down.

1.3 How to Study Cities?

The 'cooperative republic', 'cooperative town', 'red Żoliborz', political training area, utopia of a new society, housing estate regime, ideological estate and, finally, the laboratory of estate communal life and urban experiment, all these were names attached to the Warsaw Housing Cooperative founded in the 1920s. Thanks to its pioneering spirit, the WHC can be seen as an idea-driven experiment, not free of now debunked modernist ambitions to build order. It can also be analysed as a modern utopia. I do not think, however, that this begs criticism, especially when one applies today's perspective of an engaged human science. Neither should we treat this project merely as a relic of the past. No doubt, one can hold on to a harsh judgement of modernism and to a radical critique of modernity, which, rightly so, results from the abandonment of grand-narrative illusions. Yet such a critique rejects the whole cultural order with the institutional framework typical for modern societies, which provided the subject with both public and private qualities. That is why my perspective in this book, while critical, is not radical in any postmodern sense. Indeed, what seemed to me most intriguing in the Żoliborz project was the very modern faith in the possibility of the institution of a new collective urban subject, as well as this project's pragmatism, which brought the established order way beyond its borders despite its specifically local and grass-roots origins.

We can, therefore, link the proposed interpretation of the WHC with a return to a utopian mode of thinking, which has been present in the social sciences for some time, inspired by ideas of such authors as Wallerstein (1998), Jacoby (1999), Levitas (1990, 2013). These authors claim that the construction of utopian thinking results simply from our desire for a better way of life and in this sense it is a better-life design project. They thus argue for the rejection of the anti-utopia discourse that equates utopia as a road towards totalitarian nightmares. Seeing utopia as a pipe dream or fantasy leading to violence and terror disables our vision of a radically different future and dispossesses us of 'sociological imagination'. Instead, we are compelled to contemplate only certain correctives to the existing state of affairs, as if another world would not be possible. Levitas sees utopia not as a descriptive category but rather as an analytical and critical one. She offers a hermeneutic method as well as a chance to recognise 'ideal' social relations (such as those of musicians performing together) or prefigurative and transformative practices. As a method, utopia enables us notice and analyse the 'everyday utopianism' of various social practices aimed at creating new customs and alternative institutions. As a critical hermeneutic method, it focuses on systemic solutions, leaving holistic descriptions of abstractive values such as equality or justice for their concrete implementations embodied in social institutions presented systematically as an integrated totality. Such utopias consist in a stark contrast between the present and its critique. Levitas gives examples of such narratives as provided by novelists, travellers, political programmes, or social/political theories. Levitas calls the method of using the critical analysis of these integrated holistic utopias an *Imaginary Reconstitution of Society*.

This kind of convergence of the social sciences and utopias (to be precise, it is a reconvergence but of a different nature than their encounter in the nineteenth century) must certainly lead to reconfiguration of those sciences that now should focus on what they had seen as non-scientific: the normative aspects of human lives, images of good society, which are necessary for human happiness. This also means that the social sciences can no longer ignore utopian thinking as a kind of speculative knowledge, but they must now approach it as a legitimate knowledge of an imaginable future. Levitas draws on a French sociologist, André Gorz, for whom 'it is the function of utopias, in the sense the term has assumed in the work of Ernst Bloch or Paul Ricœur, to provide us with the distance from the existing state of affairs which allows us to judge what we *are* doing in the light of what we *could* or *should* do' (Gorz 1999, p. 113, quoted in Levitas 2013, p. xvii). For Levitas, this means that utopia is an instrument for the assessment of our contexts, circumstances and actions. Moreover, it allows a transformation both of the existential experience and of the social world's objective structures that generate this experience.

There are three modes of the *Imaginary Reconstitution of Society* method: analytical (archaeological), ontological and architectural. The architectural mode concerns institutional design and description of the good society. It is complemented by the archaeological mode, that is, uncovering fragments and pieces and their recombination into a coherent whole. The purpose of such archaeology is to make the foundations of the model of the good society open to control and public critique. And the ontological mode concerns mainly subjects and agents of utopia. It makes it possible to hear, in those 'political, literary and artistic' utopian 'reports', the voices of the absent or hidden actors. Although analytically these modes of the utopian method are separated, the practice of the *Imaginary Reconstitution of Society* intertwines them and leads to research on the institutional specificity of a society. Levitas writes: 'Wherever we start in the process of imagining ourselves and our world otherwise, all three modes must eventually come into play' (Levitas 2013, p. xvii). Eventually, utopia as architecture turns on the ontological mode, creating citizens who feel, desire and act differently from us. Within the institutional framework, they develop their self-productive and self-organising capacities, and they improve institutions.

Certainly, Levitas can be charged with the implementation of normative thinking to sociological discourse and with promoting utopian solutions. She argues that we should not lose 'what most of us can imagine as realistic, feasible or achievable' (Levitas 2013, p. 201). Normativity is a way of allowing one's involvement in the debate on things public, and clear scenarios for the future are fundamental for all democratic debate, from which sociology cannot refrain. Here lie the origins of the idea that inspirations from the Żoliborz project should not be overlooked when we think about contemporary ways of living in cities and of making urban cultures. They must be, however, subject to critical analysis undertaken in contravention of anti-utopia discourse in order to unravel the project's transformative quality.

The Żoliborz project was an ideological and sociological laboratory of modernity, and it can undoubtedly be analysed in the context of the history of ideas. Still more interesting seems the question of how these socialist, cooperative and anti-capitalist ideas were implemented; how they operated through concrete activities; how the estate's residents practised the rules of cooperation, rituals of labour; how these people collectively constructed day-to-day reality and their involvement in complex social relations. Was their cooperation a goal or rather a tool for putting anti-capitalist ideals in practice? Equally important is the question concerning the performative possibilities of architecture and the very role of the architect as someone convinced of his or her ability to create social relations.

1.4 The Subject of Urban Action

The subject produced in the Żoliborz project is reminiscent of Michael Hardt's and Antonio Negri's multitude, that is, 'multiplicity of singularities acting together, a class of productive subjects connected by common experience of exploitation and vocal demand of democracy' (Editorial team of Praktyka teoretyczna 2012, p. 45). These authors see metropolis as a site of biopolitical production and of 'the inorganic body of the multitude. … [T]he metropolis is to the multitude what the factory was to the industrial working class. The factory constituted in the previous era the primary site and posed the conditions for three central activities of the industrial working class: its production; its internal encounters and organisation; and its expressions of antagonism and rebellion. The contemporary productive activities of the multitude, however, overflow the factory walls to permeate the entire metropolis … The metropolis is the site of biopolitical production because it is the space of the common, of people living together, sharing resources, communicating, exchanging goods and ideas' (Hardt and Negri 2009, pp. 249–250). It is therefore a distinct realm of value production and, at the same time, a spatial object of resistance.

I will thus attempt to answer the question whether the Żoliborz project, as it was made through the networked collaboration of multiple subjects and strove for combining the uniqueness of numerous individual experiences with social roles within a community, could be seen as such metropolitan multitude. I will also ask: How was mutuality among residents possible when almost all of them were subject to top-down power mechanisms that largely initiated and oversaw their lives? How did the cooperative become real and how did it turn technical competence into prosocial experience?

Two spheres were predominantly responsible for producing citizens in the 'Żoliborz republic': the material sphere of physical, architectural or urban planning actions, and the immaterial sphere of symbols, rituals, regulations and institutions. The estate presented the type of political and ideological engagement, which emphasised civic subjectivity as its primary identity, based on the common and not on narrow group or class interests. It therefore rejected the notion of democracy as institutionalised order, in which various groups of interest strive for hegemony, compete or even struggle with each other. But neither was it a project built on the illusion of eliminating all conflict. This was not a populist idea promising a peaceful and dignified life for everyone.[5] Adam Próchnik noticed the difference and demanded that they be respected. As he wrote in 1993: 'One cannot deny that there are various issues, which divide us. We tend to hold diverse views with regard to the way in which we are heading, we opt for various methods of struggle, and we belong to diverse political current and parties. And it cannot be otherwise.... So shall we, facing these differences, refrain from disputing the matters of ideas in our field for the sake of some peace? Or shall we open the ground for fierce and passionate political struggles? Neither one nor the other. As an association attached to ideas, we cannot push for creating some kind of ideological vacuum in our institutions' (Próchnik 1933, pp. 1–2). Therefore, he encouraged disputes and ideological discussions. He warned, however, against demagogies, hot-headed and fierce confrontations and political brawling.

The cooperative was founded as a workers' estate. It was meant to meet the housing needs of the class that lived off their own labour. It soon turned out that impoverished workers could not afford to maintain their flats in Żoliborz, though this does not mean that workers did not live there. The major group of residents, however, was the so-called working intelligentsia, mostly people of liberal professions. Was the intellectual Żoliborz, then, a paradox? Should the estate's heteronomous quality be considered its failure? The neighbourhood's multitude was bound by similar positioning in social stratification. 'We all, each and every one of us, belong to the world of labour. None of us classifies as a member of the class of owners, and nobody here lives of exploitation of other people's labour. We all make our living by the 'sweat of our necks and minds' (Próchnik 1933, p. 1). First and foremost, what kept the residents together was the idea of the self-development of living labour, which became the founding concept for the estate's subjectivities. Thus, the Żoliborz multitude escapes the reductionism of particular identities: those of class, race, ethnicity or locality. It provides an example of a multitude, which 'acts and speaks directly, without intermediaries—unlike the people, who need representatives. So, whereas the notion of the people serves as a theoretical basis for representative democracy, the notion of multitude becomes the foundation of direct and performative democracy' (cf. Sowa 2015, pp. 247–268).

Hardt and Negri talk about 'biopolitical labour' when they refer to activity, which creates conditions for its own cooperation and thus becomes more autonomous and acquires resistance ability by operating through 'network forms' of cooperation. In turn, the 'self-development of living labour' is understood as the 'ability to cooperative making of social relations', naturally contrary to capitalist relations. The aim is to 'liberate living labour from the capitalist regime in order to open a possibility of creative and independent activity' (Editorial team of Praktyka teoretyczna 2012, pp. 31–32).[6]

While the ideological founders of the Warsaw Housing Cooperative, obviously, had not used the literal notion of 'biopolitical labour', they had practical understanding of it. They sought in the Żoliborz cooperative an association of citizens, residents

[5] Another interesting angle to look at the Żoliborz project would be to ask about its proximity to the agonistic and radical democracy models analysed by Chantal Mouffe.

[6] The authors refer in this context to a study by Harry Cleaver on valorisation and self-valorisation.

of the social estate founded on principles of architectural co-design, forming neighbourly relations and collaboration for community institutions. To a certain extent, they performed 'authoritative speech' acts (cf. Butler 1993, p. 17) which were supposed to normalise this very reality. That is why they appealed to all those 'who feel co-responsible for the future of the Cooperative and Estate, and who, according to their abilities and competences, deal with matters of our "phalanstère" by attending gatherings, electing delegates and formulating sober critique.... We must consolidate these fragile and still fresh forms through our solidary efforts in order to show to the capitalistic world that we are capable, without force and desire for profit, to 'create the new life ourselves' only by way of collective work and merit' (Freyd 1933). We can analyse these declarative and performative calls in terms of attempts at building 'urban biopolitics' as well as 'urban autonomous zones' based on collective 'living labour' beyond the logic of capitalism. Therefore, we need to ask if those actions were of a revolutionary or rather reformist nature. What kind of ambitions do we find in the Żoliborz citizens who desired to 'create the new life' themselves?

Today, we also observe attempts to redefine the problem of production and urban manufacturing, and, as a consequence, to take a closer look at those who create and remake urban lives. Who does 'produce cities' nowadays? The contemporary urban multitude has a potential as well as cultural and social capital reminiscent of that which emerged out of the Żoliborz experiment and which formed its basis (activists, visionaries, 'idealists', 'amateurs' and all those who live of 'their own hands and brains'). Contemporary cities are produced by those who can offer their skills in operating cultural codes and building social relations; those who, like their modernist visionary predecessors, are able to work with symbolic resources. But the production of today's urban space no longer requires a proletariat. The producers are now a precariat, i.e. all those suffering uncertain conditions of employment, working part-time, many of whom are immigrants. And it is they who have the right to reclaim that part of the city's value, which they produce.

David Harvey offers a similar diagnosis of contemporary cities. While the Left had believed that industrial workers would have been the engine of social change, now, with the shift from industrial to cognitive capitalism, we need to search for a new actor who might be able to transform the sociopolitical scene. And today we can find such actor in the urban citizen.

1.5 The Object of Urban Action

The idea of the common, which in the period of making the Żoliborz estate was neither widespread nor democratically distributed, is particularly interesting here. In times when the workers tended to dwell in dark and clammy rooms in tenement houses, the founders of the WHC demanded universal access to light and fresh air. In our times, we interpret these all-too-obvious claims as manifestations of functionalist modernism, whose ambition was merely to satisfy people's biological needs. Nevertheless, Hardt and Negri still include them in the realm of the common:

'first of all, the common wealth of the material world—the air, the water, the fruits of the soil, and all nature's bounty—which in classic European political texts is often claimed to be the inheritance of humanity as a whole, to be shared together. We consider the common also and more significantly those results of social production, such as knowledges, languages, codes, information, affects, and so forth. This notion of the common… focuses rather on the practices of interaction, care, and cohabitation in a common world, promoting the beneficial and limiting the detrimental form of the common. In the era of globalisation, issues of the maintenance, production, and distribution of the common in both these senses and in both ecological and socioeconomic frameworks become increasingly central' (Hardt and Negri 2009, p. viii). The common so defined opens new spaces of the political and of emancipation, as was the case in the 1920s. This becomes perfectly clear when we read the manifestos of the Warsaw Housing Cooperative: 'beauty and charm of this world must no longer belong only to the rich and the mighty, those lucky chosen ones, and they have to become available to the worker. We told ourselves: gardens and green spaces must no longer be a privilege' (Próchnik 1934a, p. 1). Do not the demands of contemporary urban dwellers continue to be the same?

I see the Żoliborz republic as a historical example of an attempt to replace an extensive system of representative democracy and majority rule with the principles of direct democracy and the self-governance of various subjects; an example of creating the common based on the biopolitical production of customs, norms, tastes, cooperative rules and, finally, subjectivity; the common, produced and distributed contrary to the capitalist logic, in a modern city desperately lacking common resources. This shared 'Żoliborz republic' must be filled with concrete political content embodied in experiments and prefigurative forms of self-governing. A social economy proved necessary to achieve this: it consisted in class-based (i.e. not neutral) cooperation and stressed the equivalence of social and economic objectives as well as democratic qualities in business management. All this made it a 'planned ideological and economic totality'. 'For the housing cooperative is just perfect a terrain for bringing together all spheres of socialist economy and activity, as it creates community. It concentrates territorially people characterised by homogeneous class outlook and similar ideological outlook, and it connects them through a range of economic ties…. So, along the housing cooperative we operate the consumer cooperative. We create a labour market for many of us…. We educate our children on three levels—preschool, elementary school and gymnasium. We educate ourselves—through lectures, courses, library and reading room. We provide instruction in various fields of knowledge, in languages, music, and painting. We collectively practice sports and tourism, we organise social events. We organise mutual aid engendered by the spirit of solidarity. These bonds make us into a kind of socialist mini-republic' (Próchnik 1934b, p. 4). Grass-roots production and services were practised with the purpose not to make profit but to restore the meaning of work and to build community ties. Żoliborz may also resemble, to some degree, Chris Carlsson's notion of *nowtopia* (Carlsson 2008), that is, an enclave where labour has the value of social usefulness in realising some ideal of community life with regard to the economy, residential matters, environment, etc. It is crucial to find out whether labour is an instrument or an objective in itself. Does it play a role in the struggles against capitalism or in the creation of autonomous zones (Hakim Bey[7]) on the system's fringes? Does labour mean the making and networking of multiple and often temporary areas that undermine capitalism?

I do not use here or test the concept of social capital defined by Robert T. Putnam for two main reasons. First of all, Richard Sennett's concept offers more interesting and successful 'interpretation matrices'. Secondly, Putnam's concept is incompatible with the critical and anti-system perspective proposed by Hardt and Negri, which shows how jointly created multitude disintegrates the capitalist economy from within (their categories of non-material biopolitical work, common good and multitude work more effectively on the empirical material presented here). Although I often use the category of social capital in this work, the French tradition, mainly Pierre Bourdieu's work, appeals more to me. Most often, however, I use the performative category to describe rituals, habits, housing practices and direct democracy as well as knowledge that the inhabitants of the estate created, or treat them as a common good.

One crucial inspiration for my analysis is the notion of the craft and rituals of cooperation proposed by Richard Sennett. A few of his assertions are of crucial importance: '[T]he craft of making physical things provides insight into the techniques of experience that can shape our dealings with others. Both the difficulties and the possibilities of making things well apply to making human relationships. Material challenges like working with resistance or managing ambiguity are instructive in understanding the resistances people harbour to one another or the uncertain boundaries between people. I've stressed the positive, open role of routine and practicing play in the work of crafting physical things; so too do people need to practice their relations with one another, learn skills of anticipation and revision in order to improve these relations' (Sennett 2008, p. 289). All skills (or capital and competence at our disposal, as Pierre Bourdieu would say) originate then in corporeal practices, in the 'intimate connection between hand and head' (Sennett 2008, p. 9). This brings material culture to the fore in Sennett's work. Humans are conceived as working beings, creators, who become dignified through craftsmanship, patience in shaping habits and transforming them into rituals, their reflexive modification and, finally, creating new habits. Naturally, this concerns not only the manufacturing of material objects but also dealing with others, caring about one's own body, social life and everyday acts 'in the world at hand' as Alfred Schutz once put it (Schutz and Luckmann 1973). Such an idea of the human condition and the reappraisal of materiality may be associated with some version of Marxism, although pragmatism, too, as a philosophy of concrete experience, can serve well as a theoretical foundation for studies in the perspective outlined here. Although I do not use the work of Tim Ingold intentionally in this work, I believe that in further and more in-depth analysis,

---

[7]Bey's vision of 'free enclaves', which draws inspiration, among other sources, from the eighteenth-century global networks of pirates, encourages actions that might, at least temporarily, help people detach from economic, political or cultural realities of everyday life. Bey expanded this concept into permanent Temporary Autonomous Zones or Permanent Autonomous Zones (Bey 1991). Bey's proposition is neither an academic study nor a political manifesto. According to Piotr Płucienniczak, it should rather be seen as an 'adventure story', though it still contains some interesting political observations (see his review of Polish re-edition of Bey's book: Płucienniczak 2010).

the findings of this researcher—anthropologist—could be extremely valuable. His phenomenological position inspired by Maurice Merleau-Ponty differs significantly from the tradition of German idealism (mainly in its Heideggerian version). 'Radical empiricism', close to Tim Ingold, focuses on experience (Ingold 2011). In the same vein, I deliberately gave up the Heideggerian inspiration, so popular in thinking in the categories: Bauen, Wohnen, Denken (build, live, think). I agree with the criticism of this approach, which was taken up by the American anthropologist (Jackson 1996). I admit: I made quite arbitrary choices. However, I must have done some choices so that this work does not grow exponentially, and instead of putting forward research proposals with conclusions, it turned out to be only an erudite display.

I will therefore concentrate on experience. If the biopolitical labour of the Żoliborz residents is of importance, my intention is to explore it as a collective experiment, which produces the rhythm of cooperation and mutual relations (often transformed into *communitas*), of the social construction of reality, of some intersubjective and concrete material space, collectively organised and self-governing. The study should thus focus on that, which is directed outward, not so much on sensibility, but rather on skills, patience, craft and practical proficiency.

The experience of grass-roots participation is then marked by the systematic rhythm of labour, practice, holiday rituals and daily life's repeatability. Rituals are neither individual nor spontaneous. For an activity to have the quality of ritual, it needs to be repeated, re-cited, performed routinely and also collective. For the ritual embodies ideas, norms and experiences. Ritual is a mediated (through the body) form of intersubjective experience of the social world. Ritual helps to externalise what we are not fully aware of, and it preserves activities that can only later become subject to reflection. Rituals can be seen as practising social norms on one's own body. Thus, their crucial role in educational projects aimed at 'forging the citizen and democracy'.

Jeffrey Goldfarb's idea of 'the politics of small things' convinced me to undertake a materialist analysis focused on the details of everyday life, e.g. kitchen arrangements, hygiene and interior design. As Goldfarb writes: 'As we embark on our exploration of the grounds for an alternative to the politics of despair and terror, I must clarify what I mean by the politics of small things. I do not have in mind micro-interaction in general, nor all attempts at acting locally while thinking globally. Rather, I want to highlight something built into the social fabric, by active people, a potential component of everyday life' (Goldfarb 2006, pp. 9–10).

For Goldfarb, this kind of politics can take place, for instance, by the kitchen table—in Poland or in any other country under Soviet regime—where private and free talk expands public liberties. Another example is a regular flat somewhere in a socialist apartment building, which becomes a bookstore selling illegal literature. Goldfarb simply notices the potential and the political significance of free public space present in small things and in myriads of daily interactions. He knows that the political construction of this kind of space is possible without leaving one's room. All that is needed is, in the very room, to carry on a dialogue or a struggle over the social definition of the situation, to use the possibility to reveal one's own views.

The materialist analysis can thus help to study dense networks of relations and connections between various forms of social and intimate life, of important daily arrangements, objects, architectures, flora, educational efforts, and gardening, the relationship between the emancipation of women and workers, and ways of arranging apartments by the latter.

This performative and democratic perspective, attempting to capture the essence of the political in day-to-day living, has been developed by Goldfarb in the New School for Social Research. Another scholar, Elżbieta Matynia, also draws on the tradition of this institution in her notion of 'democracy in action'. All these inspirations open the question of whether the Żoliborz initiative of the Warsaw Housing Cooperative during the interwar period possessed the qualities of a performative democracy that was stimulated by new public objectives, enabled the creation of conditions for new active subjectivity, turned tenants into citizens, encouraged negotiation and compromise in the social game and, most importantly, reduced distance between 'the society' and its 'representatives'. Inverted commas seem in place here, since the founders of the WHC, were not representatives in a literal political-scientific sense, nor were the dwellers of Żoliborz the society. While the founders of the cooperative represented the interests of the tenants, providing the latter with an institutionalised regime, dignity and the social power of decision-making, the day-to-day reality in Żoliborz developed (or merely happened) through performances of highly diverse social actors. And the eventual shape of this reality was not necessarily congruent with the intents of the project's founders.

The lesson in performative democracy, which Żoliborz gives us, bears relevance today not only for places that lack freedom, but first and foremost those where democracy is taken for granted though in fact reduced to expert-technocratic decision-making that diminishes in citizens the sense of the common and the need for open dialogue. Performative democracy is valuable not so much for its institutionalised order, but rather for being a lived experience of concrete people, experience that stimulates social imagination and enables the changes of institutions. As I have already observed, it is valuable in contemporary cities, which now tend to be seen less as urbanised spaces and more and more as active subjects.

Why do I understand all these practices and discourses as performative? Performance deals with norms, and it is embodied in actions that conserve or transform social agreements, conventions and scenarios.

Cultural performance theorists (McKenzie 2001) focus on three aspects: firstly, a social self-reflection by dramatising or embodying symbolic forms; secondly, presenting alternative agreements; and thirdly, the possibility of transgression by the challenge of efficiency. How does this relate to the Żoliborz experience? Indeed, it demonstrated reflection on acute social inequalities (housing deficit, marginalisation of cooperative and socialist ideas by the government in pre-war Poland, poverty and exploitation of the workers, disadvantaged position of women), that reflection pervaded the very founding act of the cooperative as well as various publications, economic strategies and ideological projects. The challenge of efficiency targeted the chaotic and profit-driven housing economy of capitalism. The challenge takes dramatised forms, especially with regard to housing deficit and the world economic crisis, and the efficient action will be aimed at building the smallest flats, i.e. affordable for the workers.

The project was a social housing estate, and the flats were intended for those who needed them most: working people. The cooperative was to be an alternative to capitalist reality. Indeed, it proved to be a challenge to the dominant form of ownership, manifested in cooperative housing, the narrative of solidarity and collaboration with the working class (a strong sense of identity—'we' of Żoliborz, the 'cooperative republic'—against bourgeois culture of the privileged). It was an enclave open to the world, fostering a social model based on emancipation, cooperation, equality, mutual aid and 'cultural co-dwelling', namely on immaterial biopolitical production. My view is analogous to that presented by Negri and Hardt in their trilogy: *Empire*, *Multitude* and *Commonwealth*. They analyse immaterial production as the new paradigm of ownership and subjectivity, bursting capitalist economics from within. Cooperative action and collective property rights propagated during the Great Depression in prewar Europe laid foundations for the later social-democratic politics, which for a long time prevented the development of neoliberalism, and often protected the common.

The final note is about sources. The life of Żoliborz community has been thoroughly documented. A researcher can make use of valuable studies, reports, ethnographies of material conditions and institutions, as well as historically relevant descriptions of facts and the people who made the Warsaw Housing Cooperative. Detailed statistical data on residential diversity with regard to sex, age, education, class etc., financial data on spending in social infrastructure and on rent revenue are all very important due to the WHC's pioneering nature. Economic statements and public reports were a crucial aspect of democracy and self-government in the cooperative. The transparency of annual reports had particular significance for the development of self-governing structures. We have vast and rich empirical material at our disposal.

I used monthly publications of the cooperative as well as annual reports and statements. Residents' accounts of everyday life, as editorial letters, diaries, scholarly analyses, reflections (Ossowski) as well as memoirs and retrospective narratives, were of equal value as sources. I did not ignore literary forms either: novels and even poetry, which reflected the atmosphere of Żoliborz.

I offer a research perspective based on 'cityview'—as opposed to world view—which, indeed, does not emerge as a result of general reflection about the world, humanity and society, but rather refers to concrete situations, facts, places and issues of the city, which come as effects of everyday urban struggles. The sources used in this book are concrete and local. They intertwine (thanks to the porosity and openness of institutional structures) with broader, perhaps stronger, trans-local narratives. And they contain contradictions. Yet there are no better or truer narratives, only those that may respond to the institutional problems of Żoliborz realities more effectively and more efficiently. These contradictions emerge because Żoliborz is urban space developed and changed over time in flexible and pluralistic ways. It is an example of how modernity was counterpointed by everyday life.

References

Bal M (2002) Travelling concepts in the humanities: a rough guide. University of Toronto Press, Toronto

Bey H (1991) T.A.Z.: the temporary autonomous zone, ontological anarchy, poetic terrorism. Autonomedia, New York. http://hermetic.com/bey/taz3.html#labelTAZ. Accessed 9 Sept 2018

Bollier D (2007) The growth of the commons paradigm. In: Hess C, Ostrom E (eds) Understanding knowledge as a commons: from theory to practice. Cambridge University Press, Cambridge, pp 27–40

Brenner N (2009) What is critical urban theory? City 13(2–3):195–204

Butler J (1993) Critically queer. Gay Lesbian Q 1:17–32

Carlsson C (2008) Nowtopia. How pirate programmers, outlaw bicyclists, and vacant-lot gardeners are inventing the future today! AK Press, Edinburgh, Oakland, and West Viriginia

Civic S (2013) Future possible: cooperatives from ownership to use. A conversation with sostre civic. Quaderns 265:9–12

De Angelis M (2003) Reflections on alternatives, commons and communities. The Commoner 6:1–14

Editorial team of Praktyka teoretyczna (2012) Wprowadzenie. Rzecz-pospolita i aktualność komunizmu. Genealogia i krytyki polityczno-filozoficzne projektu dobra wspólnego [Introduction to the Polish edition of Commonwealth—Commonwealth and currency of communism: genealogy and political-philosophical critiques of the common]. In Hardt M, Negri A (eds) Rzecz-pospolita. Poza własność prywatną i dobro publiczne (trans: Editorial team of Praktyka Teoretyczna). Korporacja ha!art, Kraków, pp 7–71

Freyd E (1933) Życie WSM, 1 Sept

Goldfarb J (2006) The politics of small things: the power of the powerless in dark times. The University of Chicago Press, Chicago

Gorz A (1999) Reclaiming work: beyond the wage-based society. Polity Press, Cambridge

Hardt M, Negri A (eds) (2009) Commonwealth. Harvard University Press, Cambridge, MA. http://korotonomedya.s3.amazonaws.com/Michael_Hardt_%26_Antonio_Negri_-_Commonwealth.pdf. Accessed 7 Sept 2018

Harvey D (1997) The new urbanism and the communitarian trap. Harv Des Mag 1:68–69

Ingold T (2011) Building, dwelling, living: how animals and people make themselves at home in the world. In: Ingold T (eds) The perception of the environment: essays on livelihood, dwelling and skill. Taylor & Francis, London, pp 172 –188

Jackson M (1996) Introduction: phenomenology, radical empiricism and anthropological critique. In: Jackson M (ed) Things as they are: new directions in phenomenological anthropology. Indiana University Press, Indianapolis

Jacoby R (1999) The end of utopia. Politics and culture in an age of apathy. Basic Books, New York

Lefebvre H (2014) Critique of everyday life (trans: Elliott G). Verso, London and New York

Levitas R (1990) Concept of utopia. Syracuse University Press, New York

Levitas R (2013) Utopia as method. The imaginary reconstitution of society. Palgrave–MacMillan, New York

McKenzie J (2001) Perform or else: from discipline to performance. Routledge, London and New York. https://monoskop.org/images/8/8a/McKenzie_Jon_Perform_or_Else_From_Discipline_to_Performance_2001.pdf. Accessed 9 Sept 2018

Merrifield A (2011) Magical marxism. Subversive politics and the imagination. Pluto Press, London and New York

Merrifield A (2014) The new urban question. Pluto Press, London. https://pl.scribd.com/read/314636395/The-New-Urban-Question. Accessed 7 Sept 2018

Merrifield A (2017) The amateur. The pleasures of doing what you love. Verso, London and New York

Miessen M (2017) Crossbenching: towards a proactive mode of participation as a critical spatial practice. Centre for Research Architecture, Goldsmiths College/University of London, London. http://research.gold.ac.uk/20980/1/VIS_thesis_MiessenM_2017.pdf. Accessed 7 Sept 2018

Płucienniczak P (2010) Tymczasowa Strefa Eskapistyczna [Temporary Escapist Zone]. E-splot, 2 March 2010. http://www.e-splot.pl/?pid=articles&id=479. Accessed 14 Dec 2017

Próchnik A (1933) Współżycie ideowe czy walka polityczna [Ideological coexistence or political struggle]. Życie WSM [Life of the WHC], October, pp 1–2

Próchnik A (1934a) Polityka ogrodnicza WSM [The WHC horticultural policy]. Życie WSM [Life of the WHC], July, pp 1–2

Próchnik A (1934b) Spółdzielnia czy spółdzielczość [A cooperative or cooperativism]. Życie WSM [Life of the WHC], September, pp 2–4

Schutz A, Luckmann T (1973) The structures of the life-world (trans: Zaner RM, Engelhardt HT Jr.). Northwestern University Press, Evanston

Sennett R (2008) The craftsman. Yale University Press, New Haven and London

Skórzyńska A (2016) Czy możliwe jest kulturoznawstwo aktywistyczne? Partycypacja w perspektywie filozofii praxis [Are activist cultural studies possible? Participation in the praxis philosophy perspective]. In: Matysek-Imielińska M (ed) Uczestnictwo, ruch, wspólnota [Participation, Movement, Community]. Prace Kulturoznawcze, vol 19. Wydawnictwo Uniwersytetu Wrocławskiego, Wrocław, pp 15–40

Sowa J (2015) Inna Rzeczpospolita jest możliwa! Widma przeszłości, wizje przyszłości [Another republic is possible! Specters of the past, visions of the future]. W.A.B, Warszawa

Wallerstein I (1998) Utopistics, or historical choices of the twenty-first century. The New Press, New York

Chapter 2
Modern Dreams. Modern Illusions—Ideas on Cooperativism

Abstract The modern social reformers at the turn of the eighteenth and nineteenth centuries, like Charles Fourier or Robert Owen, dreamed about healing cities. That is why they designed utopian housing estates based on tenant cooperatives. On the one hand, they believed that in the social world, one can bring order, the world can be structured and architecture would be an expression of mathematically arranged reality. On the other hand, the reformers dreamed of cooperation and self-organising citizens. This tension between the order-making ambitions of the modern state and large social structures, and the grass-roots, self-help association of citizens reveals not only the ambitions but also illusions of modernity. In Poland in total/of utter disarray which, after 123 years under Russian, Austrian and German occupation, regained its independence; these modern dreams resound with new power. The idea of cooperativism is developing dynamically. As a result, in the 1920s, on the outskirts of Europe, under conditions of brutally developing capitalism, the exclusion and exploitation of workers, and thanks to, among others, the avant-garde architects from the Praesens group and Polish socialists and cooperatives; the Warsaw Housing Cooperative was built in Żoliborz, a district distant from the centre of Polish capital city, which was to address the housing deficit, ensure a decent existence for all workers, by giving them modern, modest and cheap flats.

Keywords Modernity · Cooperativism · Polish history · Contemporary modernisation order-making practices · Housing deficit

The Warsaw Housing Cooperative is obviously not an entirely innovative project. At the turn of the eighteenth and nineteenth centuries, social reformers, driven by the then-modern ambitions of order-making, tried to 'heal' the life of large cities by designing utopian housing estates based on tenant cooperatives. Charles Fourier (1768–1830) wanted to create *phalanxes*, communities of about 1500 people each, living in *phalanstères*, i.e. huge, multi-storey buildings, designed so that people could study, work and… live there. It seems that this was the prototype of modern cities growing around large factories. Fourier, as befits a utilitarian, wanted the greatest possible happiness for as many people as possible. His dream was the abolition of mass poverty. His concept of social structure was often used as a synonym for all architectural experiments involving the design of common spaces for more people.

M. Matysek-Imielińska, *Warsaw Housing Cooperative*, The Urban Book Series,
https://doi.org/10.1007/978-3-030-23077-7_2

He came up with the idea of a total workshop with a dominant supervisor organising the phalanstère's work life. Fourier advocated for a heterogeneous composition of population (Jews, who were allowed to live there, were to be assigned the lower parts of the phalanstère, as they were to perform the most filthy work—*sic!*), and the organisation of social life was based on the collective nature of social activities structured around agriculture and the workshop. In the first half of the twentieth century, Soviet urban-planning doctrines inspired by Fourier's ideas aimed to eliminate the differences between the layouts of cities and villages in order to improve the spatial planning policy. That is why it is accurate and safe to say that Soviet industrial plans are rooted in Charles Fourier's legacy (Vidler 1987).[1] The most famous and perhaps the largest architectural object is Moscow's Narkomfin building designed by Moisei Ginzburg who implemented to some extent the ideal of collective facilities such as a laundry room, a dining room and a common room with a library.

Robert Owen (1771–1858) was more of a realist than Fourier and managed to bring his vision to life. He initially founded New Lanark in Scotland, a workers' estate with a cotton mill, an organised urban planning policy and a social programme described in *A New View of Society*. He equipped his experimental community with a special tool, 'The Institute for the Formation of Character', designed to care for children and the elderly, provide help in households, free education and the correct formation of character. Houses were to consist of a maximum of four floors with no kitchens, as they were to be replaced by a shared dining hall situated in a large building in the middle of the square. The dining hall, like many other social devices, was based on 'rational principles' developed on the basis of precise quantitative research and detailed guidelines. Owen's project, although implemented, did not last long.

Later, Owen established the *New Harmony* settlement in the USA. In 1844, Owen set out the 'Rochdale Principles' which continue to inspire many Social Democrats and cooperative societies to this day.

The principles can be summarised as follows: workshops should be open to all members, ensuring equal employment for everyone; social life should be governed by the democratic principle according to which each member has one vote; political and religious neutrality should be maintained and education of members should be promoted; the distribution of profits should be based on cash trading (departure from the principle of credit transactions). At this point, it is worth noting that the Polish cooperative movement was not political at first. The cooperative movement based its strategic goal mainly on practicing cooperation, providing organisational self-support for members of various cooperatives, including producer and consumer cooperatives. As we will see later, Owen's ideas inspired WHC activists as well, though they did meet with some criticism.

What united the solidarist and cooperative thinkers was, first and foremost, the attitude towards the industrial revolution and the idea of progress. It was not about depriving people of access to civilisation achievements but rather handing them to people and thus blocking the mechanisms of alienation. Secondly, they believed that in the social world, one can bring order, the world can be structured and architecture

[1] Roland Barthes carried out an excellent analysis of the topic (Barthes 1989), cf. also Fourier (1996).

would be an expression of mathematically arranged reality. Thirdly, and finally, modernist dreamers assumed that it was possible to manage human passions and affects and that greed and egoism could be restrained: it is enough to create, for example, the aforementioned institutions that shape human character.

It was not until later that the cooperativists realised that the essence of cooperation is above all a bottom-up desire to cooperate and be together. This desire cannot be controlled from the legislator's office or from the manager's watchtower. Neither can it be the effect of a visionary's plan. Cooperation is the willingness for mutual help, associative work, self-organisation of citizens who cooperate and manage their activities out of their own will.

Owen and Fourier became the initiators of the cooperativist doctrine. 'The tension in their vision between large capitalist structures and the modern state on the one hand and limited, grass roots, associations of an egalitarian nature on the other constitutes one of the most important distinctions on which cooperativism is based. The first socialists were thus a link between the enlightenment and future social doctrines (not necessarily of leftist provenance), among which cooperativism was to be found as an element of the cooperative movement' (Błesznowski 2017, p. 8).

Although the ideas born in the spirit of modernity often inspired various order-making social concepts, they did not focus on the issue that is central to this work—the city and the urban public space. Bear in mind that Owen's *New Harmony* was created outside of the urban context. The aforementioned authors had a traditional understanding of cities as a planned and organised space. They perceived the city as an area of human activity, the activity of urban residents, city users and as a result of these activities, as a dynamic, spontaneous process. If that process sometimes became the focus of their analyses, it was only because they saw it as an undesirable state.

2.1 Meanwhile in Poland... On Cooperativism in a Country Wiped off the Map of Europe

Although the cooperative movement in Poland was not, as I had mentioned, politically polarised at the beginning, it was extremely heterogeneous. This was mainly due to the political situation. In 1795, Poland was divided among three powers: Prussia, Austria and Russia. Until the end of World War I in 1918, we cannot refer to the Polish state, although Poles maintained an extremely strong national identity. In the interwar period, Poland enjoyed the development of the associative and cooperative communities mostly because the country had been ruined during the First World War, and cooperation and self-help were simply a necessity (Okraska 2017).

The initial self-help and association activity could not be equated with the modern cooperativism, as it was neither open nor democratic, and its members were not guided by egalitarian decision-making mechanisms. It took on a modern shape only under partitioned Poland, and at the earliest in the territories annexed by Prussia where, in 1848, the freedom of association was introduced.

In the 1870s, savings and credit cooperatives dynamically developed alongside farmers' clubs in the Prussian partition (the province of Greater Poland, Wielkopolska). Father Augustyn Szamarzewski convinced the distrustful peasants and small craftsmen to cooperate with and provide aid to the weaker social groups. Under these conditions, Father Szamarzewski tried to implement the cooperative model developed by Franz Hermann Schulze-Delitzsch, pioneer of the credit unions in Germany. Although cooperative members usually adhere to the principles of cooperativism and solidarism, they had to, due to the policy of the partitioning power, refrain from allocating the earned profit for social purposes. Any cultural, educational or proactivist activity constituted a threat that the association would be delegalised. The mainstream cooperativism in the Prussian partition may be doubtlessly called the cooperativism of smallholders and producers. It focused mainly on the owners of agricultural holdings (60–70%) and craftsmen with production properties (15–18%). The remaining members were small landowners, liberal professionals and traders. There were also craftsmen not owning their own workshops, though poorly represented. Despite this diversity, they were united by ideological bonds—national solidarism and the belief that the strengthening of Polishness is possible thanks to cooperation. What also contributed to the above was Catholic social teaching and its typical mistrust of capitalism which destroys communities and family-owned manufacturing facilities. It should also be emphasised that the best economic situation was in the Prussian partition, and the cooperativism of smallholders was one of the market mechanisms supporting this dynamic development.

The situation was quite different in the Austrian partition. There was neither ideological, political nor organisational unity. The first savings and loan banks, as well as farmers' clubs, were established here. They were established on the initiative of yet another priest–politician, Stanisław Stojałowski, who emboldened the peasants to social activeness. However, serious conflicts developed within the cooperative movement. On the one hand, the slogan of national solidarity was raised (which in the case of a country under a foreign domination played a significant bonding role), and on the other hand, the emancipation tendencies of the peasants developed strongly thanks to the idea of cooperativism which, in turn, conflicted with the interests of the conservative landed gentry. Farmers were soon joined by craftsmen, and credit unions, whose main purpose was to protect its members from usury were tainted by the Polish-Jewish antagonism. In the context of ethno-religious and class conflict, the cooperative movement in these areas was a phenomenon affected by social tensions, and it was the economic interest that prevailed over the human dimension.

In poor Galicia, however, the Schulze model did not work. A new, local model of savings banks was developed by Franciszek Stefczyk who modelled them on Friedrich Wilhelm Raiffeisen's credit unions. Stefczyk embraced the principles of free market economy and although he saw many flaws in it, such as greed, selfishness and class antagonism, he saw an opportunity for social harmony and for fixing capitalism in cooperativism and fulfilling religious duties. In addition to the consumer cooperatives which brought together wage labourers, the Galician cooperatives, quite like in the Prussian partition, focused mainly on the middle classes 'with the difference that in Galicia smallholders were poorer on account of the fragmentation of land structure in agriculture and the miserable condition of the majority of the towns (crafts)' (Okraska 2017, p. 57).

Only consumer cooperatives associating wage labourers had a strong ideological, anti-capitalist programme. Edward Milewski shared such views that 'Cooperativism is the antithesis of capitalism' (Milewski 1930, p. 59). In contrast to Stefczyk, Milewski believed that class conflict is an intrinsic element of capitalism, which could be eliminated only by a change of system and by abolishing exploitation. Undoubtedly, however, Milewski's views were the exception in Galicia. His ideas were closest to the ideological climate developed in the Russian partition.

The cooperative movement in the Kingdom of Poland (the Russian partition) developed last, at the slowest pace. This resulted from the extremely oppressive political oversight on the part of the Russian authorities, the significant restrictions on civil liberties, the distrust of all associative initiatives, as well as from the extraordinary slowness and bureaucracy.

It was the economic development which forced the establishment of credit unions, and the revolution of 1905 brought political liberalisation, paving thus the way for legal cooperative activity. In 1906, the law on associations was passed. It quickly turned out that the cooperative activists are perfectly prepared in theory, thanks to numerous bulletins published and distributed underground and thanks to the readout and popularising activities. The development of the movement was also dynamised by the social involvement of the Polish intelligentsia.

The illegal Polish People's Union was the most developed; its popularity was associated first with demands for independence, secondly, with the emancipation of the peasant class, and thirdly, with the call for radical social reforms. The Russian occupants carried out the most predatory economic policy based mainly on exploitation. The independence and emancipation emotions were most heated here, in the Russian partition, by poverty, economic backwardness and the lack of civil liberties. The instigator of this illegal workers' cooperative was Edward Abramowski, a former socialist activist, convinced that the moral revolution (which is possible thanks to mutual help, collective cooperation and solidarity) must precede the work of social revolution.

Although the cooperative was of economic and self-aid nature, some of its activists set themselves far-reaching objectives—a systemic transformation in the spirit of social justice. The leftist Polish intelligentsia joined them in their educational and cultural activities. At the initiative of Abramowski, the Cooperativists' Society was formed—whose socio-political activity focused on democratic education, preparation for self-governing work and gradual liberation from bureaucratic and governmental subjugation. The society ambitiously outlined a programme aimed at raising the residents' material prosperity and culture, as well as at promoting an economic system based on the idea of social solidarity; thus the society focused on publishing brochures and books propagating cooperativism. Abramowski's *The Social Idea of Cooperativism* [*Społeczne idee kooperatyzmu*] was published back then. In 1911, the Warsaw Union of Consumer Associations was established which consolidated all the unaffiliated cooperatives scattered across the Russian partition. Just before the First World War, there were 1200 consumer cooperatives with

140,000 members in the Russian partition. Consumer cooperatives were vastly propagated, as they were the most popular and concerned the activity of every consumer. Consumer cooperatives were also the most egalitarian, because, after all, everybody is a consumer, regardless of whether they have a workshop, savings or a land estate. They allowed the systemic change to be brought about by the most effective actions.

One of the ideologues of the consumer cooperative movement, Romuald Mielczarski, recognised the mechanism and the logic of capitalism with unbelievable intuition and research reflection. He argued that despite the fact that capitalism may appear to be an invincible fortress, it has its weak side, sales—on the one hand, and the consumer—on the other. 'Nominally the master of the situation, he is the plaything of capital, because he is not part of an organisation. But from the moment when he understands his strength and wants to organise with others, from being the servant he becomes the actual master of capital' (Mielczarski 1936). It was the consumer (and not the worker) who was supposed to 'disassemble'—according to Mielczarski—capitalism.

Despite the restrictions on civil liberties, the cooperative movement in the Russian partition was the most developed in ideological and intellectual terms. Its reformist, leftist-progressive, egalitarian and civic ideas were developed by various intelligentsia members, not always interested in economic issues, but focused on social inequalities, for example, the exclusion of women from public life and their unfavourable legal situation.[2] In the circles of savings and loan, and trade cooperatives, the role of Jewish merchants was never emphasised, and ethno-religious issues were completely disregarded. What was rather pointed out was the harmful role of trade intermediation, which was regarded as an unnecessary level that increases consumption costs. The cooperation was developing both in the milieu of National Democrats and Christian Democrats who were represented, among others, by Fr. Wacław Biziński. He worked very actively in the village of Lisków, which was commonly referred to as the 'cooperative village' because its inhabitants, in a short period of time, organised a consumer cooperative, self-aid farmers' clubs, a dairy cooperative and a mutual insurance company, which made the poverty-stricken and backward locality blossom both economically and socially (Karczewski 1939). Activists and reformers often referred to the success of this village, and the lectures and readings of Fr. Biziński were published in cooperative brochures.

After the end of the First World War, Poland regained its independence, and work began on building a new political, social and economic order. The three partitioned sectors developed different economic and legal systems, which lasted over 120 years, and had to be harmonised. Poland's independence also brought civil liberties and freedom to associate and propagate ideals. Despite considerable differences in the understanding of cooperativism, the cooperative movement enjoyed

[2]In 1912, the Warsaw Union of Food Associations [Warszawski Związek Stowarzyszeń Spożywczych] issued a brochure titled *Znaczenie kobiety w ruchu spółdzielczym* [The Importance of Women in the Cooperative Movement] which was extremely popular in this milieu (Anonymous 1912), while Maria Orsetti published, in 1933, a brochure titled *Kobieta, której na imię milijony. Rzecz o zadaniach kobiet w ruchu spółdzielczym* [Woman Whose Name is Millions: On the Responsibilities of Women in the Cooperative Movement] (Orsetti 1933).

great interest as a method of farm management (in terms of both consumption and production). Therefore, the cooperative movement enjoyed the support of socialists, National Democrats, Christian Democrats and peasant circles. The adherents of economic liberalism, who believed that cooperativism stimulates the economic activity of the poorest consumers, did not exclude it, as it did not run counter to the market mechanisms, thus supporting capitalism.

The cooperative movement therefore enjoyed the support of the newly formed government and prominent people occupying high public positions. Both members of the academic elite (Zofia Daszyńska-Golińska, Ludwik Krzywicki) and the cultural elite (Stefan Żeromski, Maria Dąbrowska) strongly supported cooperativism.

The cooperative law, which was one of the most modern legal acts in Europe regulating the cooperative sector, was adopted relatively quickly, in 1920, as for rather difficult economic and political realities. The Cooperative Council, consisting of representatives of the cooperative movement and unaffiliated cooperatives, was established under the Ministry of the Treasury. And the Minister of Agriculture established, years after Poland regained independence, the State School of Agricultural Cooperativism. The Cooperativists' movement itself created a number of supporting institutions, which were the superstructure for the cooperatives. The Cooperative Scientific Institute, a research, educational and popularising unit of cooperativism, was established in 1919. In 1935, the Cooperative Women's Guild was established, together with numerous educational establishments (such as the Cooperative High School [*Gimnazjum Spółdzielcze*]), reading rooms and publishing houses—all promoting cooperative ideas.

In addition to the most thriving consumer movement, there were also construction and housing cooperatives (both construction-investment, as well as tenant-administrative types). In 1922, there were 18 construction cooperatives and 72 housing cooperatives, and in 1930 their number increased to over 850 such associations. These were usually small tenants' initiatives involving the residents of one tenement house. But also construction-investment cooperativism was practiced, which was transformed over time into an activity consisting in renting premises to workers who lived in disgraceful conditions.

The Warsaw Housing Cooperative was recognised as a model initiative in the housing cooperative sector, conceived as 'a deliberate, holistic organism'. Although this was a local cooperative, its example shone throughout the country, inspiring other initiatives. At the end of 1927, it had 345 members, and a little over a decade later, there were 2283 of them (Okraska 2017, p. 65).

There were clear ideological differences mainly within the consumer cooperative movement, but also within housing cooperatives. Cooperative ideas were based on Robert Owen's principles of political neutrality, but in the face of poverty and extremely aggressive capitalism, bourgeois cooperatives, and the cooperatives of landowners, workshop owners and larger production plants, and especially investment cooperatives began to be perceived as blurring the cooperative movement's identity, its self-help and social character, and are not a very distinctive element of the free market economy that fits in with the logic of capitalism. There was an internal break within the cooperative movement. Proponents of neutrality (including the Union of Polish Consumer Associations [*Związek Polskich Stowarzyszeń Spożywców—ZPSS*] maintained that cooperatives should have an open character. Anyone who is a member of a cooperative can make purchases in it and thus contribute to the further development of the cooperative movement. They shied away from any political and ideological identification. On the other hand, the representatives of workers' or class cooperatives considered the principle of universality and neutrality as harmful, because they blurred the social goals of this movement. That is why they created the Union of Labour Consumer Cooperatives [*Związek Robotniczych Spółdzielni Spożywców, ZRSS*]. They included, among others, Jan Hempel, Bolesław Bierut, Aleksander Ostrowski—members of the Communist Workers Party of Poland [*Komunistyczna Partia Robotniczej Polski*] and Stanisław Szwalbe, Stanisław Tołwiński, Ludwik Libracht, Maria Orsetti—all those who later, in the winter of 1921, founded the Warsaw Housing Cooperative. Despite the fundamental differences of opinion, they promoted the social economy and not the capitalist one. They knew that the idea of self-help ran counter to the ideas of individualism and egoism developed in the capitalist culture.

As a consequence of this split within the cooperative movement, the membership universality was significantly reduced, in favour of the principle that cooperatives should be self-help movement for the people living of their own labour. The cooperative movement's goal was to emancipate this class, abolish the wage labour, socialise means of production and lay the ground for the socialist system. 'Consumer cooperativism,—wrote the founders of the Union of Labour Consumer Cooperatives in the initial programme statement—on account of its economic idealism, has nothing in common with agricultural associations of producers, nor with credit cooperatives, because they are based on the privilege of owning land or other means of production and display a clear tendency to transform into associations of larger or smaller capitalists' (Związek Robotniczych Stowarzyszeń Spółdzielczych 1919).

Adam Próchnik, who became, over time, a fervent ideologist of cooperativism among the Żoliborz activists, strongly opposed the so-called neutral or bourgeois cooperativism. The cooperative concept, according to Próchnik, consisted in the conscious struggle against capital and against the exploitation of the working class. The struggle would rely on mutual aid and placing collective interest over individual needs. 'Cooperativism thus, being a form of cooperation, is participating in the class struggle. The essence of cooperativism is replacement of the capitalist-individualist form of production and apportionment with social forms. Cooperativism thus abolishes the causes of class antagonisms in a certain area. (…) Thus cooperativism, being a form of cooperation, is simultaneously a form of struggle. The role of cooperativism depends on recreating society and therefore it is revolutionary' (Próchnik 1937a, p. 2, 1937b, pp. 279–280).

It is also worth adding that despite the very clear ideology, the representatives of a branch of the class cooperatives emphasised their autonomy. 'The consumer cooperative remains completely independent of other forms of the workers' movement—political parties and trade unions, and in no way can it be subordinated to them, aiming at the same time at the same goal as the general socialist movement' (Szymański 1989, p. 11).

Over time, it will also turn out that the split within the cooperative movement runs not only along the line of neutrality, but also ideology. The split would be the most visible between the socialists and the communists.

In spite of the great differences, in 1925, the three organisations formally united. The Union of Polish Consumer Associations [*Związek Polskich Stowarzyszeń Spożywców, ZPSS*], the Union of Labour Cooperative Associations [*Związek Robotniczych Stowarzyszeń Spółdzielczych*], the Central Union of the Cooperatives of State and Municipal Employees and Social Workers [*Związek Rewizyjny Spółdzielni Pracowników Państwowych, Komunalnych i Społecznych*]. The Union of Consumer Cooperatives of the Republic of Poland [*Związek Spółdzielni Spożywców Rzeczpospolitej Polskiej (ZSS RP)*] was established this way. However, this formal gesture did not end the ideological disputes that lasted uninterruptedly until the outbreak of World War II.

The development of the cooperative movement was hampered by the worldwide economic crisis which reached Poland in 1930. Member shares, savings deposits and trade turnover decreased dramatically, and finally some of the cooperative enterprises failed.

Cooperative members saw differently the role of the state in the face of the crisis. Some of them began to propagate state intervention in the economy. They believed that the recovery from the economic downturn and the modernisation of Poland should be part of the state's planned and rational policy, which would block the chaotic and uncoordinated private capital flows. That, in turn, was in great contradiction with Edward Abramowski's ideas—often criticised in the past—which defined, probably in the most important manner, the intellectual framework of the Polish cooperativism.

Edward Abramowski, still in the period of partitions, published, in 1905, the famous text titled *General Collusion Against the Government* [*Zmowa powszechna przeciw rządowi*] which was interpreted throughout the history of Poland in many different ways, depending on the political context. He argued in it against the supremacy of the authorities and institutions of the Russian Empire. He called upon the Poles to create their own social self-governing and bottom-up structures, not only in the name of Poland's independence, but also in the name of their personal freedom. The citizens of reborn Poland were to be prepared for independence, having arbitration courts, schools, credit unions and vigilantes at their disposal. The new country was based on a grass-roots democracy and the socialisation of the means of production. It was also possible to see in Abramowski's appeal a much later foundation for the founders of the Solidarity movement at the turn of the 1970 and 1980s, which advocated the liberation of People's Republic of Poland from the Soviet influence (its biggest impact was fully revealed during the period of martial law introduced in December 1981). It is not clear whether Abramowski was in favour of stateless cooperativism that could have resulted from the oppressive situation, or whether in the light of Poland's regained independence he saw an opportunity for state communism. After all, he strongly supported all activities aimed at regaining independence and forming a state by Poland.

Similarly to Jan Wolski and Maria Orsetti, Abramowski believed in grass-roots action of disassembling the capitalist system. He developed the idea of 'direct action'

(*action directe*) which, as a consequence, was supposed to strike at the very heart of the state organisation. These, in turn, should be replaced by cooperative institutions based on solidarity and reciprocity. Abramowski, in the period of partitions, identified the state with the impersonal form of coercion affecting the individual's social constitution, consisting in belonging to some organisation and thus eradicating individual differences. Polish contemporary interpreters of Abramowski's writings juxtapose his concepts to the interpretation patterns of Marx or Pierre Bourdieu who, when developing the category of *symbolic force*, enroling individuals in a metaphysics of power and 'abstract forms of mutual « interdependence »' (Błesznowski 2017, p. 31).

This does not mean, however, that Abramowski embraced the individualistic standpoint. In fact, the opposite is true. Snapping individuals out of the metaphysics of power, Abramowski placed them in the metaphysics of collectivism and cooperation, mutual aid and brotherhood, which is a phenomenon of a true social existence. Brotherhood is 'both a metaphysical experience of love and will' (Abramowski 1980, p. 581). Only that love is a social relation, it is a meeting with another in which will—the active creative element—is exercised. Thus, the sphere of intersubjectivity appears, in which an entity—neither individual nor social—is born. Abramowski means an entity which thanks to brotherhood, love and will is a 'social individuality'. That means that there is no humanity outside of society.

Only the entities, thus construed, create cooperatives. These, however, differ from the cooperatives having only economic character. Not only Abramowski's cooperatives reward work as the creative element and block the mechanisms of alienation. They stimulate, above all, commitment to cooperative democracy. They are experiments consisting in shaping an individual initiative that will bring a sense of accomplishment and new opportunities for self-realisation to both the individual and the community. It is therefore about civic cooperation, self-help and self-organisation. In 1912, Abramowski formulated the model of Friendship Unions, cooperatives of a brotherly nature. 'I imagine Friendship Unions as neighbourhood unions whose task is mutual aid in everything' (Abramowski 2009a, p. 218). They were to adopt various characters and functions: economic, tutelary or educational. What is the most important is that they are based on democratic cooperation and mutual responsibility, as well as on the unconditional autonomy not only from state power but also from all central or union structures. The experimental trait of the Friendship Unions consists also in the fact that they are to be primarily a space for the emergence of the new ethics of interpersonal relationships, the forging of a new man in a new, fraternal society, based on ethical economy, created as a result of moral revolution.

It is not surprising, then, that Abramowski's demands, and especially his concept of the Friendship Unions, were much criticised, also by the supporters of this vision. Both Stanisław Ossowski and Maria Orsetti discussed it with Abramowski. For Polish Communists (e.g. Jan Hempla), Abramowski's vision was not only a pipe dream testifying to his political immaturity, but above all it represented escapist ideologies. One cannot, according to Hempel, treat the Friendship Unions as real alternatives to a capitalist economy based on exploitation and profit. Hempel went even further in his criticism and pointed out that Abramowski's idealism preserves capitalism to a certain extent, by immobilising the most active individuals, depriving them of their subversive potential and pushing them to act beyond the real world order.

Therefore, cooperativism is part of a great project of modernity which gives hope that there are social instruments that allow us to shape the new social ethics and the new man who could make the world a better, more rational and friendly place. Cooperativism in this world would be understood as a rule of a new, grass-roots political and economic activity shaping modern principles of citizenship, democracy and—above all—socialisation. Bartłomiej Błesznowski writes explicitly that even if cooperativism is utopian, 'then it is a quite pragmatic utopia, whose guiding idea is the promise of hard work founded on the very basic fact of socialisation. A cooperative is thus action, becoming, within which the discord between the idea and its realisation is each time invalidated, because a cooperativist, as Abramowski wrote, « not only speaks of a new social system, of a better and more just world, but is building that world »' (Abramowski 2009b, p. 162).

2.2 Housing Situation in Poland

The tragic housing situation among Warsaw's poorest residents resulted from the tense and predatory policy under partitioned Poland. On the one hand, one could observe the negative policy of the tsar against all political and social activity. Therefore, no social housing problems were ever addressed at all. On the other hand, there was a fairly open and tolerant policy in terms of the economy. Russia needed dynamic industrial and commercial development. The Polish metal industry intensively supplied the Russian market needs and that in turn was associated with the strong industrialisation of the Polish capital. It also caused a significant migration of people from villages to cities (between 1870 and 1913, the population grew from 266.2 to 845.5 thousand, and the increase in employment in industry grew from 6000 to 41.5 thousand workers). Under partitioned Poland, Warsaw, from a provincial city, became an industrial centre (Cegielski 1968, p. 23).

In such a situation, the lack of state support quickly resulted in a fatal existential situation, housing problems, sanitation issues, bad epidemiological conditions, as well as unlawful building works dangerous to people. This also triggered the rental sector but not of apartments, but of the sublet space in someone else's flat.

After Poland regained its independence, when Warsaw became the capital of a resurgent state, the number of inhabitants continued to increase, but unemployment, in turn, rose as Russian industrialists withdrew their businesses from Poland. Those who had left the city due to warfare were coming back. In addition, there were those who thought they would find better employment in Poland's capital city. Newcomers came to look for work in institutions, which were necessary for the proper functioning of the state administration.

It was estimated then that due to warfare 1914–1918, and due to the lack of maintenance, nearly 27 thousand brick residential buildings and 0.5 million wooden ones had been destroyed. In 1927, Marcin Weinfeld claimed that one million apartments should be built in Poland (Weinfeld 1927, p. 27): 400,000 were to be built in Warsaw alone and 250,000 in Łódź. Ten years later, the Polish Housing Congress [*Polski Kongres Mieszkaniowy*] proposed the construction of 2 million dwelling premises by 1947 (Tubiasz 1938: 81). These data could be of course subject to further debate. There was no doubt, however, that the condition of housing was disastrous and that such a state was permanent. Residents were severely affected by the lack of housing.

The poorest population of Warsaw, but not the homeless, occupied one-or two-room premises (consisting of a room and a kitchen equipped only with an oven). Dwellings were often deprived of access to light (when they were located in the basement) or access to fresh air (when in the attics) and had terrible sanitary conditions. They were most often unsewered; they did not have running water, and the water had to be drawn from the well placed in the yard, near toilets—wooden sanitary facilities unadapted either for children or disabled people. Hardly ever did the dwellings have electricity. Tenants of the apartments located on the outskirts of the city, with no access to paved roads, were often deprived of efficient transportation. Many of them occupied shanties and tenement houses suitable only for demolition. A frequent practice was also to sublet space in these miserable flats to other tenants, which further aggravated the already extremely difficult housing conditions. Subletting space in one's flat, however, allowed to complement the budget destined for rent arrears. Otherwise, tenants were threatened with eviction.

In 1927, the Survey Commission for the Study of the Production Conditions and Costs and Exchanges published a report, which showed that in Poland, 95% of dwelling premises in rural areas and 83% of dwelling premises in cities amounted for the category of small (two- and three-room) dwellings. In cities, one-room flats accounted for 39% of all dwelling premises. In Łódź, which was the second largest city in Poland at that time, 59.7% of flats had one room. Statistically, there were 2.5 tenants for every room in Warsaw. The commission described in the report that many families lived in such premises at the same time. There were often six people cramping in one room. In the face of significant overcrowding and in dreadful sanitary conditions, tuberculosis spread like wildfire. In Łódź, for 10,000 inhabitants, 30 died from tuberculosis (Gandecki 1928: 12).

Although the aim of numerous construction cooperatives was to provide their members with flats, they failed to raise the funds for their construction. Neither did they have loan guarantees. Construction cooperatives, entered into court registers, waited for housing loans until finally they were wound up. Private investors perceived residential construction as unprofitable. On the other hand, the state's budget was too limited, and therefore it did not grant loans, nor did it invest itself.

The projects which were implemented were mainly those whose construction was economically viable (that is mainly flats larger than two-room ones). The Warsaw investments that were directed at meeting the needs of the most disadvantaged social strata, mainly workers, included: Workers' Housing Development Societies [*Towarzystwa Osiedli Robotniczych, TOR*], a nationwide organisation, the Warsaw Housing Cooperative (WHC) and local government (Strzelecki J 1935; Strzelecki E 1935). It is worth mentioning that TOR's flats were built on the Koło estate (19 blocks of flats in 1937, i.e. 984 flats) and on the Rakowiec estate (5 blocks of flats in 1937, i.e. 192 flats).

The 'investment' activity of the city consisted mainly in organising shelters for the homeless. In Warsaw, most of the homeless living in shelters could neither afford to rent nor to buy a flat, even on the most preferential terms. Edward Strzelecki in 1935 produced a report titled 'The Housing Issue in contemporary Warsaw', in which he wrote: 'Summing up the comments on the construction activity, it can be said that its role in improving Warsaw's housing conditions consisted in improving the housing situation of wealthy and middle-class residents, but gave nothing to the lower classes of the population' (Strzelecki 1935; Strzelecki 1935, p. 206).

Drawing a catastrophic picture of Warsaw's housing at that time, it is worth noting the scale of homelessness, which is obviously connected with the housing issue. Homelessness resulted from numerous parallel and intersecting processes. Until the early 1920s, the homeless were mainly Polish returnees who came back from former eastern Poland, post-war émigrés wandering around the country, orphans and demobilised soldiers. The first shelters were opened for them. However, with each subsequent year, the homeless came from other social groups. Housing scarcity resulted in a steady increase in the number of homeless people. In the report on the second half of 1925, the Department of Social Care of the Municipal Social Centre [*Departament Opieki Społecznej Miejskiego Ośrodka Społecznego*], while describing the situation at that time, described the processes characteristic for the whole country: 'the catastrophic housing shortage and migratory flows of people in search of work result in setting up lodging-houses; for the time being we have only 10 of them, all of which are lagging behind on sanitation and hygiene condition. At least 40 such accommodation facilities should be created in larger urban and industrial centres, with a total of 4000 beds' (Department of Social Care 1925–1926, p. 123).

The end of the 1920 and the 1930s was the time of economic crisis, which brought as a result numerous waves of evictions. A new law, allowing landlords to evict tenants, is in force since April 1924. In 1928, of 2028 homeless families living in Warsaw shelters (which amounted for 71% of the then-current homeless population of shelters), 50.5% were evicted families. The rest are returnees' and émigrés' families (12.3%), evicted from company-owned flats (9.9%), from collapsing houses (6.5%), relocated from overcrowded flats (5.5%) and located in shelters due to other reasons (12.3%) (Drozdowski 1990, p. 327).

Local authorities had neither the idea nor funds to solve the problem of homelessness. And the obligation to provide an 'adequate room with fuel and light' meant that help for the homeless was limited to building shelters and lodging-houses in the form of provisional barracks. However, it cannot be said that local governments helped the same homeless. They did not undertake any initiative to support the construction action. In the handbook for social welfare officials, it was even made explicit that the municipality [*gmina*] had no obligation to build homeless shelters, 'but only to help those who are subject to Law on social welfare'. Municipalities decided to build only profitable new residential houses, i.e. with large, at least three-room flats. Mateusz Rodak, the researcher in interwar homelessness, puts forward the thesis that, similarly to the municipalities, other initiatives, including the Workers' Housing Development Societies (TOR) and the Warsaw Housing Cooperative, although they had noble goals, were focused mainly on profit (Rodak 2013, p. 55).

Shelters—organised and run by municipalities or private associations, lodging-houses or spontaneously created squatter settlements gradually started to fit in the landscape of most Polish cities. Jan Strzelecki in his paper titled 'The Housing Situation and Needs in Poland' delivered during the First Polish Housing Congress said: 'almost every Polish city has a festering wound in terms of housing situation. What we are referring to are the makeshift buildings in which thousands of workers live in the Gdynia's poorest districts (the so-called Beijing and Budapest), terrible hovels (the so-called Madra) in Grudziądz, towns consisting of makeshift shacks in Torun on the so-called Cossacks Mountains. In Warsaw, barracks for the homeless in Żoliborz and Annopol districts, settlements of wooden shacks in Buraków and in Okęcie, flats in the 'Merry Town' and exhibition buildings in Poznań. Various elaborate mud huts and caves in Silesia. Tarnów's old warfare facilities inhabited by hundreds of people, old and dilapidated factory buildings and railway wagons in Radom. Hovels in Bałuty and Chojny districts in Łódź. Some appalling hovels-mud huts in Hajnówka. Squatter settlements in Bydgoszcz, etc.' (Strzelecki 1937, p. 10).

There were twelve homeless shelters and three lodging-houses in Warsaw at the time, when maintaining shelters for the homeless was the only strategy to deal with the problem of homelessness (1928–1932). Metal barracks created in Żoliborz, near Warsaw's Gdańs Station, eventually became a powerful symbol of Warsaw's misery. However, at the end of the 1920s, the Żoliborz shelter for the homeless is bursting at the seams and has more and more problems: epidemics, crimes and prostitution spread between the homeless. The city authorities decide to build a new shelter, this time on the other side of the Vistula River, in Annopol district. The colonies initially consist of 20 barracks, in which 1500 homeless people are planned to be settled. Wooden or brick barracks were built quickly and with the use of cheap building materials, and the terrible housing conditions did not change. Many reports and social research into these barrack 'lodging-houses' were conducted (Dembiński 1939; Dąbrowski 1965; Rodak 2010, 2011). They described severe poverty, terrible sanitary conditions and the moral decay of the inhabitants. In 1938, in 'Wiadomości Literackie' [*The Literary News*], we could read: 'Annopol is far from the city centre. It does not hurt anyone's feelings with wooden, brick, or concrete shelters. (…) This misery is avoided like the plague. Hidden in the middle of nowhere. Deprived of its teeth and claws. Such misery to watch, for demonstration, clinical misery, inscribed within the system, safe as animals in nature reserve, poisoning with its venom only itself. Misery without comparison and without contrast. Hopeless misery' (Szemplińska-Sobolewska 1938, p. 34).

Jan Starczewski was very critical of the city's activities in the area of solving the problem of homelessness. He knew perfectly well that the barrack 'gives the families a roof over their heads, but at the same time introduces them to the environment in which the population, naturally united by the location, is suffused with the characteristics of the psyche typical of people accustomed to living at the expense of society, such as a complete abandonment of efforts to base their existence on their own earnings, lack of a sense of solidarity with the world of labour, complete lack of fear of the consequences of their addictions and crimes' (Starczewski and Konopnicki 1938, p. 82). It can therefore be said that rising homelessness was the result of both the increase in poverty and erroneous construction action which, in its programmes, did not take into account the construction of small working-class flats (or—as the architects used to say—the minimum dwellings, for those who needed them most). And on the other hand, the policy of 'fighting homelessness' through the construction of barracks—shelters and lodging-houses—reproduced and deepened homelessness.

The ineffectiveness of this shelter strategy and the increasing costs of maintaining and renovating shelters led to a change in the forms of assistance. It was decided to grant financial aid to the homeless so that they can pay the rent. In the year 1934/1935, the financial aid for the rent was granted to 815 evicted families and 111 families who left shelters for the homeless (Starczewski and Konopnicki 1938, p. 93).

Although in 1936, Warsaw's shelters for the homeless did not accept new tenants, fifteen thousand homeless still remained in the barracks. In Annopol alone, where a space for 4000 people was initially planned, eleven thousand tenants were cramped in fifteen buildings. In the one-room premises, eight people were sleeping on three beds, on 12 m^2. Despite these appalling conditions, the homeless tenants were often reluctant to use state subsidies for renting flats and frequently decided to stay in the shelter. Annopol and Żoliborz, until the Second World War, served as evidence of the city's ineffective strategy in the fight against homelessness, which perpetuated misery and moral decay. Barrack estates disappeared from the map of Warsaw only during the demolition of Warsaw after the end of war.

People dealt with housing scarcity on their own, by building squatter settlements, makeshift shanties serving as shelters. Workers, who were used to physical work and had the competence to carry out construction work, built the most durable houses. They would buy cheap land, often exposing themselves to fraud. The sellers sold the same plot of land several times or the land planned for roads or municipal investments. The buyers did not know about legal complexities anyway. The price was the only criterium that mattered. Other criteria were not significant. So they would buy undeveloped plots of land, away from the city centre, and even away from the city itself. The plots were small, not going much beyond the external outline of the building. Houses were built with no plans, permits or construction designs. Even though there was a risk of high penalty (such as building demolition) or financial fine for constructing a house without a permit, the authorities turned a blind eye to this practice. After all, it was not in the city's interest to deepen the already dire homelessness crisis. The only thing that mattered was to make the house quickly inhabited. 'The construction must advance quickly so that potential inspectors can be presented with a *fait accompli*. Materials come mainly from demolition waste, as workers cannot afford to buy from a brickyard or sawmill. Window panes and sheet metal for chimney were the most expensive, as these could not be obtained from demolition waste. Building material—in order not to detect the intention to construct, but also from the fear of theft—was usually accumulated in hiding, outside the plot, but close enough so that the transport does not entail additional costs. Work would usually begin on Saturday, late in the afternoon, when it is certain that there are no inspectors or police around. Construction work would last all night and most of Sunday. The family would move into the building on Sunday evening and over the next night, they would do the finishing works and interior decoration. On Monday morning, the house would be ready and inhabited. Demolition would mean the eviction of the entire family, and few construction inspectors would be ever able to do it' (Springer 2015, p. 31). The interiors, arranged by families, are one-room rat-holes ranging from 9 to 16 m^2 with one window, a single-pitched roof, equipped with stoves without a kitchen. They withstand eight to ten years. After that time, they start to break and fall into ruin. Many of them survived practically unchanged until the outbreak of World War II, which wiped them out physically and from the memory and history of Warsaw. In order to imagine the conditions in which these 'cavemen' lived, look at the description published in the *Dom. Osiedle. Mieszkanie* [House. Estate. Flat] magazine, edited by Teodor Toeplitz. In the issue describing the Werkbund exhibition and the second CIAM conference, there is also a report on how these inhabitants organise their living space. 'Outside, in front of this "flat", they wash, dress, cook food, work, and undertake various activities (…). Children spend time outdoors; they spend time inside the 'flat' only when it's cold or when it rains. They even sleep outside, in their beds and cradles placed in front of the house, whenever the weather is good. This settlement extends into one long street, 'built-up' with houses on one side and a square on the other side, around which there are hovels or mud huts. In the middle of the square, there is an iron stove, which is used by many families (all those who do not have a stove in the 'flat'); the social life of these people is centered around this place. Women come here to talk, men—to cut their own deals (arrange casual employment, as most of them are 'unemployed') and children – to play together. Residents are usually those who '*voluntarily became homeless*—[the original italics]. These are former subtenants who could not live under the current housing conditions. The hell they had at home forced them to change their fate (…)'. The reportage has a staggering end with the following sentence: 'one can kill another person with the flat like with a bludgeon' (L. B. 1929, pp. 17–18).

In 1933, three shelters were closed. In 1934, Warsaw's municipal council decided to restrict the admission of additional homeless people. With time, more shelters for the homeless were also closed, especially those in which the worst sanitary conditions prevailed.

It is impossible today to estimate how many homeless people there were in Poland in the interwar period. Undoubtedly, however, neither the problem of homelessness nor the housing deficit had been solved. It is safe to say that under the Second Republic of Poland, housing scarcity and homelessness were permanent elements in the functioning of the Polish cities. They concerned a large group of pauperised workers and, as a rule, they were rarely the result of pathological phenomena (i.e. alcoholism, crime or drug addiction).

This is why Warsaw activists, social activists, architects and urban planners worked extremely actively on the definition of 'the minimum dwelling', while giving practical shape and form to the ideas of both the International Congresses of Modernist Architecture, the International Housing Congress and the Polish Society of Housing Reform. The minimum dwelling is, under Polish conditions, a flat with an area of up to 42 m^2 for the lowest-paid wage labourers, and for the better-paid wage labourers—a flat with an area of up to 56 m^2. However, it is not about the square footage of the flats. It is known that the flat has to be small and cheap so that a labourer can afford it. But such were both rat-holes and tenement houses with subtenants. The minimum dwelling shall first and foremost ensure the citizen's right to housing, which thus protects their family from disintegration and moral decay. The minimum dwelling provides decent housing conditions. These are determined by the category of the existential minimum (die Wohnung für das Existenzmi-nimum) (Tołwiński 1939).

References

Abramowski E (1980) Metafizyka doświadczalna [Experiential metaphysics]. In: Abramowski E, Metafizyka doświadczalna i inne pisma [Experiential metaphysics and other writings]. Państwowe Wydawnictwo Naukowe, Warszawa

Abramowski E (2009a) Związki przyjaźni [Friendship unions]. In: Abramowski E, Braterstwo, solidarność, współdziałanie [Fraternity, solidarity, cooperation]. Stowarzyszenie 'Obywatele Obywatelom', Łódź

Abramowski E (2009b) Kooperatywa jako sprawa wyzwolenia ludu pracującego [The cooperative as a matter of freeing working people]. In: Abramowski E, Braterstwo, solidarność, współdziałanie [Fraternity, solidarity, cooperation]. Stowarzyszenie 'Obywatele Obywatelom', Łódź

Anonymous (1912) Znaczenie kobiety w ruchu spółdzielczym [The importance of women in the cooperative movement]. Wydawnictwo "Społem", Warszawa

Barthes R (1989) Sade, Fourier, Loyola (trans: Miller R). University of California Press, Berkley and Los Angeles. https://thecharnelhouse.org/wp-content/uploads/2015/04/roland-barthes-sade-fourier-loyola1.pdf. Accessed 9 Sept 2018

Błesznowski B (2017) Utopia in the service of modernity: on the sources of cooperativism. In: Błesznowski B (ed) Cooperativism and democracy: selected works of Polish thinkers. Brill, Leiden, pp 1–46

Cegielski J (1968) Stosunki mieszkaniowe w Warszawie w latach 1864–1964 [Housing relations in Warsaw in the years 1864–1964]. Arkady, Warszawa

Dąbrowski J (1965) W barakach na Żoliborzu [In the shanty barracks of Żoliborz]. In: Dąbrowski J, Kosowski J (eds) Niepiękne dzielnice. Reportaże o międzywojennej Warszawie [Unpretty districts: stories of Warsaw between the world wars]. PIW, Warszawa, pp 124–129

Dembiński H (1939) Z zagadnień urbanistycznych i społecznych Żoliborza [Urban and social issues of Żoliborz]. Życie WSM, May 1939

Departament opieki społecznej. Drugie półrocze 1925 roku. (1925–1926) Praca i Opieka Społeczna, 1–2 [Department of Social Care, the second half of 1925 (1925–1926). Work and Social Care, pp 1–2]

Drozdowski MM (1990) Warszawa w latach 1914–1939 [Warsaw in the years 1914–1939]. Państwowe Wydawnictwo Naukowe, Warszawa

Fourier C (1996) The theory of the four movements. Charles Fourier. Edited by Gareth Stedman Jones, Ian Patterson, translated by Ian Patterson, Cambridge University Press, Cambridge

Gandecki A (1928) Budownictwo mieszkaniowe. Społem [Housing construction. Together] 18:11–13

Karczewski W (1939) Lisków: Dzieje jednej wsi polskiej [Lisków: the history of one Polish Village], Warszawa

L. B. (1929) Ucieczka. Mieszkanie. Osiedle. Dom [Escape. Flat. Estate. House] 6:17–19

Mielczarski R (1936) Cel i założenia stowarzyszenia spożywców. Mielczarski R, Pisma, vol 2. Związek Polskich Stowarzyszeń Spożywców R. P, Warszawa, pp 225–226

Milewski E (1930) Sklepy społeczne (rzecz o kooperacji spożywców) [Social shops (On consumer cooperativism)], 2nd edn. Związek Spółdzielni Spożywców Rz. P, Warszawa

Okraska R (2017) From self-help to the vision of new system: an outline of the material and ideological development of the cooperative movement in Poland to the year 1939. In: Błesznowski B (ed) Cooperativism and democracy: elected works of Polish thinkers. Brill, Leiden, pp 47–79

Orsetti M (1933) Kobieta, której na imię milijony. Rzecz o zadaniach kobiet w ruchu spółdzielczym [Woman whose name is millions: on the responsibilities of women in the cooperative movement]. Skł. Gł. 'Dom Książki Polskiej', Związek Spółdzielni Spożywców R. P. w Warszawie, Kraków

Próchnik A (1937a) Ideologia spółdzielczości robotniczej [The ideology of labour cooperativism]. Życie WSM, May, pp 1–4

Próchnik A (1937b) The ideology of labor cooperativism. In: Błesznowski B (ed) Cooperativism and democracy: selected works of Polish thinkers. Brill, Leiden, pp 277–283. Accessed 2017

Rodak M (2010) O "Cyrku", który nie był cyrkiem – rzecz o warszawskim międzywojennym domu noclegowym. Praca Socjalna [On the 'Circus' which was not a circus—on the Warsaw lodging-house in the interwar period. Social work] 4:86–102

Rodak M (2011) Kolonia dla bezdomnych w Warszawie na Żoliborzu (1923–1939). Praca Socjalna [Colony for the homeless in Warsaw's Żoliborz (1923–1939)] 1:106–113

Rodak M (2013) Zjawisko bezdomności w Drugiej Rzeczypospolitej (ze szczególnym uwzględnieniem Warszawy [The phenomenon of homelessness in the Second Polish Republic (with particular emphasis on Warsaw)]. In: Grata P (ed) Od kwestii robotniczej do nowoeczesnej kwestii socjalnej. Studia z Polskiej polityki społecznej XX I XXI wieku [From the workers' issue to the new social issue. Studies in the Polish social policy of the 20th and 21st century]

Springer F (2015) 13 Pięter. Wydawnictwo Czarne Wołowiec

Starczewski J, Konopnicki J (ed) (1938) Materiały o organizacji i działalności WOSiZP Zarządu Miejskiego w m. st. Warszawie Warszawa [Materials on the organization and activities of the WOSiZP in the capital city of Warsaw]

Strzelecki E (1935) Zagadnienie mieszkaniowe współczesnej Warszawy [The housing issue in contemporary Warsaw]. In: Anonymous (ed) Zagadnienia gospodarki samorządu Warszawy, Warszawa

Strzelecki J (1935) Organizacja i działalność Towarzystwa Osiedli Robotniczych [Organization and activity of the workers' housing development societies (TOR)]. Dom Osiedle Mieszkanie 2

Strzelecki J (1937), Sytuacja mieszkaniowa i potrzeby mieszkaniowe w Polsce (materiały Pierwszego Polskiego Kongresu Mieszkaniowego) [The housing situation and needs in Poland (Materials from the first Polish Housing Congress)], Warszawa

Szemplińska-Sobolewska E (1938) Annopol. Rezerwat nędzy [Annopol. Reserve of misery]. Wiadomości Literackie 52–53:34

Szymański JA (1989) Warszawska Spółdzielnia Mieszkaniowa 1921–1970. Zarys dziejów [Warsaw Housing Cooperative 1921–1970. History Outline]. Wydawnictwo Spółdzielcze, Warszawa

Tołwiński S (1939) Definicja mieszkania "społecznie najpotrzebniejszego". Dom. Osiedle. Mieszkanie [Definition of the 'minimum dwelling'. House. Estate. Flat]. no 6, pp 2–30, June 1939. http://bcpw.bg.pw.edu.pl/Content/7229/05dom39_czerwiec.pdf

Tubiasz S (1938) Sprawozdanie z Polskiego Kongresu Mieszkaniowego. Praca i Opieka Społeczna [Report from the Polish Housing Congress. Work and social care] 1:80–92

Vidler A (1987) The writing on the walls. Princeton Architectural Press, Princeton

Weinfeld M (1927) Kwestia mieszkaniowa w Polsce. Praca i Opieka Społeczna [The housing issue in Poland. Work and social care] 3:27–29

Związek Robotniczych Stowarzyszeń Spółdzielczych (1919) Jak powstał i czym jest Związek Robotniczych Stowarzyszeń Spółdzielczych [What the union of labour cooperative associations is and how it arose]. Wydawnictwo Związku Robotniczych Stowarzyszeń Spółdzielczych, Warszawa, pp 4–5

Chapter 3
Progress Through Architecture. Two Modernisms?

Abstract In recent decades, contemporary modernisation practices, as well as modernist architecture itself, have been the subjects of a severe critical narrative, mainly due to their radical ambitions of implementing an imagined social order and attempts at introducing social change. Their utopian visions have been exposed. In my analysis of the Warsaw Housing Cooperative, I ask whether modernity is, after all, worth defending. Modernist architects believed that they had the tools for implementing great social change by shaping the urban ways of living. When analysing the International Congresses of Modernist Architecture, and especially the first Congress in La Sarraz, I show how the architects' ambitions and goals differed. The discrepancies between them make us seek answers on whether architecture is a goal in itself or a tool for achieving social goals. The answer to this question leads to another one: is the idea of social housing estates, developed during the period of architects' great faith in the agency of architecture and urban planning, worthy of a contemporary redefinition and analysing in the context of today's developing cities? Here, I reconstruct the idea of a social housing estate proposed by Barbara Brukalska, the modernist architect forgotten even in Poland. Only a few copies of her 1948 brochure, *Zasady społeczne projektowania osiedli mieszkaniowych* [*Social rules of housing estates design*] have survived (Poland's communist authorities demanded its entire print run to be destroyed) and it has never been reissued. Still, it can be an extremely inspiring source for contemporary urban studies researchers, animators of public life, or urban activists.

Keywords International Congress of Modernist Architecture · Social housing estate · Neighbourhood unit · Residential housing · Modernist architecture · Urban planning

Modernity has marked its social ambitions also in the area of architecture. Since the end of the nineteenth century, Europe witnessed the emergence of projects based on the idea of urban planning as a social activity (today we would rather call it social intervention), shaping not only the space, but also the city dweller, not only the individual, but also the entire complex system of social relations, neighbourly ties, housing habits, urban strategies of coexistence and coshaping reality. Not only Ebenezer Howard, the founder of the English garden-city movement, or the Scottish

M. Matysek-Imielińska, *Warsaw Housing Cooperative*, The Urban Book Series,
https://doi.org/10.1007/978-3-030-23077-7_3

sociologist Patric Geddes, who saw the city as a social institution, had known that the city is the people, and not the space itself; Le Corbusier, Wolter Gropius, Siegfried Giedion, Pieter Oud, Mies van der Rohe and Ernst May were also aware of that. At the beginning of the twentieth century, and especially in the interwar period, modern urban planners had already adopted their vision of social coexistence and interpersonal relations which could be tried in practice thanks to their urban-architectural and social projects. It can be said that Europe in the 1920 and 1930s was committed to the idea of social housing and established a new role for the architect. The Germans, Dutch, Belgians, Swiss and English were pioneers in the construction of workers' estates. They had great expectations of the industrialisation of the housing sector. It was precisely the industrialisation of the housing sector that would solve all of the social, economic, technical and also artistic problems (Conrads 1970, p. 81). Most often, however, these estates were designed as ready-made 'products', places to settle in, modern, innovative, comfortable, but not yet fully tamed. Such houses built by municipalities or affordable housing offices (e.g. in Italy by the 'popular dwellings' institution) were technically, administratively and financially managed by public officials. They also belonged to and remained the property of the municipality or some other administrative body. Company towns were also being built by large industrial companies, inhabited by employees as long as they worked at the factory. Housing estates seemed somewhat resemblant of employer's patronage. However, flats bound employees with the factory, making them dependent on the employer. When a worker lost his job, he would also lose his flat.

If the above-mentioned modernist architects were indeed so socially, reformistically and emancipatively involved, then it is worth asking what events had to transpire for modernism to ultimately become considered synonymous with a technocratic and liberal-capitalist order preserving the existing class system. Modernist housing estates have been identified with social exclusion and degradation, and Le Corbusier, who is widely recognised as an undisputed champion and visionary of modernism, ended up designing palaces for the world's elite. The above question does not mean, however, that residential buildings were not being built. After all, the architect's profession also involves designing 'machines for living' for the rich, and modernist architecture is meant for everyone, regardless of social status. However, already in the 1970s critics of modernism (initially mainly Jane Jacobs, then Charles Jencks) claimed that modernist trends in architecture turned out to be the main cause of the movement's failure. Charles Jencks triumphantly proclaimed that 'Modern Architecture died in St. Louis, Missouri on July 15, 1972 at 3:32 p.m. (or thereabouts) when the infamous Pruitt-Igoe scheme, or rather several of its slab blocks, were given the final *coup de grâce* by dynamite. Previously it had been vandalised, mutilated and defaced by its black inhabitants, and although millions of dollars were pumped back, trying to keep it alive (fixing the broken elevators, repairing smashed windows, repainting) it was finally put out of its misery. Boom, boom, boom' (Jencks 1977, p. 9).

Jencks used the spectacular demolition of the Pruitt-Igoe housing estate to make extremely reductionistic assumptions concerning the reasons for its demise. The first being environmental determinism based on the belief that space and form govern human behaviour, that, for example, badly designed spaces breed crime. The second assumption is based on the modernist myth of the architect who designs interpersonal relations and can thus solve burning social problems.

Critics of modernist architecture, such as Newman (1972), Rowe and Koetter (1978), Jacobs (1992) announced the twilight of architecture feeding on the naïve utopia of the twentieth century. Whereas Jencks' notion of modernist architecture suffering crushing defeat, meeting its sorry end in St. Louis, in the horribly designed and flawed beyond any doubt Pruitt-Igoe project, can be regarded as the final, rhetorical argument and, at the same time, a gross simplification on which this myth of defeat is based. However, such a categorical argument is deceptive. Some researchers very often ignore the convergence of economic, political, social and cultural determinants in the formation and functioning of housing estates (personified by Robert Moses who is discredited today). Whereas others, in their attempt to eliminate such simplifications, look for new interpretative paths, go beyond the history and theory of modernist architecture and engage in critical ethnographic, sociological and urban studies reflections looking at the broader perspective. A reverse narrative about the defeat of modernism was proposed by Bristol (1991), according to whom holding the designers, i.e. architects, accountable for the failure of the Pruitt-Igoe housing project would mean disregarding the institutional and structural sources of the problem. Those included mainly ghettoising the poor and unemployed African Americans within the complex. An extraordinary film by Chad Friedrichs, *The Pruitt-Igoe Myth* (2011) was released twenty years after the publication of Katherine Bristol's article; it has clearly shown that the Pruitt-Igoe complex was ultimately a tool of racial segregation, an instrument of the helpless social welfare system which forced black men to get out of sight of public authorities. The authorities predicted that the housing estate would be a social tool designed to alleviate the pain of social exclusion, poverty and injustice. Social policy had to first prepare a profile of the typical resident of the Pruitt-Igoe housing project: a black single mother, which left no room for men. Friedrichs' film presents the memories of Pruitt-Igoe's inhabitants who, on the one hand, have confirmed that at a certain point the housing estate 'stopped functioning properly', losing the battle against crime and moral demise. On the other hand, however, they have indicated the reasons for this decay. Undoubtedly, architecture was not to blame.

What is most surprising today are not the arguments put forward by Bristol or Friedrichs, but rather the fact that the critique of modernism and the resulting death warrant on the movement have been accepted so easily and with such certainty. We are talking, of course, about the American context here. The above bears mentioning, as the migration of modernist architecture to America has largely deprived it of its ideological dimension, revealing mainly its technocratic character. It should be noted, however, that in North America, urban planning and architecture were understood as one of the innovations of modernity, anticipating the rational vision of the social world, and aimed at strengthening social ties.

For Clarence Arthur Perry, the basic unit of urban planning was the notion of the 'neighbourhood-unit'. Perry had been developing his concept from 1923 till the formulation of the method of the theoretically elaborated 'neighbourhood unit formula' in 1939.[1] The neighbourhood unit's main function, apart from providing satisfactory housing conditions, was to recreate local communities, typical for small towns and villages, in an urban environment. Perry's theory, therefore, was not only about spatial concepts, but also about social ones.

The first harbingers of the depleting social potential of modernist architecture appeared in Europe even before the transcultural transfer of modernism to North America. In fact, the divergence of conceptual architectural approaches took place in the late 1930s, but the symbolic split had occurred basically at the very beginning, going back to the first CIAM conference in La Sarraz, where modern and engaged architects had met.

Now, let me take a look at the history of the two modernisms (modernism as a style and modernism as a trend in social housing) from a peripheral, local perspective. A Polish designer, who followed new developments in architecture and whose husband took part in the CIAM conference, after many years recreated the history of the 'split' in modernism which today is seen as a reason for the demise of modernism.

At the International Congress organised by avant-garde architects (it was the first CIAM conference—*Congrès Internationaux d'Architecture Moderne*) in La Sarraz in 1928, the term 'The Minimum Dwelling'—'*habitation minimum*'—was coined, and architects and urban planners, for the first time ever, turned their attention to social, educational and cultural problems. They declared that they would teach people how to live. At that time, the International Residential Reform Union was established in Frankfurt am Main, aiming to radically improve housing conditions in European cities. Europe was bustling with new social and architectural ideas, and the concept of cooperativism was born. The conditions to carry out modernist projects were favourable.

The CIAM conference in La Sarraz had set two paths for the coming decades: social housing and modernist architecture. Helena Syrkus, while investigating the social causes of the CIAM programme in 1970, outlined the atmosphere that was brewing in Europe in the early 1920s. She wrote about the fundamental differences between the views of architects centred around Ernst May on the one hand, and Le Corbusier on the other, and these differences were already apparent in the initial period of CIAM's activity.

The programme statements on social housing were published in 1924–1928 by the radical 'ABC. Beiträge zum Bauen'[2] magazine from Basel. Swiss architects protested against the production of goods that do not have a social meaning and whose costs must be borne by society. 'We will give up—they wrote in 1926—on idealism based on false reckoning, on the artistic taste of bygone treasures, on practicing art at the

[1] It was not until 1939 that C.A. Perry finally clarified and defined the concept of the neighbourhood unit and published *Housing for the Machine Age* (Perry 1939).

[2] The author reports the activities and programme manifestos of these young architects-editors in great length (Hannes Meyer, Hans Schmidt, Mart Stam who took part in the CIAM conference in La Sarraz).

expense of social wealth. We have been thinking in terms of historicism, idealism and aestheticism for far too long. Today, we need a clean and transparent slate: materialistic thinking'.[3]

However, the most expressive forum for socially engaged architects, taking into account the economic and technical realities, was the 'Das Neue Frankfurt' monthly, edited by Ernst May, who at the time headed the largest European 'urban planning laboratory'[4] and mass industrialised construction, i.e. the development of Frankfurt am Main. In the foundational article entitled *Das soziale Moment in der neuen Baukunst* (May 1928), May outlines the role of an architect whose social sensitivity does not allow him to be indifferent to people in need. Thus, the proposed programme of new construction was presented broadly in the context of the vision of European cities, May opposed the 'imperialism of the cities', 'delusions of grandeur' reflected by great representative avenues and unparalleled dimensions' (May 1928, cited in Syrkus 1976, p. 42). The grandeur of a city is to be evidenced by the quality of life that it can guarantee to all of its residents, not only those belonging to the elite, to the upper classes. May advocated for an egalitarian city that takes into account the needs and dignity of the poorest.

The manifesto of May's brigade sounds extremely modern, especially when compared to the writings of contemporary representatives of the younger generation of critical urban studies (including Andy Merrifield) on parasitic and planetary urbanisation. The interwar urban condition was diagnosed by young modernists as a 'human migration which drove the inhabitants of rural plains to large cities where they succumbed to degeneration. In the name of profits, land prices skyrocketed, and the push for profits gave rise to cramped tenements. Profits banished light and sun from the big city deserts'. Architects protested against this type of economy, as they believed that it simply would spell the end of big cities. However, at the same time they declared: 'we are not going to call for the end of the big cities as the utopians are doing'. Instead of asking whether to build metropolises, they asked how one should go about building them. Thus, they imagined a city that the inhabitants would want to live in: over there, on cheaper plots, 'we will create vast housing estates with low-rise buildings, with gardens in the immediate vicinity of flats. Not because of cost-efficiency, not for the purpose of eliminating the efficiency of food production through the cultivation of home gardens, but first and foremost for reasons dictated by social economy which requires creating the best possible conditions, allowing residents to rest the body and soul after long days of nerve-racking work. There, in a natural environment, our children will grow up and become healthy and joyful citizens' (Syrkus 1976, p. 43). As it would turn out later, the authors of the Żoliborz housing estate had been implementing the declarations of 'Das Neue Frankfurt' architects. It is worth adding that May also had an extremely experimental experience in the new urban cost-efficiency and construction.

Helena Syrkus brings up the left-wing nature of these Swiss-German social and architectural manifestos in order to show the most influential and visionary ideas of

[3]Syrkus does not provide a detailed bibliographic record. As cited in Syrkus (1976, p. 42).

[4]Term coined by H. Syrkus.

Le Corbusier's modernist architecture (cocreated at that time by Pierre Jeanneret). Although they met with universal applause and were the highlight of international exhibitions (such as the Pavillon de l'Esprit Nouveau displayed at the Paris EXPO exhibition in 1925, or a housing estate presented at the exhibition in Stuttgart in 1927), their sheer enormity and originality led to their becoming luxury houses for wealthy clients. We must not forget that Pavillon de l'Esprit Nouveau was demolished after the exhibition, although Le Corbusier wanted it to be inhabited by one of the many families who visited the exhibition. The Pessac housing estate near Bordeaux, on the other hand, where workers' families were to live, 'an act of sabotage, unprecedented in the history of construction, took place' (Syrkus 1976, p. 49).[5] Le Corbusier was a perfect visionary of forms, but he lacked the social imagination and sensitivity typical for architects associated with the 'ABC' and 'Das Neue Frankfurt' magazines.

This lengthy introduction was necessary to better understand the context of the two architectural concepts formulated during the CIAM conference in La Sarraz: new social housing and modernist architecture. And although it is obvious that the goal was to establish and promote the idea of architecture that would satisfy the most common needs of the broad masses, that the main problem was housing (*habitat*), it should be said that in the first case—architecture was an instrument of social change, and in the second—architecture itself was the goal. Helena Syrkus demonstrates this by analysing the differences between the names of the organisation founded in La Sarraz. In German, it was: Internationale Kongresse für Neues Bauen (International Congresses of New Construction), in French *Congrès Internationaux d'Architecture Moderne* (International Congresses of Modern Architecture). The German name, therefore, indicates the goal, whereas the French name refers to the form. A similar discrepancy accompanied the events that followed. The second CIAM conference focused on the topic of 'Die Wohnung für das Existenzminimur' (according to social radicals from Switzerland, Germany and the Netherlands), and around 'Habitation minimum' [*The Minimum Dwelling*], according to the Frenchmen. Those two names reflect the two architectural concepts and their social and political role. Those differences made CIAM 'avoid, at least in the initial phase, the risk of becoming a passive team subjugated to a single leader' (Syrkus 1976, p. 50).

Did Le Corbusier have, in the CIAM's initial period, such a large impact and ambitions as to shape the character of architecture? 'In a conversation with architecture students (in 1943) (…) [Le Corbusier—author's note] mentions that he came to La Sarraz with a print-ready draft resolution, but he met with an unexpected reaction mainly from the young Swiss and Dutch who meticulously analysed and criticised his text' (Syrkus 1976, p. 59). Le Corbusier undoubtedly referred to Meyer, Schmidt, Stam and Steiger; but it was Ernst May who proved to be his main opponent (probably due to his well-established position within the group). Helena Syrkus sees both architects as continuators of nineteenth-century utopian socialism. 'Le Corbusier was a utopian raised on Fourier's philosophy—his concept of 'Unité d'habitation de grandeur conforme' stemmed directly from Fourier's concept of phalanstère.

[5]The housing estate degraded both physically and morally because the workers were unable to live in the houses designed by Le Corbusier.

Ernst May—a disciple and follower of Raymond Unwin—transferred many of Robert Owen's ideas to the satellite systems he designed for Wroclaw and Frankfurt am Main' (Syrkus 1976, p. 67). Helena Syrkus' claim is of course debatable. However, we can without any doubt agree that 'the German housing estate concepts show a clear preference for radical social motivation' (Heyman 1976, p. 214). Ernst May willingly cooperated with specialists in other fields (hygienists, psychologists, sociologists) and suggested conducting consultations with residents and even advised designers to engage in participant observation.

However, when May, together with Meyer, Stam and Schmidt, decided to design new socialist cities in the Soviet Union, the balance of power was increasingly difficult to maintain because they were not taking part in European discussions. Hence, CIAM became dominated by visionaries of modernist architecture. The Athens Charter, as a result of the fourth CIAM conference, announced the beginnings of a technocratic and functional vision of architecture. The social programme, although present of course, was less visible, and over time became less important for architects. The last CIAM conference took place in 1956 in Dubrovnik and although there were attempts to continue the activities of the organisation (by establishing CIAM II), these efforts ended with failure. In the 1960s, the greatest dreams of the supporters of the Athens Charter came true with the construction of the largest housing estates ever. Starting in the 1970s, however, they became subject to fierce criticism in the architectural discourse which also predicted the downfall of architectural avant-garde.

The discrepancy described here seems to be very important for the defence of modernism. It was not architecture that failed. The sources of the failure, collapse or crisis of modernism may be found in the lack of social programmes and political imagination, in the architects distancing themselves from the residents and in the lack of housing education. The fall of the Pruitt-Igoe project (in St. Louis, USA) was inevitable, because the estate was designed as a tool for racial and class segregation. While designing the Bijlmermeer estate (in the Netherlands), on the other hand, the living habits were not taken into account. Although the Bijlmermeer estate was designed according to Le Corbusier's idea of an 'ocean liner sailing through a sea of green', what counted more than light, greenery and open space was providing car access to the high-rise buildings. Inclined ramps directed car traffic to multi-storey parking garages, with another set of ramps directing the traffic outside. The 'sea of green' stretched beneath the umbrella of streets and parking lots. Thus, the building's ground floor was a dead zone: dark and unfrequented corridors have not even been conceived as a commercial or office part intended for the so-called social devices. The housing estate was not well connected to the city centre at that time. It seems that the estate's designers did not think about the social relations or neighbourly ties, as the Bijlmermeer estate consisted of 31 high-rise buildings with a total of 13,000 flats.[6] Perry's neighbourhood unit had been forgotten while designing the Bijlmermeer estate.

---

[6]Perhaps the attempts to rescue the modernist public housing estates were the reason for the 1980s renaissance of interest in the manifestos of radical architects of the 'ABC' magazine. Helena Syrkus wrote that 40 years after 'ABC' folded, assistant lecturers at the Building Department of the Technical University in Eindhoven in the Netherlands reprinted all issues of the magazine. I think that

Le Corbusier's project built in 1965–1967 in Firminy, as the results of research conducted by Nöel Jouanne show, is an example of optimal solutions ensuring the social functioning of this type of buildings. The mayor of Firminy, on the other hand, is aware of the fact that 'if we wanted to fill Le Corbusier's building with immigrants, it would not be a problem. But it would become a ghetto' (Jouenne 2005, p. 120). Firminy Vert has thus become a socially diversified housing estate in which not only low-income families but also teachers, architects, young people and the middle-class found their homes (Jouenne 2005, p. 120). The flats were designed in a flexible manner to enable residents to adapt the space to their changing needs over time.

Thus, the criticism of modernism in architecture and the narrative of its downfall proposed by Jane Jacob and Charles Jencks do not take sufficiently into account the many complex factors that result in failures or successes of modernist housing estates. It is therefore important to reverse this narrative and re-examine—this time from the perspective of engaged social sciences and contemporary critical urban studies—more complex issues, not limited only to architecture and urban planning, but also take into account social, political, economic, and, finally, cultural aspects. It is worth telling the story of the Warsaw Housing Cooperative using this reverse narrative.

The designers and architects of the Warsaw Housing Cooperative strongly advocated for building housing for 'people in need'. Architects joined the battle against the 'housing deficit' and created a Polish concept of social housing estates with a detailed cooperative, educational and ideological programme. Łukasz Heyman argues that in terms of its programme and implementation of the socio-spatial structure, 'the WHC team is akin to the concepts of the 1920s German residential reform movement which materialised to a certain degree in the architectural activity of functionalism of Bauhaus and Ernst May's brigade, who was the architect of the New Frankfurt and city planner of numerous Soviet cities' (Heyman 1976, p. 214).

3.1 Beginning. The Establishment of the Warsaw Housing Cooperative

The Warsaw Housing Cooperative was founded on 11 December 1921 in the premises of the Folk High School located at 6, Oboźna Street and lent gratuitously for a short period of time. Before that, however, three workers' activists feverishly worked on the cooperative statute in a rented room located at the address Nowy Zjazd 6/7.

Filip Springer, a contemporary reporter who analyses source materials with diligence, builds an interesting narrative about the establishment and origins of the Warsaw Housing Cooperative. '*The room is very modest. Floor-to-ceiling pinewood*

bringing back the social housing programme could indeed be a result of the crisis of faith in modernist architecture which, at that time, had suffered multiple social failures. In the Netherlands, a possible urban renewal project of the failed modernist Bijlmermeer social housing estate was being discussed at the time.

bookshelves, two iron beds covered with thin jute blankets, and between them—a paper-covered work table made of wooden table legs and a tabletop made of planed boards. Three men sit at the table. They rub their hands and stamp their feet. It is the end of November 1921; through the draughty windows of the apartment at 6, Nowy Zjazd Street, blows the wind, heralding the onset of winter. A tall man dressed in an exotic, short, sleeveless sheepskin coat, who drinks a hot herbal infusion out of a strange bottle with a steel tube, assumes the role of a host. His name is Jan Hempel. He has recently moved to Warsaw, and has lived in this cold room for less than a month. He spent some time abroad, in Japan, working as a teacher in a Polish school and as a surveyor in Brazil. His correspondence was published in 'Kurier Lubelski', among others. He went through a long ideological evolution—from a pagan mystic to a militant vegetarian and then an anarchist with a disastrous tendency towards radical communism. (...) The impression that Hempel is walking in the clouds is enhanced by the way he dresses—eccentric tailcoats and vests, too big a black bow-tie (...). The second tenant occupying the cold room and co-host of the meeting is a young, almost thirty-year-old communist staring at Hempel blankly, a certain Bolesław Bierut. They met in Lublin where they had founded a consumer cooperative before the war. They also worked together in the 'Książka' Publishers Cooperative. They become inseparable. Bierut embraces most of Hempel's opinions. He is called the 'cooperative model' by the leftist Warsaw-based intelligentsia. An elegantly dressed man, polite, somewhat shy, maybe even a bit reserved, who often blushes and behaves sometimes exaggeratedly gallantly towards women. When he gives a speech, he gains countenance and energy. He can be smart and witty at that time. The third man is a guest here, but his presence will soon turn out to be the most important. The other two will soon completely disappear from the horizon of his actions. It is Stanisław Tołwiński. He is younger than Bierut, barely twenty-six years old, but has a lot of experience. Right after finishing his studies in Saint Petersburg, he became involved in the secret activities of Polish socialists. He fought in the October Revolution, wrote articles for the underground press under the alias Szymon Judym. He came to Poland in 1918. (...) Currently, he works at the Central Statistical Office, and after working hours, he is actively involved in an independent cooperative society—that is where he met Hempel and Bierut; together they founded the Union of Labour Cooperative Associations'[7] (Springer 2015, pp. 46–48).

The work on the Union's statute took a very long time. The authors of the statute studied cooperative models developed in England, France and Italy (where the associations were run by one-man management). Eventually, they adopted the German model, according to which the cooperative was managed by two bodies: the Management Board (Vorstand) and the Supervisory Board (Aufsichtsrat). They probably also used the just-published work of Władysław Dobrzyński, PhD, titled

[7] The Polish-Soviet War erupted in 1919. Poland, defending itself against the Soviet invasion, on the one hand fought against communism, and on the other hand, tried to maintain its newly regained independence. And the victorious fight in 1920 was hailed as the 'Miracle at the Vistula'. The Bolshevik coup that threatened Poland caused mistrust towards the communist world. Under the pretext of fighting against Bolshevism, workers' cooperatives and associations were liquidated, the communist party was banned and the most active members were put under surveillance or arrest.

Kooperatywy mieszkaniowe [*Housing cooperatives*] (Dobrzyński 1921), in which the author analysed in detail the draft statute of a housing cooperative prepared by the Organising Commission of the Cooperative Housing Association of Government Officials. They also imagined that solely people living of their own labour could become members of the cooperative established by them. Although the statute did not specify the members' property rights to the flat they received, Jan A. Szymański believes, when reconstructing the history of the Warsaw Housing Cooperative, that the fact that property rights were never mentioned shows that flats were to remain socially owned property (quite like the English tenant societies), and the cooperative member—the user of the flat for as long as they pay contributions and rental fees. This model rapidly became common in Germany, and the Berliner Spar und Bauverein cooperative served as a model for Polish activists.

After a few days, the final document was sent to several dozen cooperative activists throughout the city. Szymański writes that all of them were members of the elitist progressive intelligentsia, and most of them were members of the Polish Socialist Party. However, I do not agree that one can describe them this way. They were, after all, very young people who, due to their age, could hardly be immediately assigned to the elite. And Szymański's strategy can be interpreted as a historical projection. After all, the icon of the Polish cooperativism had to have, according to Szymański, a noble genealogy. Undoubtedly, however, the cooperative members were linked by their progressive faith in a better future. Springer writes: *'many of the signatories are not more than thirty years old. They work mainly as petty officials, accountants and journalists. Their heart is definitely on their left side. After hours, they work for the Polish Socialist Party (PPS), organise trade unions or local publishing and consumer cooperatives. They are here, there, and everywhere, seething with energy. They want to change the reality which affects them deeply. They live in rented rooms or sublet space in someone else's flat. Every day they are faced with ever-increasing costs, inflation and fear of losing their jobs'* (Springer 2015, p. 49).

Those who signed the document on 11 December 1921, at the seat of the Folk High School, became the first members of the Warsaw Housing Cooperative. On that very day, a nine-member supervisory board was also created. The management board was composed of: Bolesław Bierut and Stanisław Szwalbe. The Warsaw Housing Cooperative was entered in the court register on 1 February 1922. The Polish mark exchange rate was completely unpredictable during this period. The inflation was rampant, and the capital accumulated by the founding members was rapidly losing its value. Therefore, WHC members decided to pay up their shares not in cash, but in bricks which they bought in a brickworks near Warsaw and stored in the company's stockyards in which Bierut worked (Tołwiński 1970, p. 250).

It was time to find a suitable construction site. Over 90% of Warsaw lands belonged to private owners. The purchase of a plot of land was out of the question, as it would raise the cost of the future rent, while it had to be affordable to workers. The only land available to housebuilding was in the northern part of the city, the area of the Warsaw Citadel, where Russian soldiers were stationed (Figs. 3.1, 3.2, 3.3, and 3.4). The area remained undeveloped at the time. Tołwiński and Szwalbe, therefore, went to the urban councillor, Teodor Toeplitz, who was quite influential within the city

Fig. 3.1 Map of great Warsaw from 1932, 8500024, p. 45 (Wacznadze, D. Red. Weiss, R. Rys. Zakład Graficzny W. Cukrzyński (Warsaw) (1932). Warsaw 'Samopomoc Inwalidzka' [*Self-Help for Handicapped*], 1932 (Warsaw: (Zakł. Graf. W. Cukrzyński). Public Domain. Scale 1:30 000. Brochure: In print. Z. Frączkowski, Warsaw public access by: National Library)

authorities. They also knew Toeplitz' views; he had published in 1920 a report entitled *Klęska mieszkaniowa i próby jej usunięcia* [*The housing disaster and attempts to fix it*]. Toeplitz understood the housing problems perfectly well and identified himself strongly with the cooperative movement. He pursued consistent activities aimed at increasing urban lands and starting construction of housing for working people covered by public funds.

On 4 June 1923, as a result of enormous efforts and unprecedented commitment on the part of Toeplitz and the entire management board, the Warsaw Housing Cooperative received from the Regional Directorate of Public Works [*Okręgowa Dyrekcja Robót Publicznych*] the land for the construction of multi-family houses in Żoliborz.[8] Toeplitz advised also young cooperative members to use the 'Robotniczy Przegląd Gospodarczy' [*Workers' Economic Review*] biweekly to promote the plan for the construction of multi-family houses as part of the WHC project among the activists

[8]Żoliborz, similarly to Rakowiec, is one of the districts of Warsaw, the capital of Poland, which at the beginning of the 1920s was still mostly empty, and the conditions for obtaining plots for the development of socially most necessary houses from the city authorities were favourable.

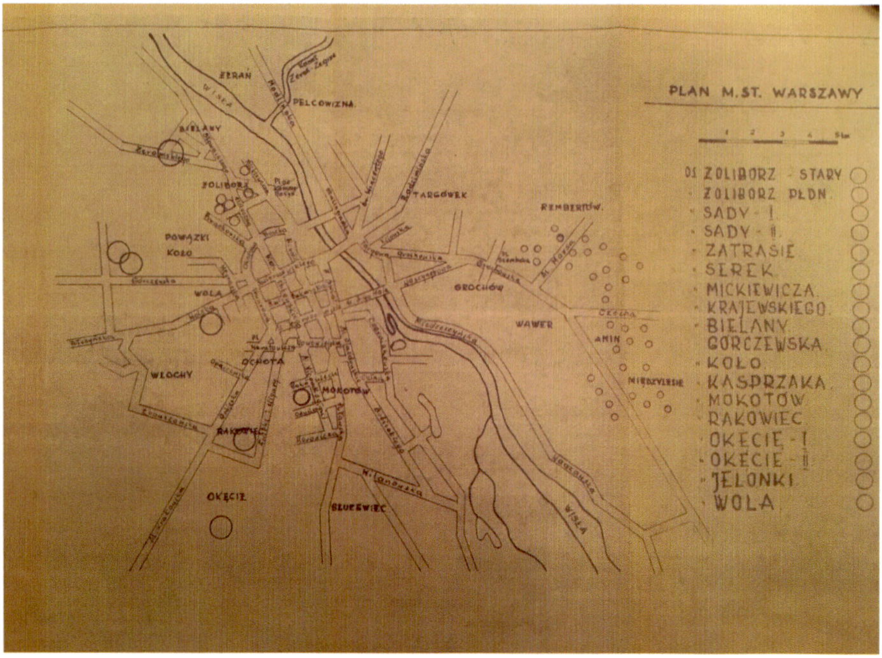

Fig. 3.2 Map of the Capital City of Warsaw with marked districts—WHC History Chamber, issuance date and author unknown. 3891, p. 45

of trade unions and workers' movements. This is where they met Antoni Zdanowski and Marian Nowicki.

However, the cooperative did not have the money for the construction of houses. The cooperative members submitted to the President of Warsaw, Władysław Jabłoński, a memorandum concerning the subsidies and municipal loans for cooperatives, thus involving the representatives of the Trade Union Council of the capital city of Warsaw into the WHC project. The Construction Committee awarded a small loan to the Warsaw Housing Cooperative—a sum sufficient to cover the costs of construction works for a few months. It was well-known back then that the cooperative founded by socialist activists was not to the liking of Warsaw's right-wing, conservative authorities.

At the beginning of 1924, a currency reform was adopted which introduced changes to the WHC statutes (the registration fee had to be paid in the Polish zloty, and not the Polish mark). The economic situation of Poland was very bad at the moment and the reform itself did not improve it. *'For that, however, the Polish Republic needs cash other than the worthless Polish mark. And here comes to the fore Józef Toeplitz, brother of Teodor, CEO of one of the biggest private banks—Banca Commerciale Italiana. The bank is ready to support the currency reform by granting a loan secured by the Polish Tobacco Monopoly. Negotiations with the Polish government last several months. Not all contractual terms are*

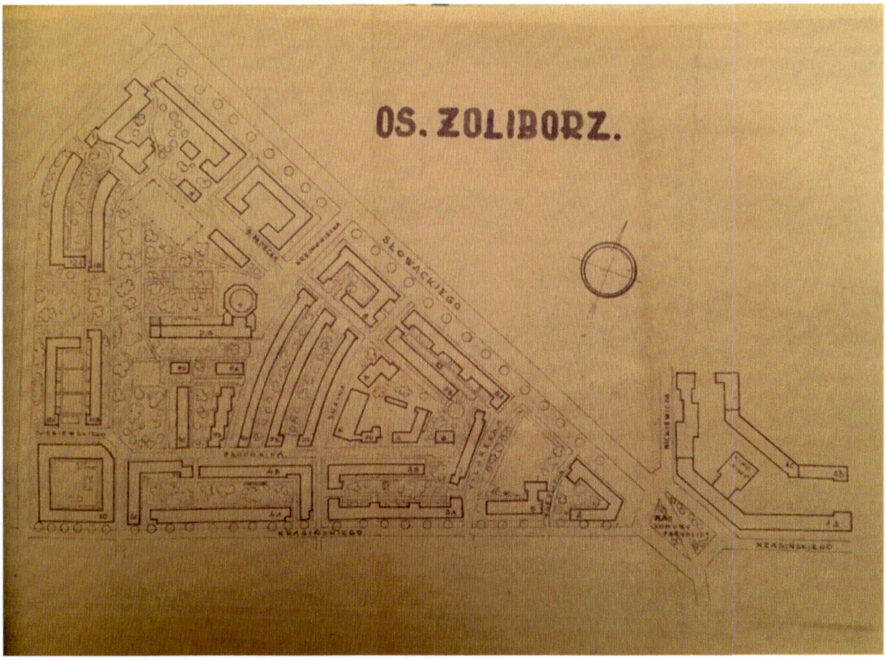

Fig. 3.3 Map of Żoliborz—WHC History Chamber, issuance date and author unknown. 3897, p. 45

included in the document signed by both parties. In addition to the above, the Italian bank, and in fact its Polish president, demand the release of Leon Toeplitz,[9] *the allocation of five hundred thousand dollars from the "tobacco loan" to the Warsaw Housing Cooperative by the newly created Bank Gospodarstwa Krajowego, whose aim is to provide money for housing purposes, and the guarantee that the Polish Tobacco Monopoly, which was seized as security for the carrying out of the entire contract, would buy flats for its employees from the WHC'* (Springer 2015, p. 54).[10]

Although it may seem that the Italian loan would solve the cooperative's problems, it is worth noting that for the WHC the loan represented a significant compromise. The interest rate on the loan was quite high, which would translate later on into a high cost of future rent for the cooperative members. The signing of the contract with the Polish Tobacco Monopoly was quite a big concession. It meant for the Warsaw Housing Cooperative authorities building company towns, whose idea conflicted with the public housing ideological assumptions. Company towns made the employees

[9]Leon Toeplitz was the son of Teodor Toeplitz. He was arrested for his activity within the Young Communist League of Poland. Unofficially, however, it was rumoured that his arrest was aimed at removing Teodor Toeplitz from the Polish political scene. The City council demanded his dismissal from public functions to no avail.

[10]There are some sources available in Poland which describe this complicated financial operation. However, I am quoting this excerpt after Filip Springer because of its general description and fairly faithful presentation of its financial consequences.

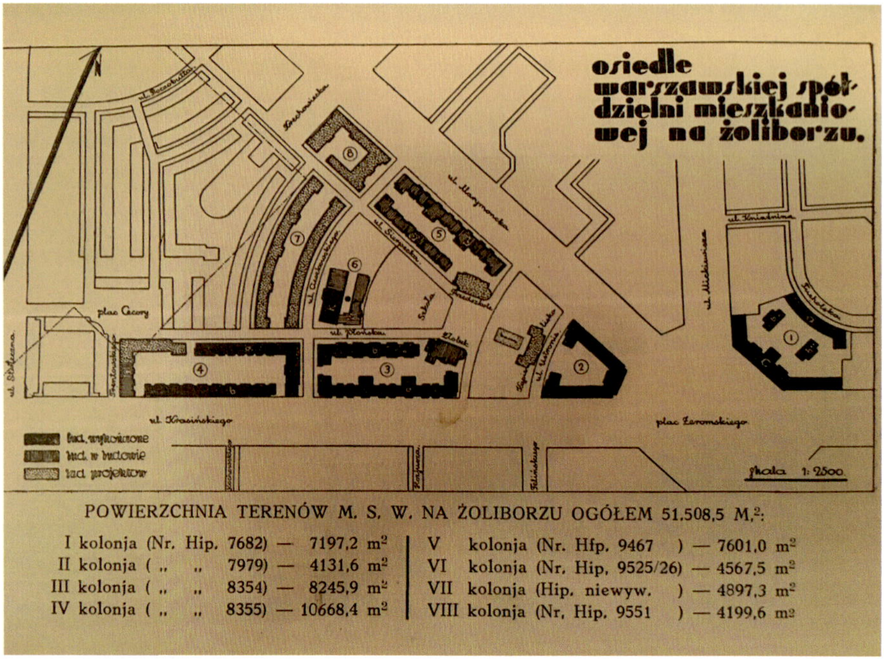

Fig. 3.4 Plan of the Warsaw Housing Cooperative estate in Żoliborz. 3459, pp. 45, 89

dependent on the employers: when a worker loses his job, he would automatically lose his flat. And this was difficult to accept for the idealists sensitive to the workers' situation.

Finally, on 12 December 1925, the foundation stone was laid for the construction of the first cooperative house of the Warsaw Housing Cooperative in Żoliborz. The slogan 'The liberation of the working class can solely be done by the workers themselves' was included in the foundation act. This Marxist motto had already had a long tradition in the history of the cooperative movement in Poland. In June 1917, during the Warsaw Congress of the Union of Polish Consumer Associations, Romuald Mielczarski gave a lecture during which he argued that 'only in cooperation shall the motto "The liberation of the working class can solely be done by the workers themselves" be materialised. For cooperation does not only transform social relations, but at the same time transmutes souls and educates people who are able to direct this transformation. In cooperation, the previously unorganised crowd, becoming acquainted with the economic needs, learns to practice solidarity, begins to understand, duly assess, and what is more—to love the common good. The cooperative is also a school in which administrative, technical and organisational talents are educated and developed' (Mielczarski 1917, p. 8).

The Warsaw Housing Cooperative started its proper construction activity in 1926, and on 9 January 1927, the first 28 apartments were put into use. The ceremony was attended by the representatives of the government, city authorities, Bank Gospodarstwa Krajowego, trade unions and the Institute of Social Economy.

In conclusion, it can be said that in Poland, the still-developing working class needed housing, though the struggle against the so-called housing deficit was not a priority for the Sanacja government. The cooperative movements were developing dynamically in the capitalist system, gaining ground in Poland after World War I. However, the cooperative movements were significantly marginalised politically, mainly due to being anti-capitalist and class struggle in spirit. Looking at this phenomenon from a modern and critical perspective, one can say that the WSH initiative was a cultural and economic practice of urban subversion within the then capitalist system, with interesting strategies and tactics of resistance, and even an attempt to fight for the right to the city.

3.2 New-Build Construction Projects or Modernist Architecture in Żoliborz

In 1925, Szymon Syrkus wrote *The Architectural Foundations of Residential Housing* (Syrkus 1925). This article proved that the young architect knew exactly how and what to build. However, Syrkus had a vague idea for whom to build and who was the addressee of modern buildings. 'He is a left-wing architect—wrote his wife—looking for allies and recipients of his work among bourgeois democrats and its elite—professional intelligentsia. He became a sternly left-wing architect and social activist only after 1927, when Teodor Toeplitz responded to the appeal, published for the second time in 'Praesens', for the industrialisation of residential housing.'[11] From then on, Syrkus was associated with the Warsaw Housing Cooperative, which was established in 1921.

It was in Toeplitz' luxurious house in Otrębusy that in January 1929 those who eventually turned out to be the most creative and innovative in implementing the social experiment in Żoliborz and Rakowiec had a chance to meet: the milieus of public authorities and WHC architects (including Bruno Zborowski, Barbara Brukalska and Stanisław Brukalski), avant-garde architects from the Praesens group, Helena Syrkus and Szymon Syrkus, who could accurately and specifically relate the CIAM conference which had taken place in La Sarraz in 1928. At the meeting, it was pointed out that the inability to formulate important housing missions leads to faulty architectural solutions. These, in turn, generate costs to be borne by society. One of the points of CIAM's La Sarraz Declaration assumed that 'elementary housing

[11] Syrkus (1976, p. 19). The article that drew Teodor Toeplitz' attention was entitled Preliminarz architektury [The Architectural Estimate] (Syrkus 1926). Toeplitz was extremely open-minded and had multiple global contacts, as well as unprecedented knowledge in the field of social housing, architecture and urban planning.

principles could be effectively disseminated by introducing them into the curricula of schools and educational institutions. We should focus on explaining the principles of sanitation, the effects of air and sunlight, essential hygiene, the practical use of household appliances and tools' (Syrkus 1976, p. 63). It was all about teaching people to live, to use the city's public space, but also to supplant academicism which 'urges the state to incur great expenses on monumental building undertakings and thus preserves the already bygone luxury' (Syrkus 1976, p. 63). The role of the state is, therefore, to shape the architect's attitude 'in relation to new buildings, so that it corresponds to the overall economic and cultural missions of society' (Syrkus 1976, p. 63), modern society. And all of this sounds like resistance to Haussmanisation.

While Szymon Syrkus discussed the La Sarraz declarations, Teodor Toeplitz presented the programme of activities of the International Residential Reform Union. Thus, thanks to the meeting in Otrębusy, a group of activists was formed: Polish avant-garde architects, engineers and social activists 'and the Warsaw Housing Cooperative became its patron. [Szymon] Syrkus could present a new model of a national group, related to the social institution which builds workers' housing estates, at the [CIAM—author's note] delegates' meeting in Basel. It was considered to be an exemplary model worth replicating by other groups. It was also the first Polish contribution to the joint work of the International Congresses of Modern Architecture' (Syrkus 1976, p. 78).

Let us recall that the WHC began building the socially most necessary houses as early as in 1926, two years before the CIAM conference in La Sarraz. The dwellings were to ensure a hygienic, comfortable, intimate life to the working classes, while preventing the natural tendencies of the inhabitants to separate themselves from others and maintain anonymity. 'Social urbanism' gave its initiators the possibility of rational social coexistence, based on mutual help and solidarity. It was a kind of radical left-wing programme[12] taking its roots in the ideological unity of the cooperatives' initiators who dreamed that such unity would grow among the inhabitants, but it seemed that this wish seldom came to fruition.

The modernism of Polish ideological artists and architects from Żoliborz reflected the ideological, socially radical, left-wing character of social housing. Although the rational and modern architectural solutions perpetuated the myth of a designer striving to strengthen social ties, the technocratic vision of architecture and environmental determinism were avoided. The Żoliborz project was primarily emancipational and educational in character. The architecture was meant to teach

[12]Jan Strzelecki defines this radical leftism as follows: 'I derive the notion of the radical political left from a dictionary about pre-war relations in Poland. It described the type of socio-political orientation of people who recognise the need for a fundamental social reconstruction in the spirit of socialism and co-creating centres of thought and action, proclaiming and justifying the need for such a reconstruction. It does not coincide with any of the political circles of that time, embracing people from both factions of the organised labour movement and those with no political affiliation, expressing their connection with the values leading to the ideas of radical social reconstruction in a different way than through party membership and the types of actions and responsibilities resulting therefrom. Let us not forget that they included people like Krzywicki or Broniewski, which has been cited as yet another proof of how rich the tree of life is' (Strzelecki 1970, p. 220).

people how to live and thus 'produce a new man in a new housing estate'. This was possible mainly due to the idea of housing cooperativism devised by Teodor Toeplitz and Adam Próchnik based on strongly ideological argumentation. 'The ideology of real housing cooperativism—wrote Toeplitz in 1928—may and must be a socialist ideology—all its characteristic features logically stem from it: its democratic management board, the allocation of apartments based not on the privilege of capital, but taking into account housing, family and social determinants of the cooperative's members—and eliminating the possibility of speculation and subletting. And further on—social settlement of suitable life matters, such as: in the field of education and culture—nurseries, preschools, reading rooms, meeting rooms, etc'. (Toeplitz 1928). This means that housing cooperativism greatly differs from company towns, as well as from the housing estates built and administered by municipal or state authorities. It is based on the bottom-up organisation of residents and on satisfying their needs in a self-governing manner.

3.3 The Social Housing Concept

The social housing concept[13] is based on the assumption that its purpose is to foster the creation of neighbourly ties and that many of the needs of residents are satisfied on a group, rather than individual, basis, relying on so-called social amenities: cafeterias, libraries, laundries, nurseries, preschools and, last but not least, schools. Reform movements, which had been started by progressive, left-wing architects, social activists and physicians, led to the construction of new neighbourhoods, such as Sweden's Välingby and England's Radburn and Harlow.[14] During the interwar years, the social estate movement took hold in Germany and Poland, with sociologists and social activists, but also socially aware architects, spreading the idea. Polish architects of the 1920 and 1930s had engaged in creating new types of estates, such as WHC, which met with criticism during the war and the post-war period. There is a wealth of literature on the subject from that time. Szymon and Helena Syrkus wrote about 'experimental urban planning' in 1946. They were aware that the 'social content of a typical democratic urban unit, i.e. estate, is rich and diverse, consisting of a variety of functions with deep and extensive roots planted in the needs of a large group of people. An estate's social composition will, therefore, not be a decoration meant to obscure its internal emptiness, but rather a lively form reflecting a concrete social content. The content will grow as all of the diverse aspects of social ties develop and deepen' (Syrkus and Syrkus 1946). Social estates allow 'social content' to emerge, but also lead to social ties and relationships taking on new forms, e.g. civic and democratic, going beyond just housing needs. All of Poland's social housing

[13]More information on the concepts described herein in detail can be found in Wenderski (2019) and Crowley (1992).

[14]Examples of housing estates built according to social principles and the concept of the neighbourhood unit were described by Czarnecki (1960).

projects and the theoretical discussions surrounding them can be attributed mainly to the efforts of Szymon and Helena Syrkus and Barbara Brukalska. In 1942, Helena Syrkus wrote a brochure entitled *Social Estates as Part of a District, City, Region*. All of the above-mentioned architects wrote numerous articles on the subject of social urban planning and the concept of social estates for the *Dom. Osiedle. Mieszkanie* magazine (Brukalska 1946; Syrkus and Syrkus 1947). In 1948, Brukalska's *Social Rules of Housing Estate Design* was published, considered by Janusz Ziółkowski to be the most complete theoretical works on the concept of social estates (Ziółkowski 1964, p. 187). In her book, Brukalska defined a precise hierarchy of urban units, including a family homestead, a residential building (housing from 100 to 150 people) and a colony (consisting of residential buildings with a total population of less than 800 people). A set of colonies is considered a housing estate and should not house more than 5000 people (pioneering estates may house up to 8000 people). A housing district has between 25 and 40 thousand inhabitants. The hierarchy was designed in such a way as to allow satisfying group needs, which in the case of social estates (bigger than Perry's neighbourhood unit) are related to schooling. Perry claimed that the population of each neighbourhood unit should be served by a single school, whereas Brukalska envisaged an elementary school and a gymnasium per housing estate. Despite the differences that will be discussed later, representatives of the Polish social estate school were very consistent in their conviction that architecture and urban planning shape social life. It was not, however, a naive belief of modernist radicals. Brukalska was aware that 'each building or group of buildings enforces certain lifestyles and forms of coexisting on its inhabitants. Estates that were designed mechanically, that is taking into consideration only geometric composition, had a social life that developed spontaneously, whereas today we strive to create housing estates with forms that would reflect the social content that we consciously inject into them' (Brukalska 1948). Housing cooperatives which, as Próchnik had written several years earlier, were necessary to curb the impact of capitalistic production based on exploitation and rivalry were the desired social content. Therefore, establishing an educational-emancipative and political programme, for which the WHC's creators were responsible, proved to be a prerequisite for the development of social estates.

Many urban sociology publications on the topic were released in the 1960s, placing the ideas of WHC architects in the broader context of global urban planning solutions. Wallis (1964), Ziółkowski (1964, 1965), Jałowiecki (1966) all analysed the interrelations between sociology and spatial planning, treating estates as complex social systems—an approach that was by no means new among sociologists. What is interesting, however, is their view of the achievements of Polish architects in furthering the development of enormous cross-class estates that had been erected under communist rule in the 1960s. Their thoughts prove that a curious school of urban studies was emerging in Poland at that time, which is growing relevant today as it enriches the image of the social impact of architecture and makes it more profound. They have also somewhat negatively contributed to pre-war social estates now being associated with the post-war urban development projects. That, in turn, leads some to equate the emancipative ideas of Barbara Brukalska with socialist-realist doctrine.[15] The history and development of the sociology of housing in Poland have been analysed by Maciej Cesarski, providing the historical contexts of the creation of specifically understood urban studies within its scope (Cesarski 2015). Subsequent sociologists have also benefited from the tradition of this school (Kaltenberg-Kwiatkowska 1982, Karpiński, 1975). It can also be assumed that the entire issue of 'Sociological Studies' from 1965, representative of the Polish school of urban sociology, is devoted entirely to the city in the context of very broad sociological analyses (Wallis 1965; Piotrowski 1965; Pióro 1965), centred on social relations in local communities and power relations (Jałowiecki 1965). Some of them are of course obsolete and anachronistic today. The contemporary analyses of urban public spaces and their social meanings, apart from cultural and humanistic studies, are conducted today in Poland by, among others: Bierwiaczonek (2018), while the question of neighbourly relations in Polish cities is analysed by Kotus (2006, 2009), Kotus and Rzeszewski (2013), Kotus and Hławka (2010).

3.4 Modern Social Content of the Żoliborz Estate

The Cooperative's goals were not only related to 'building and renting low-cost and healthy apartments through collective self-help' and 'joining together to satisfy the cultural needs of its members' (Warszawska Spółdzielnia Mieszkaniowa 1930b, p. 3). Something bigger was at stake—creating a new societal culture, forming an 'enlightened citizen', ensuring full participation in cultural life through an attitude of social engagement and individual moral sensitivity. With such lofty goals, collaboration was paramount. But how do you encourage a group of very different people to collaborate with each other? Activists focused on the day-to-day activities of residents: raising children, teaching, shopping, recreation. If social relations are based on specific experiences and not political forms, then they should be built on the principles of informality and loose exchange. The cooperative, however, introduced an institutionalised order into the residents' social life, relying on regulations and detailed guidelines of coexistence and living. Did that threaten the collaboration between residents? Can we say that the management of the cooperative had the inclination to forcefully introduce order and oppression? These questions beg asking because many of today's interpretations often take on the form of critical analyses seeking to unmask the underlying intentions of the cooperative. They wish to expose the hypocrisy of the pioneers of the modern order, the inherently repressive institutions and the ambivalence of social life that they had created. This modern perspective is sensitive to all kinds of attempts at dominating and controlling society. It allows for examining historical examples of establishing institutional forms of modern organisation of society. Such an approach can be tempting, as proven by the meticulous analyses of Michel Foucault.

[15] Ziółkowski, for example, associates the collective consumption and emancipation of women with the draft project of the CPSU (Communist Party of the Soviet Union) published in 1966 in *Trybuna Ludu*.

Fig. 3.5 Kindergarten of the Workers' Society of Children's Friends (RTPD), surrounded by a garden, Report on the WHC activities for 1938. 3866, p. 53

The Warsaw Housing Cooperative is a perfect example of modernist ideals introduced in practice: (hygiene, fresh air, access to sunlight, parks and green areas) achieved through avant-garde architecture, designing apartments in accordance with Taylor's principles of scientific management, a holistic, if not total, approach to organising social life. There is no doubt that WHC may be considered as a prime example of modernism. Today's analyses, however, criticise the Żoliborz estate for those very modernist solutions.

The author of a certain critical, unmasking analysis shows that the modernist idea of Glass Houses found its reflection in Żoliborz's architecture and that the building housing the 'Szklane Domy' [Glass Houses] Tenants' Mutual Assistance Association was highly glazed (I should also add that the first Żoliborz preschool commonly referred to as the 'Little Glass House', which was designed by Janina Jankowska, had a similar design) (Fig. 3.5). In that author's opinion, the Glass House design raises some doubts. 'This ambivalence has obvious and legitimate causes—the glazing not only offers a wider view to individuals inside of the building, but also makes them more exposed to those looking in from the outside' (Gmochowska 2008, p. 290). She went on to draw parallels with the thought of Michel Foucault, Jeremy Bentham, George Orwell and arrived at the following conclusion: 'the idea of a glass house is a thoroughly intriguing phenomenon, demonstrating how a single cultural theme may be interpreted and assessed in different ways. Bauman's, Orwell's and Foucault's writings have instilled a certain sense of apprehension to all kinds of projects seeking to impose order' (Gmochowska 2008, p. 292).

Although attempts at fulfilling the modernist ideal of Glass Houses were undertaken all across Europe, and despite modernist avant-garde being considered a stark and obsessive modernist order-making imperative, it is the modernity of Żoliborz's architecture that becomes 'exposed'. One important detail should be taken note of however. None of the creators of WHC wanted to glaze apartments and intimate, private, personal areas. The above-mentioned examples of Glass Houses were public buildings, which may be considered the agora of Żoliborz's social life. Where then does the mistrust towards those 'modernist utopias come alive' come from? Maybe the issue is not with modernism but with socialism? The author is right when she writes that Bauman, Foucault and others (many others) taught us to identify ambivalence, expose oppression and dominance, allowing us to analyse social life. I am under the impression, however, that the author's analyses are somewhat tainted with certain prejudice resulting from the historical experiences of Poles, which is totally understandable. The phrases 'Red Żoliborz' and 'socialist utopia' still bring to mind associations with the communist regime, socialist realism, the system's absurd and distorted character, all of which were part and parcel of daily life in Poland from the late 1940s almost up to the year 1990. Such associations lead to interpretations which view Barbara Brukalska's 'socialised individualism' as an invitation to adopt a collectivist lifestyle as part of a community (commune), a life in which being forced to live alongside others is inherent, and not as an attempt at finding balance between the freedom of social relationships and individuality (but not narcissistic individualism typically involving one's withdrawal from society). The rules of collaboration and coresponsibility for cooperatively-managed spaces are perceived as a regime of collective voluntary work, and not as quality teamwork (cooperation) or biopolitical work, as defined by Hardt and Negri. 'The conviction that good work moulds good citizenship became distorted and perverted in the course of modern history, ending in the hollow and depressing lies of the Soviet empire'—writes Sennett (2009, p. 269). Such distortions are often injected into the various narratives surrounding WHC. For instance, when talking about the creators of WHC experts feel the need to strongly emphasise that they were 'non-communist socialists' or 'secular socialists', or simply Żoliborz intelligentsia.

Teamwork is the cornerstone of a city understood as an objective organism, a city that works. When discussing a-androgynous cities, Krzysztof Nawratek underlines the importance of cooperative activities and collective voluntary work (e.g. time banks), which have become necessary for citizens to 'plug into the city' (understood as a common good, a political idea, not as a space of consumption) (Nawratek 2011, p. 149).

Therefore, today's analyses of Żoliborz's social estate concept should not rely solely on a critical approach to modernity (and especially modernism). A change of paradigm would allow us to see how we can build the cities that urban culture studies academics of the 1970s had envisioned. I am of course aware that they each had different visions of what a city should be. I will look to their works for solutions and arguments which will allow me to test and analyse Żoliborz's 'city within a city'. They will also help me to bring down the myth of the failure of modernist architecture, as I plan to analyse the social and political debates that accompanied the Żoliborz project from the very beginning. Therefore, my approach may be described as an attempt to deal with the claims that modernist architecture has failed. I also wish to switch around the narrative concerning the oppressive and order-making ambitions of modernity. In that sense, my work will be a critical defence of modernity.

References

Bierwiaczonek K (2018) 'Miejskie przestrzenie publiczne i ich społeczne znaczenia – próba systematyzacji' [Urban public spaces and their social meaning—an attempt to systematise]. Przegląd Socjologiczny [Sociological review], LXVII (67), 1:25–48
Bristol K (1991) The Pruitt-Igoe myth. J Arch Educ 44(3):163–171
Brukalska B (1946) Urządzenia społeczne w osiedlach [Social devices in housing estates]. Dom, Osiedle, Mieszkanie 8–10:37–38
Brukalska B (1948) Zasady społeczne projektowania osiedli mieszkaniowych [Social rules for the design of housing estates]. Developed at the Institute of Social Economy. Wydawnictwo Ministerstwa Odbudowy vol 22, Warszawa
Cesarski M (2015), Sociology of housing in Poland—genesis, development and future prospects. Wars Forum Econ Sociol (6), 1(11), Spring:141–164
Conrads U (1970) Programmes and manifestoes on 20th-centruty architecture. MIT Press, London
Crowley D (1992) National style and nation-state: design in Poland from the vernacular revival to the national style. Manchester University Press, Manchester, New York
Czarnecki W (1960) Planowanie miast i osiedli [Planning of cities and housing estates], vol 2. Państwowe Wydawnictwo Naukowe, Warszawa-Poznań
Dobrzyński W (1921) Kooperatywy mieszkaniowe: wskazówki praktyczne [Housing cooperatives: practical tips]. Instytut Wydawniczy 'Bibljoteka Polska', Warszawa
Gmochowska K (2008) 'Szklane domy'. Żoliborska WSM. Miejsce uwikłane w historię ['Glass Houses'. Żoliborz WHC. A place entangled in history]. Konteksty: Polska Sztuka Ludowa 62(3–4):286–294
Heyman Ł (1976) Nowy Żoliborz 1918–1939. Architektura – urbanistyka [New Żoliborz 1918–1939. Architecture—urban planning]. Zakład Narodowy im. Ossolińskich, Wrocław
Jacobs J (1992) The death and life of great American cities. Vintage Books Edition, New York
Jałowiecki B (1965) Władza w społeczności lokalnej a procesy urbanizacyjne [Power within the local community and the urbanization processes]. Studia Socjologiczne 4:89–102
Jałowiecki B (1966) Socjologia stosowana jako instrument planowania przestrzennego [Applied sociology as an instrument of spatial planning]. Studia Socjologiczne 23:79–103
Jencks Ch (1977) The new paradigm in architecture: the language of post-modernism. Yale University Press, New Haven
Jouenne N (2005) La vie collective des habitants du Corbusier. L'Harmattan, Paris
Kaltenberg-Kwiatkowska K (ed) (1982) Mieszkanie: analiza socjologiczna [Flat: sociological analysis]. PWE, Warszawa
Karpiński J (1975) Ludzie i zbiorowości (zagadnienia „poziomu analizy" w socjologii) [People and communities (the issue of the 'levels of analysis' in sociology)]. Studia Socjologiczne 3:131–151
Kotus J (2006) Changes in the spatial structure of a large Polish city—the case of Poznan. Cities 23(5):364–381
Kotus J (2009) Fragmenting and isolating neighbourhoods: a way of creating flexible spaces and flexible behaviours? GeoJournal 74(6):551–566
Kotus J, Hławka B (2010) Urban neighbourhood communities organised on-line—social ephemeron or a new form of self-organisation in the Polish city? Cities 27(4):204–214
Kotus J, Rzeszewski M (2013) Between disorder and livability. Case of one street in post-socialist city. Cities 32:123–134

May E (1928) Das soziale Moment in der neuen Baukunst. Das Neue Frankfurt 2(5):77–83

Mielczarski R (1917) Cel i zadania stowarzyszenia spożywców [The aims and tasks of consumer associations]. Społem [Together] 5–6:8

Nawratek K (2011) City as a political idea (trans: Pyzik A). University of Plymouth Press, Plymouth

Newman O (1972) Defensible space. Macmillan, New York

Perry CA (1939) Housing for the machine age. Russell Sage Foundation, New York

Pióro Z (1965) Ekologiczna interpretacja układów osadniczych (założenia i problematyka badawcza) [Ecological interpretation of settlement systems (assumptions and research topics)]. Studia Socjologiczne 4:21–23

Piotrowski W (1965) Z problematyki przystosowania ludności napływowej do środowiska miejskiego [On the issues of adaptation of the migrant population to the urban environment]. Studia Socjologiczne 4:57–73

Rowe C, Koetter F (1978) Collage city. MIT Press, Cambridge

Sennett R (2009) The craftsman. Penguin Books, London

Springer F (2015) 13 Pięter. Wydawnictwo Czarne Wołowiec

Strzelecki J (1970) Mój nauczyciel – Stanisław Ossowski [My teacher—Stanisław Ossowski]. In: Anonymous (ed) Żoliborz wczoraj, dziś, jutro [The Past, Present and Future of Żoliborz]. Książka i Wiedza, Warszawa, pp 220–224

Syrkus S (1925) Architektoniczne podstawy budownictwa mieszkaniowego [Architectural foundations of residential housing]. Odbudowa gospodarcza 2:21–26

Syrkus S (1926) Preliminarz architektury [The architectural estimate]. Praesens: Kwartalnik modernistów [Praesens. Modernists' quarterly'] 1:6–16

Syrkus H (1976) Ku idei osiedla społecznego 1925–1975 [Towards the idea of the social housing estate 1925–1975]. Państwowe Wydawnictwo Naukowe, Warszawa

Syrkus H, Syrkus S (1946) Budownictwo doświadczalne [Experimental construction]. Dom, Osiedle, Mieszkanie 6–8:46–52

Syrkus H, Syrkus S (1947), Budowa II kolonii osiedla WSM na Kole [Construction of the colony II of the WHC estate in Warsaw's Koło district]. Dom, Osiedle, Mieszkanie [House, estate, flat] 9–12:18–34

The Pruitt-Igoe Myth (2011) [film] directed by Chad Freidrichs. Unicorn Stencil Documentary Film, Columbia, MO

Toeplitz T (1928) Robotnicza spółdzielczość mieszkaniowa [Workers' housing cooperatives]. Robotnik 10

Tołwiński S (1970) Wspomnienia 1895–1939 [Memoirs 1895–1939]. Państwowe Wydawnictwo Naukowe, Warszawa

Wallis A (1964) Socjologia miasta a planowanie urbanistyczne [Sociology of the city and urban planning]. Studia Socjologiczne 13:211–230

Wallis A (1965) Hierarchia miast [A hierarchy of cities]. Studia Socjologiczne 4:75–87

Warszawska Spółdzielnia Mieszkaniowa [Warsaw Housing Cooperative] (1930b) Statut i regulaminy [Statute and regulations], Warszawa

Wenderski M (2019) Cultural mobility in the interwar avant-garde art network. Poland, Belgium and the Netherlands. Routledge, New York

Ziółkowski J (1964) Osiedle jako system społeczny [The housing estate as a social system]. Studia Socjologiczne 4(2):204, 183–209

Ziółkowski J (1965) O socjologicznej wyobraźni urbanisty [On the sociological imagination of the urban planner]. In: Ziółkowski J, Urbanizacja, miasto, osiedle. Studia socjologiczne [Urbanisation, city, estate]. Państwowe Wydawnictwo Naukowe, Warszawa, pp 227–251

Chapter 4
How Does the Space Perform?

Abstract Since I am adopting a performative perspective here, following John McKenzie I ask how Żoliborz's estate founders challenged architecture. Architects turned out to be performers who, through residential decorum, initiated new urban lifestyles. From the kitchen design (inspired by the Frankfurt kitchen model) to the furnishing of the flats, they shaped new rituals that over time have evolved into social habits. The residents were taught how to live in a modern, hygienic, rational, cultural, social and neighbourly way. By means of materialist analysis, I study the dense network of relations and connections between the various forms of social and intimate life, significant everyday devices and objects, architecture and flora, educational and gardening efforts; I analyse the relationship between the emancipation of women and workers and the ways in which the latter arranged their homes. Not only did the emancipation of women—who wanted to work professionally—turn out to be very important here, but mostly the emancipation of domestic servants, assumed by the founders of the Warsaw Housing Cooperative. In addition to workers and women, children turned out to be a significant actor for whom space in flats was arranged. Jeffrey Goldfarb's concept of 'the politics of small things' convinced me to carry out a detailed analysis focused on the details of everyday life.

Keywords Interior design · Frankfurt kitchen · Urban lifestyle · Rational interior arrangement · Hygiene · Housing habits

4.1 What Does It Mean to Inhabit a Place?

Following the ideas of Barbara Brukalska, one of the most active architects of the Żoliborz district, and recognising in them the profound proxemic sensitivity and concern for the entanglement of the individual in the political system, it can be boldly said that she was a social activist writing scripts of new changes made possible thanks to the architecture. Brukalska openly expressed her opposition to social injustice and humiliation of the poor. Her strategy, therefore, relied on socially engaged architecture, but always understood as a measure leading to the achievement of social goals, in contrast to Le Corbusier's modernist ambitions.

M. Matysek-Imielińska, *Warsaw Housing Cooperative*, The Urban Book Series,
https://doi.org/10.1007/978-3-030-23077-7_4

Barbara Brukalska had written down her critical thoughts during the Nazi Occupation of Poland, and therefore in the 'in-between' period, a time yet unspecified by any political system. She knew perfectly well that the liberal-capitalist system in Poland was unfair, but she also looked with fear upon the approaching totalitarianism. Reflecting on her involvement in the design of the Żoliborz housing estate, she tried to identify—in the then-current problems of social housing—the influence of the systemic phenomena on the contemporary concepts of architects and urban planners and, in doing so, differentiate between the spheres of influence of capitalist liberalism (treating the resident as a passive tenant, an individualised contractor), 'totalism' (under which the resident is subordinated to intrusive social activism) and social democracy. Nowadays, these words can be read as an anti-capitalist declaration of a strategy of resistance against social divisions, exclusion and humiliation brought on by capitalist economy.

Although Barbara Brukalska had always shown staunch support for the construction of workers' housing estates, she acknowledged that company-sponsored dwellings were an offence against human rights, 'A flat that binds a worker with the workplace creates a far-reaching dependency. Because this form of bondage is not immediately apparent, it is that much stronger' (Brukalska 1948, p. 175).

By adhering to the principles of socially engaged architecture, which in the interwar period, were introduced throughout Europe under the slogan of 'socially most necessary houses', Brukalska wanted to build flats that essentially would not restrict or impede the tenants from changing their jobs freely so that they could be independent of their employers. Cooperative housing was also meant to enable the 'organisation of mutual help' (Brukalska 1948, p. 175). Housing was not defined as a goal in itself but rather as a tool for social change. Hence, Brukalska's thoughts, written down with the benefit of hindsight, can be regarded as a kind of independent and critical development of the cooperative's current activity which, in her opinion, often diverged from its original assumptions.

Today, the prospect of occupying a 30 m^2 two-room flat, with a kitchenette, without a bathtub, but equipped with a toilet and a washbasin, does not impress anyone. But let us remember that we are talking about the 1920s. It is easy to imagine the housing conditions of the workers who migrated from the countryside to the city: a basement flat, living in single room of a tenement house with the entire family. In order to give an adequate idea of the housing conditions, it is enough to mention the words uttered by Ludwik Krzywicki, a sociologist and the leading ideologist of the social left in interwar Poland: 'So, in terms of housing issues which will be decided today, I care only about the most disadvantaged tenants, those who sublet some space in someone else's flat, and even those who engage in the so-called 'usury of space' in their own flats… There must be approximately 70,000–80,000 such itinerant tenants (…) in Warsaw. We need not discuss the negative effects of such overcrowding. (…) But these inconveniences, contentions, more frequent occurrence of epizootics, devoiding people of their modesty and a chance for clandestine promiscuity are coupled with extremely high usury fees for such a miserable place to stay' (Krzywicki, cited in Syrkus 1976, p. 99).

At a conference in 1930, during which housing issues were debated, Teodor Toeplitz described this problem as 'the matter of living conditions of the low-income strata of people who make their living with manual labour' (Syrkus 1976, p. 97),[1] and it should be added that this concerned only urban populations. On the other hand, at the exhibition entitled 'The smallest possible dwelling', he said: 'The necessity of providing each family with their own flat, and every family member with their own bed, given today's economic conditions and the current state of technology, has to be addressed using the most economically viable solutions' (Syrkus 1976, p. 96). Thus, it can be seen that the smallest possible dwelling in Polish political and economic conditions was equated with the socially most necessary house. Whereas in the architectural discourse, it took the form of 'the smallest possible dwelling' and was implemented as a political demand. The maximum use of limited resources is, after all, an everyday practice of an architect-activist.

The challenge posed to architecture was immense. It was not so much about designing a cheap flat as a form, but as a process. The avant-gardist, Le Corbusier, saw architecture as a goal, while the Żoliborz activists, including architects, saw it as a means leading to broader social and cultural projects defined in the Cooperative's constitution. The flat as a process made apparent the need for building a connection between the architecture and the tenants. In 1926, Szymon Syrkus wrote in the 'Praesens' modernists' quarterly: 'the use of new inventions changes the character of cities—we live differently and we must dwell differently, and therefore the flat layout must also be different' (Syrkus 1926, p. 7). But did the designers of the new social order from the Żoliborz district respect the rhythm of workers' lives, their experiences and their lifestyle? They undoubtedly tried to build relationships between objects, bodies, space and local authorities. However, dwelling as a process challenges the practice of cooperation, which was understood very broadly in this context. It did not concern solely architectural co-design, i.e. a dialogue in which the architect listens to the residents' voices. It is the architecture itself which stages the performance, making the act of dwelling more conscious (although it can also make it impossible). It allows residents to interact, communicate, to consolidate neighbourly ties, and is a form of political activity at the same time, transforming a passive tenant into an active citizen taking part in direct democracy. That means that the architect-activist must win people's hearts to make them susceptible to these transgressive and educational experiences, to make them develop new living habits, new rituals of purity, intimacy, relaxation, the freedom to furnish and equip their own place, and finally—to develop their neighbourly rituals. So just how far should the architect go? What limits of interference could have been imposed by the cooperative authorities? The modernist ambitions of the avant-garde were aimed at building 'machines for living', thus completely subjugating the tenant and disciplining him. The fact that no building was devoid of ideological meaning within the Żoliborz cooperative did not go unnoticed by the cooperative authorities; the architecture only allowed residents to shape their habits and new style, but did not order them to do so. The demands of urban and architectural perfection started to diminish both for economic reasons

[1] The detailed housing situation of workers has been described by Cegielski (1968).

and because of the trust in the natural, spontaneous, cooperative possibilities of the residents and their ways of using and applying 'living devices'. In order for a resident to be able to cooperate with the architecture, they had to understand it, know how 'it works'. Hence, it was necessary to shape 'housing culture' through an extensive counselling system. Therefore, contrary to the adopted and well-established critical interpretations of the Żoliborz estate, I would like to propose a different approach consisting in examining the relations between the WHC authorities and its residents with particular focus on the different kinds of tricks and tactics undermining the dominant ways of organising everyday life. This works both ways: the authorities used various types of 'rhetorical tricks', thus maintaining their dominant position, while the residents maintained their housing strategies, organising their own and unique space.

Let us illustrate this performative, experimental, cultural-civic character of the WHC housing estates by comparing it with an interesting example.[2] An industrialist from Bordeaux, Henri Frugès, asked the famous Le Corbusier to design a large housing estate. Henri Frugès ensured that this design would be implemented using cutting-edge construction methods. The 'Pessac housing estate was meant to be a laboratory. I authorise you to break with all conventions, to reject any traditional construction methods' (Le Corbusier and Jeanneret 1965, cited in Syrkus 1976, p. 49). While visiting this estate, the architects rhapsodised about the richness of polychrome, the balance of structures, the interpenetration of inner and outer space. In 1929, workers, who received long-term, low-interest loans, moved into the housing estate. However, no one taught them how to organise their everyday life in flats which were significantly different from what they were used to. Nobody had ever taught them how to live in a housing estate. Helena Syrkus wrote: 'after just a few years, the entire estate was "morally and physically worn out". Some of the windows which ensured this interpenetration of inner and outer space had been bricked off in order to allow storing all kinds of old junk; various sheds had been added, the polychrome had faded or been covered with a layer of paint—sometimes of neutral colour, sometimes defying the original colour scheme with its gaudiness; villas appealing to the petty-bourgeois tastes had mushroomed in the neighbourhood, leading to the degradation of the entire estate. And yet, both Frugès and the architects had imagined that the Pessac housing estate would be a great contribution to the construction of socially most necessary houses on behalf of the French progressivists' (Syrkus 1976, p. 50).

There was no housing counselling in France at that time. Such institutions had been introduced in Poland by the WHC, in the Netherlands—by Jacobus Johannes Pieter Oud and in Germany—by Ernst May. Polish modernists were very familiar with the concept of housing counselling. In 1926, they published an article in the 'Praesens' monthly entitled *Wychowanie przez Architekturę* [Education Through Architecture] (Oud 1926).

[2]I referred to this example by pointing to the educational dimension of the WHC housing estate in: Matysek-Imielińska (2016a, b).

Since the very beginning, the Warsaw Housing Cooperative has been an experimental endeavour and, at the same time, a showcase of the 'Praesens' group on the international scene. The cooperative was an embodiment of CIAM's ideas. It was within the area of the WHC estates that the Frankfurt exhibition entitled 'The smallest possible dwelling' was presented, including the presentations of new flats in the Żoliborz colonies. The flats had been appointed with furniture designed to fit well in the modestly sized rooms and be cheap enough as to not overburden the equally modest financial standing of the residents. Among the presented projects was Barbara Brukalska's kitchen design, inspired by Grete Schütte-Lihotzky's Frankfurt kitchen, Szymon Syrkus' fold-away bed and even the modern 'Glasshouse' preschool designed by Nina Jankowska, Brukalska's friend, and run by the Workers' Friends of Children Association (Robotnicze Towarzystwo Przyjaciół Dzieci), one of many WHC self-help organisations (Fig. 4.1).

The Żoliborz estate, as befits a 'cooperative republic', had a number of institutionalised forms of counselling and contacts with its residents: it had, for example, its own press, at first in the form of an information bulletin (one-day issues), eventually published as a monthly. The 'Życie WSM' [The Life of the WHC] magazine informed tenants about the activities undertaken by the cooperative's institutions and housing associations, about the prepared projects, and encouraged active participation in the cultural life of the communal house; it also published messages from the local self-government. The magazine was used to resolve disputes, discuss controversies, debate various topics in writing and publish critical articles (Stanisław Ossowski, Julian Hochfeld, Adam Próchnik, and Maria Orsetti, among others, wrote for the 'Życie WSM' magazine). It was a kind of forum for the res-

Fig. 4.1 Kindergarten of the Workers' Society of Children's Friends (RTPD), Report on the WHC activities for 1934. 3793, p. 60

idents of the housing estate. It also provided advice concerning housing issues (how to arrange one's own flat), but also broader problems important for the entire housing estate (how to organise the estate's daily life in cooperation with others). The magazine published standardising speech acts, i.e. specific performative acts. 'Stefan Zbrożyna was punished because his son had vandalised the housekeeper's shed, Wanda Skupiewska—for having left her dog unattended, Henryk Schpiner—for unlawful appropriation of a washtub' (Życie WSM 1936, p. 15).[3] 'The Disciplinary Board, having considered the case of Kazimierz Petryka, son of Ms. Weronika Petryka, tenant of the Colony V, for causing a brawl in the courtyard of the Colony III, decided to administer punishment to the tenant Ms. Weronika Petryka amounting to one penalty point and to publish the following statement in the 'Życie WSM' [The Life of the WHC] magazine: 1. Mr. Kazimierz Petryka admits to having caused a scandalous incident in the courtyard of the Colony III and wishes to apologise to the housekeeper, Mr. Wisniewski, for publicly insulting him. 2. Mr. Stanisław Wiśniewski takes back his insulting words aimed at Mr. Kazimierz Petryka' (Życie WSM 1937, p. 213). Such messages conveyed more than just information about the fact that someone had been punished, but first and foremost demonstrated the authorities' power to enforce order, discipline and rules of social coexistence. Such press reports acted as preventive measures, manifesting the social control system. Quoting norms and standards, formulas and behaviours by repeating them in the practice of everyday life served to maintain the dominant order through a system of punishments and admonitions, thus creating a certain housing *habitus*, as well as workers' and cooperative *habitus*. Such performative speech acts, however, were not used exclusively by the authorities. Statements against the authorities' dominant character rested with the residents and the cooperative's workers (who were moved and outraged by this), appealing to their conscience and integrity. 'The Cooperative management published a notice in the last issue of our magazine about Ms. Szczepańska's dismissal. The notice was accompanied by a comment that, although the investigation into the robbery of the Cooperative's office was in progress, before the Court would issue its judgment, the Cooperative management should dismiss Szczepańska because she had lived with a man who turned out to be a conman. Ten months before, the Cooperative management published an article that the Board had dismissed the housekeeper having gotten drunk on the New Year's day and being unable to work for the following two days. (…) Various institutions would address such offenses in different ways. In the worst-case scenario, when the breaches committed by a worker interfered with the performance of their duties, the employer would dismiss such a worker. However, the worker in question would not be stigmatised in the press, and the real reasons would not be mentioned in the certificate of dismissal. This was not the case here.

[3]Stefan Zbrożyna was an active member of the WHC. In 1930, he joined the board of the 'Szklane Domy' [Glass Houses] Tenants' Mutual Assistance Association. He participated in a lecturing committee headed by Adam Próchnik. In 1932, together with Próchnik, he was elected vice-president of the board of the 'Szklane Domy' [Glass Houses] Tenants' Mutual Assistance Association, headed by Stanisław Szwalbe. His community actions were appreciated by the President of the Republic of Poland who, on 22 December 1938, awarded Zbrożyna with the Gold Cross of Merit. In the following year, he became the vice chairman of the WHC Supervisory Board.

Not only was he dismissed from work, but his dismissal was announced publicly. The worker to whom I am referring apparently was not so bad, since the tenants had demanded his return and he has since been returned to his previous job. He was downgraded, which he accepted with humility, and eventually he continued working flawlessly. We are already familiar with the case of Ms. S., but let us consider whether the Cooperative management's proceedings in her case were just and equitable. (…) We, the WHC workers, strongly and categorically protest against treating us this way' (Święcicka 1933, p. 11). This peculiar struggle, which set the rules of what is 'normal' and what is not, took place at the level of the *habitus* and took the form of a public 'press spectacle'. It was a continuous public strategic conflict between various entities. There was no unilateral misappropriation of power nor did the use of rights to govern others persist in the Żoliborz estate; on the contrary, mutual relations of dominance appeared in various areas of social life in which disputes had arisen. Unlike Foucault's dramatic clash of an individual with the forces engaged in power struggles based in rationality, here we have disputes, resistance, conflict, tactics and strategies.

The 'My dwelling' counselling centre (whose purpose was also to display and rent various types of furniture and equipment), affiliated with the 'Szklane Domy' [Glass Houses] WHC Members' Assistance Association,[4] was founded in 1932 and included architectural, horticultural and hygienic departments. Competitions for the best-arranged small dwelling and the most beautiful courtyard garden were organised. The WHC educational experiment consisted in shaping the aesthetic needs of the residents, among them new tastes, based on modern design, simplicity and functionality which were associated with progress and a departure from backward, petty bourgeois and parochial tastes and routine habits. It was about creating a new urban lifestyle which brought on… 'humanities in action'.

The nationwide 'Dom, Osiedle, Mieszkanie' [House, Estate, Flat] monthly was first published by a group of avant-garde architects in 1929. In 1930, the magazine became the press organ of the Polish Society of Housing Reform [*Polskie Towarzystwo Reformy Mieszkaniowej*]. It was edited by, among others, WHC

---

[4]The name of the association was obviously not a coincidence. First of all, the idea of glass houses has been well established in Europe at the beginning of the twentieth century. For example, in 1914, Paul Schreerbart published a book entitled *Glasarchitektur*, while Bruno Taut presented his *GlaShaus* at the Werkbund exhibition. WHC designers based their work on Stefan Żeromski's *Przedwiośnie* [Early Spring] novel and its main character, Cezary Baryka, became the role model as an avid social activist and philanthropist. Cezary Baryka, Doctor of Medicine and constructor of glass houses, was an expression of distancing oneself from filth, backwardness and exploitation and the approval of a proactive attitude to health, progress and active participation. And the attitude of 'Żeromszczyzna' [*sensitivity to social injustice and the suffering of the Polish nation*—translator's note] was quite popular among WHC designers. Discussing the idea of glass houses in great length is beyond the scope of this work, this topic has been widely discussed in Polish literature. Cf. Mencwel (1990, 1998). The 'Glass Houses' action programme was based on Edward Abramowski's thoughts described in *Związki przyjaźni* [The Friendship Bonds]. It was to be a mutual help system based on the cooperation of small groups of people who know each other personally. The idea was to avoid the official form of charity and put an emphasis on mutual help. Ossowski, however, was aware that Abramowski's proposal was doomed to failure, because it was difficult to have friends among people who had nothing in common with us. Cf. Ossowski (1967, p. 361). See Szymański (1989, p. 91).

architects, interior designers and socially engaged members of the cooperative authorities. The magazine featured a discussion on the functional model of housing for workers and the intelligentsia (as the number of intelligentsia tenants in Żoliborz was growing). The intricacies of designing everyday objects, furniture, wall decorations and adornments were debated.

This monthly served a very important function of 'authoritative speech', which also had the character of performative acts that were to be educative and exert a disciplinary effect (Austin 1962). In doing so, the monthly designated the space of authority (of taste and aesthetic power), thus provoking the residents to act. Interestingly, the monthly was emancipatory (promoting the liberation of women, workers, the working class, and thus creating new urban lifestyles), but on the other hand, it was the 'authoritative speech' and constituted a normative force.

4.2 How to Observe Housing Decorum?

Both the 'Życie WSM' magazine and the above-mentioned professional press criticised the petty bourgeois style: 'There is a turntable on the *étagère*, which itself is covered with a plushy Art Nouveau-style bedcover adorned with tassels and fringes. Next to it is a table which is not in fact a table but a four-legged monstrosity with bulging, spherical tumours scarring its surface every few centimetres—the height of creative absurdity and infallible proof that the creators of Vienna's Art Nouveau furniture had similar bulging tumours in their heads... (...) Beds were used for sleeping. Well, what if there is no room for beds in a new flat, as all aunts had died long ago... The couch has a massive backrest, also covered with plushy fabric, enclosed in a wooden frame and adorned with a frolicsome shelf, a tiny gallery on which useless shells, buttons, horns of happiness, and terracotta figurines are arranged. (...) A pattern here and there, covered with a picture frame, portraying a forest, a path, a hunter in a red coat on the path' (Wohnout 1933, pp. 2–3).

Such discursive practices were aimed at stigmatising petty bourgeois tastes and their aesthetic habits. The authors wanted to engage readers and make them want to change, make them feel embarrassed and provoke them to reflect on life. They were not merely simple critical acts formulated in a declarative mode, but rather an imperative battle cry: 'This cannot continue!' This performative utterance triggered a heated discussion in the 'Życie WSM' magazine. Maria Kownacka, for example, emphasised that the presence of 'rubbish from bourgeois houses' does not result from conscious styling decisions, but is rather an expression of necessity (Kownacka 1933). Elżbieta Mazur, a contemporary researcher, admitted that 'the furniture, apart from those made of light ash or pine, was either given to the residents or they brought it along from their previous flats: cupboards, wardrobes, padded chairs and armchairs. No one wanted to leave good furniture behind. Only when the need for something new arose, modern and functional pieces of furniture were purchased' (Mazur 1993, p. 124).

Articles similar to Kownacka's were numerous, and the green, plushy sofa appeared in them as a symbol of backwardness and bad taste. One can have the impression that such articles were multiple regular repetitions from different parties or different actors: the cooperative's founders, architects, artists, and finally the residents themselves who took part in numerous neighbourly contests and exhibitions, putting their flats and gardens on display. It was therefore a negotiation process in which the 'norms' of aesthetic judgements were being established as the expression of socially arranged forms of social division distributed on a class basis.

The articles—or the appeals—promoted modern residential *decorum* in a systematic and very suggestive manner: tips regarding artificial light, stoves and bedrooms were being published. Szczęsny Rutkowski, for example, in an article entitled *'Every room fulfils two functions'*, promoted a modern lifestyle in which listening to music, resting, relaxing and reading books became an activity of an emancipated and educated working class, even though the authors were perfectly aware that their performative speech acts would not have an effect upon the working class. Instead, they shaped the modern habits of the underprivileged intelligentsia who lived off of their own labour.[5] *Cieślewski and son* designed a modern design setting: 'Let us hang a modern picture inside a modern interior, and we will see that it is no longer bizarre, and that, on the contrary, a naturalistic picture of flowers or a 'field landscape' would look quite strange and simply bad when hung on a wall in such an interior' (Cieślewski 1930, pp. 19–20). The Żoliborz style of the working class (mainly, however, of the intelligentsia) was to be introduced through active transformation and repair. It was advised that 'if we decided to introduce old furniture to a new place, it should be upgraded (simplified, unnecessary elements should be removed)'. The authors of an article entitled *Jak ze starych mebli zrobić nowe?* [How to turn old furniture into new?] (Dom, Osiedle, Mieszkanie 1934a, p. 11), described in detail do-it-yourself carpentry techniques, provided schematics and tips. They designed a 'home workshop' and encouraged people to work manually, to repair and transform their property. They not only promoted being active, but also told people 'how to do it'. 'The dining table should be as simple as possible, cheap, comfortable and suitable for everyday use without a tablecloth (…). The trend of making the legs of a dining table look like the dumpy legs of a piano or a billiard table was gone forever. Likewise, chairs should be as simple as possible, but at the same time as comfortable as possible, smooth, devoid of solemnly carved armrests' (Dom, Osiedle, Mieszkanie 1934a, p. 6).

The rhythm of this physical work and the flat design also transformed identity of the Żoliborz housing estate's residents. Repairing objects led to committed participation and the desire for 'productive work', which was of special value in the Żoliborz housing estate, carried out not only by workers themselves, but also by the intelligentsia (Igor Abramow-Newerly, a writer and a journalist cooperating with Janusz Korczak on the editing of the magazine for children and youth, *Mały Przegląd* [Little Review], was an interesting figure in this regard. Newerly worked

[5](Dom, Osiedle, Mieszkanie 1929a). Toeplitz describes, for example, 'mass construction' giving an example of a housing estate near London and the Becontree housing estate (Dom. Osiedle. Mieszkanie 1929b).

as a glassman in the Żoliborz housing estate, repairing and building canoes, among other objects) (Abramow-Newerly 2000).

The one-and-a-half-room flat was really difficult to arrange. The Brukalskis suggested that flats be designed with furniture already in place, thus guaranteeing the rational and hygienic use of space. Andrzej Pronaszko, member of the 'Praesens' avant-garde group, wrote: 'Every piece of my house is constantly working. (…) The work performed by my house translates directly into the convenience of my flat' (Pronaszko 1929, pp. 4–5).

Architects and creators of the *decorum* noticed a new actor who had not yet been considered, namely the child. Children's needs were attended to with particular care in the Żoliborz estate. Hygienists and paediatricians had been involved in the architectural design process.[6]

It was also proposed that in private spaces, even in very small flats, a child should be provided with such conditions as to make it possible for it to develop its physical and mental needs. It is worth emphasising that such care devoted to and the attention paid to the child's mental needs was not commonplace at the time, especially in the working-class environment, not only because of the world view, but also due to financial reasons. Thus care for the child's needs was, after all, distributed on a class basis.

The flat was therefore an educational process and an identity-building project. Although it could be argued that people continued to operate on an individual basis, furnishing and arranging their private interiors, living in their private spaces, but because these interiors had been modelled on the Żoliborz *decorum*, their actions took on a public feature, making the flat a social issue, and a public act. Thus, residents implemented Goldfarb's 'politics of small things' which makes the border between the private and the public domains disappear simply by creating spaces of interaction, and creating a cultural meaning. The modernist idea of functional constructivism turned the flat into a laboratory of the rational organisation of space, of orderly and non-chaotic activities linked to work, relaxation, and kitchen-related matters. And although it disciplined the residents, by shaping in them new habits with the use of some ideological legitimations, it was not oppressive. Maria Kuzańska put it this way: 'The designs of individual colonies and flats of the WHC estate had been carefully prepared and thoroughly and critically analysed. Prior to their implementation, numerous discussions had been organised, and modifications proposed by its future residents had been introduced. The concern for human beings, for the comfort of all family members, and especially for the best possible conditions of life, work and play for children was behind the concept of the Żoliborz housing estate. The flats were small (usually consisting of one and a half room), but sunlit and modern. All of the flats were equipped with central heating, electricity, water supply and sewage systems, as well as with gas infrastructure, large windows, niches, lockers and handy cabinets. Brightly painted walls, white doors and windows, level flooring made it easier to keep the flat clean. (…) All this obliged the residents to respect

[6]A separate subchapter within Chap. 5 has been devoted to the question of children within the Żoliborz district and their importance within the Żoliborz housing estate.

and take care of their flats, encouraged them to maintain cleanliness and order, and led them to organise their family life and consolidate new neighbourly ties based on friendship and cooperation in the new housing environment (without subtenants and 'itinerant tenants')' (Kuzańska-Obrączkowa 1966, p. 75). Of course, we can see that this statement from 1966 was pure wishful thinking. It is difficult to believe now that the finished flats that the tenants had moved into made them suddenly start treating them with care. It seems that the care that eventually grew was due to the standardising speech acts contained in the regulations and guidelines (e.g. the housing savings books contained a set of principles concerning rational behaviours described in point 16, or habit changes, e.g. regarding hanging out laundry on a clothesline installed in the flat's windows—point 22), but also stemmed from the cooperative's atmosphere. Undoubtedly, however, the architecture of residential interiors and of the entire housing estate fulfilled a performative function, directing the residents towards a change, developing in them a sense of commitment and a strong sense of identification, but it also led to the transformation of their social status. One of the residents reminisced: 'My daddy, who used to read "Gazeta Codzienna" daily (he bought it for 5 groszy), had read once that new and wonderful houses were being built in a beautiful district in Żoliborz. They were appointed with fantastic equipment, complete with central heating, bathrooms, hot and cold water—a true paradise on Earth! My father got such a flat, on Krasinskiego Street at number 20, and we moved into this flat in 1932. Compared to the Old Town, where we had lived before, it was a luxury flat'.[7]

The arrangement of flats and its aesthetics was also the subject of an extremely interesting study carried out during the winter months of 1938–1939. Irena Reicher, a researcher, was interested in how people lived in the 9th colony of the Żoliborz estate and in the Rakowiec estate. She wanted to cognise the housing problem among the workers 'from the point of view of order and aesthetics, workers' interests, cultural and housing needs, and their opinion on the project usefulness for the Furniture Cooperative. I wanted to observe at the same time—wrote Irena Reicher in an unpublished manuscript of her research report—the influence of the Warsaw Housing Cooperative on the development of residents' aesthetic culture' (Reicher 1938–1939, p. 1). Responding to the tenants' needs and at the same time wishing to influence their tastes and housing habits, the WHC decided to establish the Furniture Cooperative which would design and sell furniture for the expanding WHC housing estates. However, we do not know who inspired this project, if Barbara Brukalska was its initiator or someone else. What we know is that the Furniture Company was never established, probably because of the outbreak of World War II. We can, however, follow the evidence of outstanding, pioneering empirical research studies developed by a woman of extraordinary research knowledge. Irena Reicher's research studies shed light on the aesthetics of flats, the effects of educational and propaganda activities, as well as on the open-mindedness of the WHC residents who seem to be used to various forms of interviews, surveys and questionnaires gauging their self-awareness. It is especially visible when Reicher writes, for example, 'In general, after entering the

[7] Janusz Jarzęcki's recollections were written down as part of the activity of the Oral History Archive and published (Jarzęcki 2010, p. 58).

flat, the contact with tenants was quite quickly established, and I was often shown and told more than it was necessary for the questionnaire. (…) Tenants, when their curiosity about the purpose of the visit was slaked, said willingly and showed great interest in the Furniture Cooperative project' (Reicher 1938–1939, pp. 2–3).

Reicher emphasises that tenants complained about the narrow kitchen layout, the insufficient number of kitchen furnishings and their poor arrangement. The residents of the Rakowiec estate complained about the poor layout of ovens placed in the wrong place.

Irena Reicher drew up a report on the condition of the flats evaluated on the basis of the following factors: cleanliness, order, aesthetics and the size of the furnishings. Reicher analysed 46 flats (i.e. 15.33% of the total number) in the Rakowiec estate, and the analysis shows:

- there was a general excess of large and bulky furniture; 28 flats (60.8%) were cluttered, including:
  - 3 flats were dirty,
  - 25 flats were clean and tidy;
- a moderate usage of furniture (both the amount and the size) could be seen in 18 apartments (39.2%), including:
  - 3 flats were dirty,
  - 12 flats were clean and tidy,
  - aesthetic order reigned in 3 flats.

In Żoliborz, in 54 flats covered in the questionnaire, which represented 56.25% of all the flats of the 9th colony (A and B):

- there was insufficient amount of furniture in 8 apartments (i.e. 14.8%), including:
  - 2 flats were dirty,
  - 6 flats were clean and tidy;
- there is a general excess of furniture; 9 flats (16.2%) were cluttered, including:
  - 1 flat was dirty,
  - 8 flats were clean and tidy;
- a moderate usage of furniture (both the amount and the size) could be seen in 37 apartments (69%), including:
  - 3 flats were dirty,
  - 24 flats were clean and tidy,
  - aesthetic order reigned in 10 flats.

The largest number of flats with an excessive amount of furniture (too big for the flats), characterised by cleanness and order, was found in the Rakowiec district—25 flats (54.8%) out of a total of 46.

The largest number of flats with adequate furniture in terms of size and amount, characterised by cleanliness and order, was in Żoliborz—24 flats (44.4%) out of a total of 54.

Also, 10 flats (18.1%) in Żoliborz are distinguished not only by moderation in furniture but also by aesthetic order. In the layout of furniture (of simple and pretty forms) and in the decorations, the selection of curtains, napkins, pictures, photographs, etc., these flats demonstrate the developed sense of aesthetics of its residents. Only 3 flats (6.2%) in the Rakowiec estate demonstrate the similar taste. 'One of the flats worth noting belongs to an artist, a graduate of the Academy of Fine Arts. It is arranged and decorated in a folk manner, maybe even exaggeratedly. The Hutsuls' tradition reigns here (table, shelves, ceramics, kilims, etc.) and the rest is complemented by folk products from Polesye and Kurpie' (Reicher 1938–1939, p. 11).

Almost all flats in Żoliborz have been qualified by the researcher as in aesthetic order, but equipped with banal pieces of mass-produced modern furniture. However, the amount of furniture is limited, and the type and dimensions are well adapted to the size of the flats. Reicher also notices moderation in the selection of curtains, napkins and small ornaments (vases, figurines). She regards this as a proof of 'the proper understanding of the modern flat's character, sense of proportion and a certain sensitivity to the form'. In terms of the 'hue culture' (*kultura barwy*)—as the researcher often describes it—things have actually gotten worse, 'hence bedspreads, cushions and kilims not always harmoniously arranged. (…) However, a large number of poorly furnished and cluttered flats, characteristic for the Rakowiec estate, could indicate that the majority of Rakowiec residents do not yet have a modern sense of proportion, and they have no understanding of the proper function of the furniture (…). The flats in the Rakowiec estate are usually cluttered with tables and shelves. Shelves and tables, in turn, are overloaded with plants, vases with artificial flowers, and gypsum statues, sometimes quite large. These are the signs of longing for old dwellings, petty bourgeois living rooms. In Żoliborz, there is an elephant sculpture in almost every flat. In the Rakowiec estate, napkins play an important role in decorating the flat' (Reicher 1938–1939, pp. 11–12). It is worth emphasising that the issue of space dedicated to children captured Irena Reicher's interest the most. In Żoliborz, 42 families (out of a total of 54) have children (71 children in total). Among them:

- 20 families occupy two-room flats
- in 12 families, children go to school and have separate work tables, including 2 desks
- in 9 families, children go to school and have their own separate cabinets and bookshelves
- in 3 families, children have special kids' chairs
- in 1 family, there is a cot for an infant
- 36.6% of all children in families covered by the questionnaire are sleeping separately on a bed or a sofa. In the Rakowiec estate, this percentage amounts to 53%.

In addition to the description of flats and their state of cleanliness and aesthetics, the researcher wonders how the Cooperative influenced the tenants. Has the prolonged residence in the WHC estates affected their sense of aesthetics and taste? I notice, however, that in the Rakowiec estate, the desire for change is associated with fashion (residents buy new furniture, thus cluttering up flats). I shall therefore conclude: 'In the Rakowiec estate, the Cooperative's authority in cultural matters goes unnoticed' (Reicher 1938–1939, p. 20). On the other hand, Żoliborz residents were largely influenced by the 'Smallest Flat' exhibition which interested 76% of the respondents from Żoliborz and 52% of the respondents from the Rakowiec estate. Żoliborz residents admitted that it was necessary for them to see this exhibition and, when talking about it, they often used the term 'our exhibition'. The researcher draws our attention to the statements given by the respondents, in which the effect of housing education can be easily seen. They say, for example, 'I like modern furniture because of its easiness to arrange the space, and it is nice and hygienic' or 'I like modern furniture because it is light, simple and without decorations, so I do not waste time clearing the dust' (Reicher 1938–1939, pp. 24–25). In the Rakowiec estate, however, books and magazines on interior design were more often used to enhance the aesthetic tastes, education and housing counselling.

Irena Reicher summarises her research in a precise table (Table 4.1).

4.3 Emancipatory Potential of Modernist Kitchen?

Owen's worker flats had no kitchens, as he decided that they will be replaced by communal cafeterias. The kitchen proved to be the key element of modernist WHC designs, reflecting class differences and, in the opinion of Marta Leśniakowska, ideological performances. 'How then did the kitchen, being one of the main "themes" of modernist architecture,—she asks—start the process of toppling the traditional hierarchy of living quarters and changing their meaning, which was equal to, or more precisely—considered equal to, changing/modernising the social structure and customs?' (Leśniakowska 2004, p. 191). One could also wonder if opposing the modernist and progressive form of rationality of Żoliborz's designers was even possible, given that the people were so thoroughly educated by the state.

Let us therefore examine the kitchen performances and the ways in which the kitchen issue had challenged modern customs. The kitchen struggle marks a very important point in the conflict between the cooperative's top-down approach and the bottom-up habits of the residents. It also makes apparent the ongoing class conflict (which Leśniakowska called 'modernising the social structure') and differences in customs that could not have been disregarded given the cooperative's emancipatory and educational goals.

WHC buildings forming part of the 1st Colony located next to Wilson Square had three-room flats (60–65 m^2) with a big kitchen outfitted with a bathtub, two bedrooms and a pantry. The flats were designed by Bruno Zborowski who drew inspiration from

Table 4.1 Employment of WHC residents (Żoliborz 9th colony and the Rakowiec estate) and their interest in interior design)

		Number of people employed				Positive assessment of modern furniture				Proposed changes in flat furnishings				Number of flats		
		Rakowiec		Żoliborz		Rakowiec		Żoliborz		Rakowiec		Żoliborz				
		Father	Mother	Father	Mother	Father	Mother	Father	Mother	Father	Mother	Father	Mother	Rakowiec	Żoliborz	Total
Industrial workers	Tobacco industry workers	5	3	3	3	–	8	2	5	1	6	1	5	8	6	14
	Alcohol industry workers	2	–	2	1	–	2	2	1	–	1	2	2	2	3	5
	Metalworkers	2	–	4	–	1	2	1	3	1	1	2	3	2	4	6
	Gasworks, waterworks and power plants	–	–	2	–	–	–	1	2	–	–	–	1	–	2	2
	Construction workers	2	–	4	–	–	2	–	3	1	2	1	2	2	4	6
	Typographers	–	–	1	–	–	–	–	1	–	–	–	1	–	1	1
	Other food industries	1	–	1	–	–	1	–	1	–	1	–	–	1	1	2
	Other industries	4	–	2	–	1	4	1	2	1	2	1	1	4	2	6
Communal workers	Railwaymen	2	–	3	–	–	1	–	2	–	1	1	2	2	3	5

(continued)

Table 4.1 (continued)

		Number of people employed				Positive assessment of modern furniture				Proposed changes in flat furnishings				Number of flats		
		Rakowiec		Żoliborz		Rakowiec		Żoliborz		Rakowiec		Żoliborz				
		Father	Mother	Father	Mother	Father	Mother	Father	Mother	Father	Mother	Father	Mother	Rakowiec	Żoliborz	Total
	Tram drivers	8	–	9	–	6	6	6	7	2	6	9	6	8	9	17
	Other/e.g. drivers	5	–	2	–	1	4	–	2	2	1	2	2	5	2	7
Other workers	Employees of social institutions	–	–	1	–	–	–	1	–	–	–	–	–	–	1	1
	State officials	3	–	5	–	2	2	4	3	2	2	4	4	3	5	8
Employees of social institu-tions	Workers' and employees' institutions (trade unions)	–	–	3	–	–	–	–	1	–	–	–	1	–	3	3
	Social security institutions	–	1	–	–	–	1	–	–	–	1	–	–	–	1	1
Other white-collar workers	State officials	2	1	1	1	2	–	–	1	2	2	1	–	3	2	5

(continued)

Table 4.1 (continued)

		Number of people employed				Positive assessment of modern furniture				Proposed changes in flat furnishings				Number of flats		
		Rakowiec		Żoliborz		Rakowiec		Żoliborz		Rakowiec		Żoliborz				
		Father	Mother	Father	Mother	Father	Mother	Father	Mother	Father	Mother	Father	Mother	Rakowiec	Żoliborz	Total
Craftsmen, liberal profes-sions and others	Craftsmen, independent	2	–	5	–	2	2	–	5	1	2	1	4	2	5	7
	Pensioners	3	–	–	1	2	2	–	–	1	1	1	–	3	1	4
Total														46	54	100

The compilation prepared by Irena Reicher

Fig. 4.2 1st Colony of the Warsaw Housing Cooperative, general view, Report on the WHC activities for 1928. 2461, p. 71

Vienna's Gütrel estate, an embodiment of left-wing ideals (Fig. 4.2).[8] Kitchenettes were seen as a means of accommodating the needs of big worker families with peasant roots, whose family lives were usually centred around the kitchen (Figs. 4.3 and 4.4).

Initially, WHC creators and architects made attempts to tailor the architecture to the lifestyle of peasants who were migrating to the city to become the new working class. However, the economic crisis that ensued shortly thereafter required a revision of their approach. It soon turned out that 60 m^2 apartments were equally unaffordable for blue-collar and white-collar workers alike (Figs. 4.5, 4.6 and 4.7). Thus, further colonies were designed by the more modernist Brukalskis, who decided to make all of the rooms small and revise the design of the kitchens. Barbara Brukalska designed a standard kitchen model in 1927, which was widely discussed and deemed 'revolutionary' (Fig. 4.8). Brukalska was inspired by Greta Schütze-Lihotzky's Frankfurt kitchen. The kitchen adjoined the living room, had an area of 2.20 m $\times$ 1.37 m, a window and numerous useful cabinets and shelves. 'As a living-room kitchen,

[8] Brukalska was inspired by Rotterdam's Tusschendijken estate, which was designed by J.J.P. Oud in 1919. The common areas of the Żoliborz estate were based on Rotterdam's Kiefhoek Estate, which was also designed by Oud and built in 1925. It should be noted, however, that WHC's common areas were more socially oriented and included two stores, two newsstands, two playgrounds and a central heating boiler room. Cf. Heyman (1976, p. 90).

Fig. 4.3 Kitchen in the WHC 1st Colony—Report on the WHC activities for 1928. 3410, p. 71

Brukalska's design was suited for servant-less residences. In his leftist social estate programme, Stanisław Towiński, WHC's chief ideologue, expressed his desire to root out servants as a social and professional group' (Leśniakowska 2004, p. 192). He claimed that servants were a remnant of bourgeois customs and served as proof of the inequality and exploitation of women. Marta Leśniakowska is convinced that Tołwiński's emancipatory slogans were only a facade as, in reality, dissolving the institution of domestic workers would lead to women having to work harder at home or suffer unemployment. In my opinion, however, even when women got rid of servants (and, in return, had access to a communal cafeteria and grocery deliveries handled by the Cooperative Inn) or lost their jobs as servants (they could find a new job at the cafeteria, laundry, cooperative stores, settlement house, communal house and many other WHC institutions offering jobs for unemployed former servants[9]), they gained a lot more than they had lost—a new way of life. I am not claiming that emancipation was victimless. Among the victims was Brukalska herself whose activity in the interwar period, as noted by Leśniakowska, was marginalised and treated merely as implementing the architectural concepts of either her husband,

[9]In 1933, for example, WHC and its auxiliary institutions employed 103 white-collar workers and 308 blue-collar workers, including seasonal workers.

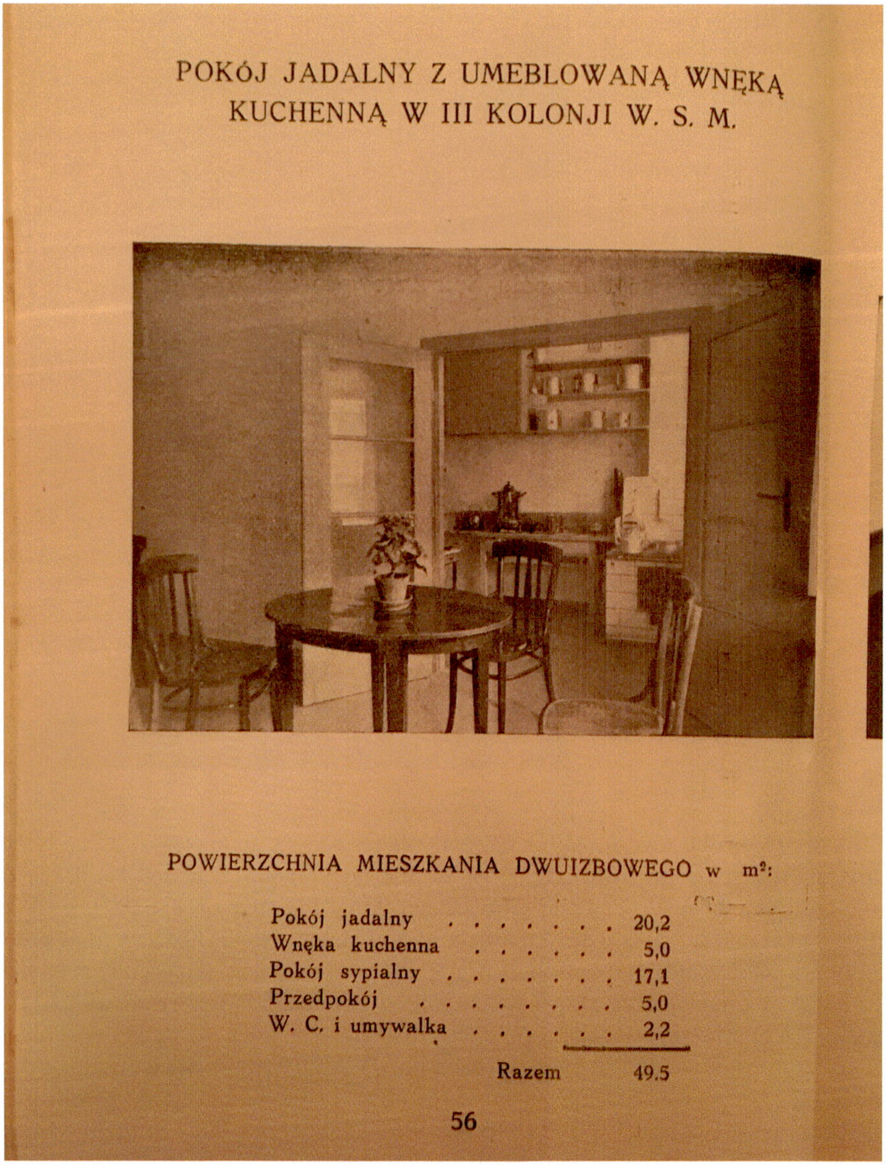

POKÓJ JADALNY Z UMEBLOWANĄ WNĘKĄ KUCHENNĄ W III KOLONJI W. S. M.

POWIERZCHNIA MIESZKANIA DWUIZBOWEGO w m^2:

Pokój jadalny	20,2
Wnęka kuchenna	5,0
Pokój sypialny	17,1
Przedpokój	5,0
W. C. i umywalka	2,2
Razem	49.5

56

Fig. 4.4 A dining room with a furnished kitchenette in the Colony III, Report on the WHC activities for 1929. 3447, p. 71

Stanisław, or Bruno Zborowski.[10] Furthermore, it soon became apparent that the efforts to abolish domestic work would ultimately prove unsuccessful.[11]

The 'Życie WSM' [The Life of WHC] magazine published opinion pieces as part of an open debate, where one female 'servant' decided to voice her views. Staging conflict and discourse in such a way is exciting, stimulating and creates new spaces for agonistic public debate. The *Servant's Letter* was a projection of thoughts on the social rules of conduct; therefore, I would like to present some excerpts of the letter that will allow me to examine the arguments and fears of women who worked as servants. 'The state that thought up this cooperative may be wise, but does not care for us servants, only for itself. They have built big flats, the rent is high, so we know that they are not for workers, but for the state. Servants have little space to do anything; the kitchen is tight and packed with fancy conveniences, like gas, heating, so that the madams would not need our help. Back in the day, you had to be poor to send your children to a children's shelter, and now you hear madam say that it is not a children's shelter but a preschool, and that they have no need for us working girls when the child is away from home for half a day. Now they are saying that they plan to build a nursery as well, and the cooperative wants to build an inn which, as I have read, means that the state wants people to eat at the inn instead of at home' (signed: Servant living in 1st Colony) ('Servant' 1933, p. 3).

The 'servant's' letter caused quite an uproar. The paper published a rebuttal from a female 'Resident' which, it seems, was in fact the Cooperative responding in a way allowing for reorganising social rituals. The rebuttal combined economic, political, legal and cultural strategies. Its goal was to work through the social rules that were brought up and discussed by the 'servant'. 'The Warsaw Housing Cooperative has indeed included only small kitchens in the flats, but that was only because we had predicted that the first buildings will house working-class families, who would not be able to afford having servants. It turns out that we were mostly correct. Out of the one thousand families living in our Estate, eight hundred do not employ servants. The WHC management was convinced that all women who require help with housework will receive assistance from the 'Work Cooperative', consisting of free women living in the Cooperative, who chose this type of work by their own free will, and who would be treated as equals. (If you wish to learn more about this topic, I would advise reading J. Wolski's brochure entitled *Podajmy sobie ręce* [Let Us Reach Out to One Another]). Convincing families to eat at the 'Community Inn' should not be viewed as a slight

[10]To emphasise the extent to which women were marginalised in that seemingly progressive community, Marta Leśniakowska writes that Barbara Brukalska and her husband collaborated on the consultations and design of Colony IV. She also mentions a very important meeting between Barbara Brukalska and Grete Schüttze-Lihotzky that took place in Frankfurt am Main, where Brukalska learned about the Frankfurt kitchen design, which was based on Taylor's rational work principles. Leśniakowska notes that Tołwiński makes no mention of Barbara Brukalska's participation. Instead, his very general recollection reads as follows: 'a very heated debate took place over the arrangement and furnishings of kitchens in working-class flats' (Leśniakowska 2004, p. 194).

[11]In the flats—which were meant initially for blue-collar workers but in the end were occupied by intelligentsia families—a domestic worker's room became a necessity. In the 1930s, housekeepers constituted 20% of WHC's total population. Cf. Szymański (1989, p. 167).

PROJEKT II KOLONJI W. S. M. NA ŻOLIBORZU.

OPRACOWANY PRZEZ ARCH. BRUNONA ZBOROWSKIEGO

ROZWÓJ WARSZAWSKIEJ SPÓŁDZIELNI MIESZKANIOWEJ

Rok	Liczba członk.	Udziały	Wkłady mieszk.	Rezerwy	Suma bilan.	Wartość budowli
		w	z ł	o t	y c h	
1924	29	3069	—	70	3442	2464
1925	87	4918	1241	345	172860	15993
1926	109	13745	9810	535	683835	601158
1927	345	50560	52570	17837	3095301	2.363.624

17

Fig. 4.5 Building design of the Colony II developed by architect Brunon Zborowski, Report on the WHC activities for 1927. 3384, p. 71

Fig. 4.6 View of the completed Colony II. Chimney of the Boiler House visible in the background, WHC History Chamber. 2487, p. 71

against domestic workers. The truth is that eating out is already common in many families. The difference is, however, that the 'Inn' employs the very same women that used to work as servants, thus they continue working, albeit for a different employer, and, what is more, the state often offers better employment conditions than their former madams did' (signed: 'Female Resident') (Życie WSM 1933a, pp. 3–4). As an aside, I should also mention that the push for emancipating women and workers was motivated by the notion of ending exploitation. As the 'resident' points out, 'all of the benefits brought on by modern technology and replacing humans with machines are reaped by capitalists, who rake in enormous profits, allowing them to lead lavish lifestyles that no worker can ever dream of. (…) While we do tolerate the existence of machines (…), we cannot allow only a small fraction of people to profit from their work. Therefore, workers unite in labour unions and declare war on the classes of landowners and capitalists' (Życie WSM 1933a, pp. 3–4).

That strategy was a local attempt to dismantle the inequality-based capitalist model at the estate level. It made possible the shaping of new working conditions and lifestyles, but also led to the emergence of a network of cooperative institutions on which WHC's economic self-sufficiency was based.

Such performative discursive acts prove that each participant in Żoliborz's rituals could take a position and manifest it, which was then discussed, giving rise to interactions and consequences. The estate's press used such critical practices to capitalise on misunderstandings and a proactive approach to conflicts, treating them as an entry

Fig. 4.7 Staircase of the WHC 1st Colony, WHC History Chamber. 2479, p. 71

point for further experimentation and ongoing modification of social rules. We can go even further and say that the estate's actual goals (contrary to what was stated in the estate's constitution) were in fact set on-the-go as a result of public reflection on aspirations and emerging meanings. Therefore, instead of treating Żoliborz as an attempt to build a social utopia, it may be worthwhile to examine the interrelations between action and reflection, i.e. shaping goals using the available means. Goals are not set ahead of time or imposed arbitrarily. Instead, they are created in reaction to current events. It was what Hans Joas called 'acting in a situation' (Joas 1996, pp. 153–164). When the availability of a certain resource becomes apparent, the goals that we strive to achieve also become clearer. Thus, they are not general objectives, but concrete goals that have to be ordered and systematised.

However, if we were to follow Leśniakowska's train of thought, who considered the aforementioned ideological projects to be an embodiment of male domination ('a disguised quote drawn from cultural norms created by men'), we would have to admit that those projects had already been deconstructed by Brukalska. She turned out to be a modernist through-and-through, creating a kitchen that no longer was a woman's domain or the hearth of a home, but had made possible a new lifestyle.

Fig. 4.8 Kitchen design in the WHC Colony IV presented in 1927 by architect Barbara Brukalska; Warsaw Housing Cooperative Report for 1927, pp. 71, 75

Brukalska created kitchens with a consistent set of equipment: 'the kitchen area is minimalist in size (2.2 m × 1.37 m). Kitchens come fully furnished, as no one would be able to fit all of their tables and cupboards, often of an irrational design, in such a small space. If residents do not have their own furniture, we may be sure that they will not install even a single additional shelf to make use of the very limited space available in the kitchen' (Brukalska 1929, pp. 8–9).

Of course, Brukalska's kitchen was no match for Greta Schütze-Lihotzky's shiny, chrome- and nickel-plated Frankfurt kitchen. The kitchens in Żoliborz were made out of wood and painted white. The inbuilt furniture, characterised by neutral colours and simple forms, was designed in accordance with the principles of viewing psychology (research into the physiology and psychology of colours and work effectiveness was particularly popular at the time) and the functional principle of rationally organising creative activities.

Brukalska's modernist and avant-garde inspirations drove her to design kitchens in line with the scientific principles of households. To that end, she made use of the rigid Taylorist framework (e.g. relying heavily on Erna Meyer's *Modern Household*[12]). The kitchen was a modern laboratory allowing the architect to rationally organise work in a manner reminiscent of manufacturing plants: all working processes ran in a single direction, using as few movements as possible to complete tasks, thus eliminating chaos in an already cramped space. 'The sink was mounted under the window, and beside it was a table with a small shelf for keeping spices. The design adhered to the principles of rational work. The dishes were cleaned first, then food was prepared on the table and cooked on a coal-gas stove. The best possible kitchen design could be fit into such a confined space because the workflow and working area were basically the same regardless if the kitchen was of an open or closed design' (Mazur 1993, pp. 111–112).

With that said, we are only beginning to address the kitchen issue. What about the furnishings and their arrangement, which are far from an innocent decorum? Leśniakowska wonders whether Brukalska was aware that her modernist hope and faith in the scientific rationalism of homesteads may have potentially led to transforming humans into machines. It turns out that Brukalska's creativity led her to challenge interior design and engage residents in a new way of living and doing housework. Not only did she design the kitchen, but she also placed it in a scalable environment. That modernist scaling stage featured a modern woman actor with short hair (*à la* Josephine Baker), wearing a sporty outfit and a knee-length dress. It was, in fact, Brukalska herself (Figs. 4.8 and 4.9).[13] That model also sets the role of the modern woman as professionally active, emancipated from housework, able to handle her tasks in the kitchen quickly, easily and rationally. In the modern lifestyle of the working class (an equivalent of today's middle class), the kitchen was reduced to the bare required minimum. Brukalska creates 'activity in a situation',

[12]Meyer (1928). Women's press covered the so-called Taylorised home in great length at the time. Cf. Pogorzelska (1929) and Kobieta w Świecie i w Domu (1929).

[13]The kitchen of a 3-room apartment in Colony IV at 18 Krasińskiego Street. Designed by Barbara Brukalska, 1927–1928 (final design completed in 1929), built in the years 1929–1930.

directs events and stages them through the means of pictures. They are purposely exaggerated in order to emphasise the importance of stage decorations (modern interior design, ventilated pantries and functional closets), treating them as a part of the social decorum. The kitchen is presented in a way that makes readers want to adopt 'Brukalska's style'. Brukalska invokes an 'induced effect', a reaction to the stage performance. She becomes a performer who consciously shapes her behaviour in order to invigorate the public and keep their interest. She is critical, innovative and determined. Such characteristics speak to the readers.

In her efforts to liberate women from the kitchen, Brukalska envisioned estate cafeterias modelled after similar establishments in New York: 'you take a card, listing all of the prices, from a machine that is standing by the door. You then take a tray and walk along the counter; (…) Such solutions could be introduced here as well. The cafeteria should have access to a terrace, arcade or garden, allowing patrons to rest outside during the warmer months of the year' (Brukalska 1948, pp. 74–75).

Apart from her efforts to improve the life of hitherto neglected groups (mainly women and children), Brukalska also adopted a very modern approach to the city as a common public space for collective action. Szymon Syrkus had very similar views, which he expressed in his young modernist manifesto: 'architecture follows the economy of everyday life and, thus, does away with all of the luxuries of yesteryear. No longer will private living quarters have ballroom, salons, enormous dining rooms or guest rooms. The collective shift in lifestyles makes all of that obsolete. Vast indoor areas will only be used for collective purposes: communal houses, schools, universities, theatres, cinemas, hotels, clubs, cafeterias, game and sports halls (…) Making proper use of collective institutions relies on organisation and, more importantly, education' (Syrkus 1926, p. 9).

'The kitchen issue' and Brukalska's other projects introduced, or only prepared, in the Warsaw Housing Cooperative aimed to provoke and promote a modern way of life, modelled after democratic societies where professionally active women ate lunch during the day, rested and relaxed at home, and were able to quickly and efficiently prepare dinner for the family. In no way were the flats meant to resemble a working-class dive bar with its ghastly kitchen smells. The designers of the Żoliborz estate assumed that not only men, but also women and children, have their psychological needs and have to take time off to relax and be sometimes alone with their thoughts.

Kitchen designs presented by the Housewife's Association are interesting in this regard.[14] They were very similar to Brukalska's ideas in both furniture arrangement and layout, featuring multi-functional furniture (including a folding table with numerous drawers or a 'cooker' table) and walk-in, ventilated pantries. If we look at 'working photographs', however, many differences become easily apparent. The model, the 'kitchen actress', is no longer a slim, sporty, contemporary woman, but rather a focused, level-headed housewife with a white scarf on her head. Janina Jankowska's kitchen was separated from the living room, but there was a window in the wall allowing dishes to be easily passed into the living room. Although the kitchen was

[14]The Housewife's Association also presented their proposals for solving the 'kitchen issue' (Dom, Osiedle, Mieszkanie 1934b, pp. 9–13).

KUCHNIA MIESZKALNA Z WYODRĘBNIONĄ CZĘŚCIĄ GOSPODARCZĄ W IV KOLONJI W. S. M.

POWIERZCHNIA MIESZKANIA DWUIZBOWEGO w m^2

Kuchnia mieszkalna	19,4
Część gospodarcza	3,5
Pokój sypialny	15,0
Przedpokój	1,9
W. C.	0,9
	40,7

57

Fig. 4.9 Residential kitchen with a separate storage zone in the Colony IV Report on the WHC activities for 1929. 3446, p. 75

modern, efficient and included many solutions that made housewives' lives easier, it was a different kind of 'event' altogether. Jankowska offered a new interior, but it was not accompanied by a new lifestyle. She was not critical, innovative or determined. Unlike Brukalska, she was not an activist; her manner was unprovoking and unengaging. She simply explained that her idea reflected the modern way of life (Fig. 4.10).

Under pressure from tenants, Brukalska's open kitchens started being separated from the living room: initially, with twin sliding doors, and later with a single door. The window that connected the living room and kitchen, making it possible to 'pass dishes' to the family, was an inherent part of working-class flats. Flats meant for members of the intelligentsia, on the other hand, had the kitchen doors in the hall, making the kitchen a completely separate room. Thus, it could also be used as servants' accommodations, to which end folding beds were often installed there. Such modifications show that interiors are transformed to suit the needs of users, regardless of what the designers had intended or planned. Estates are living organisms where tenants are free to experiments with new ways of living.

Tołwiński's emancipatory efforts did not eradicate the intelligentsia's need for servants. Therefore, the Housewife's Association campaigned for setting a minimum size of servant's quarters (2 m × 2 m). 'The kitchen niche does not solve the problem, as it would force the housekeeper to wash, groom and dress herself in the kitchen. It would be advisable for the housekeeper's quarters to be furnished with a sink

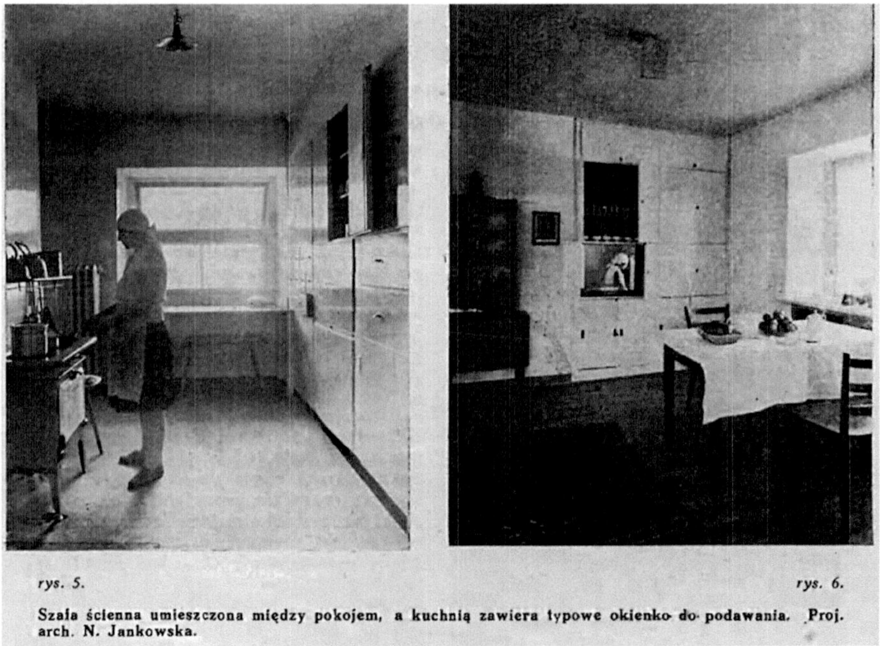

Fig. 4.10 Kitchen by N. Jankowska—Kitchen designed by Nina Jankowska, 'Dom. Osiedle. Mieszkanie', 1934, no 1, p. 76

and wardrobe' (Dom, Osiedle, Mieszkanie 1934b, p. 13). Much was done to solve the 'kitchen issue' and 'servants' quarters problem', but tenants nevertheless devised various tricks to be exempt from the uniformisation imposed on them by the principles of rational space management (which turned out to be, in a sense, irrational, as they did not account for the needs and habits of residents). They looked for various 'gaps' and mistakes made by the organisers of the estate's cultural life and took advantage of them to oppose the prevailing discourse of technological functionality and aesthetical judgements.

20% of Żoliborz flats were meant for members of the intelligentsia and had two, three or three and a half rooms with a separate kitchen, doubling as servants' quarters, and bathroom (colonies IV, VII and VIII). The important and widely discussed issue of cleanliness and personal hygiene was solved in various ways: initially, a big bathtub was installed in the kitchen, and short bathtubs were added in the bathroom later. All of that, however, was too expensive for members of the housing cooperative. Therefore, it was decided to build public baths. With time, however, tiny bathrooms, containing a shower, sink and toilet were installed in the intelligentsia flats.[15] The two types of Żoliborz flats (meant for the intelligentsia and working class) did not take their final form until 1930. Servants' quarters were not included in the designs until 1932. In practice, however, kitchens doubling as servants' quarters became an integral part of intelligentsia flats (cf. Dom, Osiedle, Mieszkanie 1936).

We may assume that those guidebook initiatives were aimed mostly at members of the intelligentsia, who were thought to be more malleable and less entrenched in their ways. Working-class residents were allured by the bourgeois way of life, which they associated with luxury and a conservative world view. The intelligentsia, on the other hand, were expected to abandon their king-sized beds in favour of daybeds and fold-away children's beds. 'A small, tidy flat with simple wooden or rattan furniture can be beautifully decorated with fabrics and ceramics. Shelves and walls may be adorned with Hutsulian bowls, for example. A white linen tablecloth with an elaborate pattern may be used to cover the table. A Bolimów milk pitcher or flower vase would nicely compliment the interior, as would Vilnius tapestries' (Dom, Osiedle, Mieszkanie 1932, p. 34). The flats were meant to be tidy, bright and comfortable, but also modern with some folk motifs. Alas, even the intelligentsia found it difficult to adopt this functional constructivism. Such interior design trends emerged during the post-war period and were slowly losing their avant-garde spirit. Most thought of them as overly folksy. Although the "kitchen debate' had introduced political experiments with modernist utopias into areas previously unknown' (Leśniakowska 2004, p. 197), in practice it mainly allowed residents to experiment with new ways of living.

It had been assumed that everyday objects and modern equipment could be used to shape the identity of residents. It had been believed that flats are the source of social identification, invoking a feedback loop that would allow tenants to lead

[15] When work started on Colony IX of the Żoliborz estate, Jan A. Szymański noted that 'earlier buildings did not have separate bathrooms with showers' (Szymański 1989, p. 55).

a modern lifestyle, consciously shape their habits, resting patterns, approach to cooking, hygienic rituals and intimate life.

Such 'authoritative speech acts' were no single occurrences, but rather a long-term educational process based on repetition (in national and local press, housing cooperative regulations, constitution, housing savings books, tenant flat exhibitions and countless speeches). Even during the war, Stanisław Ossowski wrote about the necessity of popularising a different type of working-class accommodations. He was aware that 'when thinking of climbing the social ladder, workers most often see themselves living a bourgeois lifestyle, complete with plush furniture and figurine-filled glass-fronted cabinets' (Ossowski 1970, p. 140). Class-based differences were evident in the architectural styles of buildings and their interiors. Thus, Żoliborz was considered an intelligentsia estate, while Rakowiec was associated with the working class.[16] We should, therefore, consider Michel de Certau's writings in which he states that 'It is that which happens beneath technology and disturbs its operation which interests us here. This is technology's limit, which has long since been noticed but to which we must give a significance other than the delimitation of a no man's land. This is a matter of actual practices. Conceptual engineers are familiar with this sort of movement, which they call 'resistance' and which disturbs functionalist calculations (…). They cannot fail to notice the fictive character instilled in an order by its relationship to everyday reality' (Certeau 1984, p. 200).

4.4 How Do the Architects Work and Who Do They Appeal to?

The Cooperative's ideological objectives were emancipatory in nature, but we must not forget that the smallest flats were also meant to satisfy the psychological needs of the intelligentsia, and shape such needs among the working class. The apartments were, therefore, designed with the following psychological needs in mind: intimacy, separateness (embodied by the 'separate beds for all tenants' slogan, or the notion that there are separate rooms for parents and children), as well as more intellectual needs, such as: reading, music, meeting friends and after-work activities. 'Regardless of that—writes Brukalska—we see the need for a social environment that would satisfy the tenants' psychological needs on a bigger scale or in more specialised areas through libraries, events, clubs, socialising opportunities and other means' (Brukalska 1948, p. 23). Those are luxury common goods, forming a foundation of 'live work', which Michael Hardt and Antonio Negri enthusiastically referred to as 'biopolitical work'. Żoliborz's activists were, also in this case, aware that such goods

[16]This issue has been covered extensively by M. Leśniakowska, who provides a detailed description of Brukalska's struggles with the 'kitchen issue', concluding that it was one of the most important problems that modernists had had to tackle. E. Mazur presented an interesting approach to documenting the arrangement of worker's flats by reconstructing the critique of bourgeois accommodations based on the 'Życie WSM' [The Life of WHC] magazine and 'Dom. Osiedle. Mieszkanie' [House, estate, flat] (cf. Mazur 1993).

cannot be distributed on a class basis. That is why the collective ran surveys among tenants to determine the desired arrangement of flats, e.g. the number of beds. The rationale given was that 'whether a child sleeps in their own bed or has to share it with its siblings is not without impact on the health of the children and their parents, the teachers and the community that we are striving to create' (Życie WSM 1933b, p. 6).

In order to ensure unbridled development and allow children to concentrate, the architect has to create conditions in which the need for solitude may be satisfied. Victory in the fight for silence may be secured on several levels: flats have to have at least one non-connecting room, allowing those who seek peace and quiet to find it. As regards public spaces, social buildings had so-called quiet rooms (Fig. 4.11), there were secluded places in the community centre, courtyards and green areas were designed in such a way as to allow finding peaceful spaces that were ideal for doing activities requiring concentration. Even today, many of Żoliborz residents (including those that never had a chance to see its pre-war form) describe the estate as 'easy-going', 'people-friendly', where one can 'feel free'.[17]

That is how Żoliborz was built. At the same time, construction started on a residential estate in Rakowiec which would provide affordable housing for working-class families. The project was led by the WHC management, Praesens architects and tenant representatives. Unlike Żoliborz, with its ever-growing population of 'working intelligentsia', Rakowiec was intended to be a purely working-class estate. Thus, flats were designed to be affordable, meaning that they were not as well-equipped and comfortable as those meant for the intelligentsia.[18] Helena Syrkus, who together with her husband designed the entire estate, recalls: 'our whole team was shocked when the future tenants told us just how little they needed in terms of furniture and equipment. It was then that we understood just how impoverished working-class families were, having to get by on two hundred zlotys a month. It was one thing to hear that from a sociologist, but it really hit me hard when I heard it from a bitter father of malnourished children…' (Syrkus H 1976, p. 102. cf. Syrkus and Syrkus 1937, pp. 22, 27). For example, Rakowiec tenants insisted on central heating not being installed, preferring old-fashioned stoves instead. 'Even a small rent increase would be too much for the already strained home budgets of the tenants. If we cannot

[17]These are characteristic expressions describing Żoliborz; they appear in many residents' statements. At this point, I shall quote Leszek Szkutnik's statement, '*Rozmowa z anglistą, autorem podręczników i sztuk dramatycznych – Leszkiem Szkutnikiem'* [A Conversation with an Anglist, Author of Textbooks and Dramas—Leszek Szkutnik] (Szkutnik 2009, p. 202).

[18]'Życie WSM' from March 1933 presents, in this context, an approximate estimate of earnings which perfectly reflects the proportions of rent paid to the Cooperative: 'Approximately 75% of blue collar and white collar workers in Poland currently earn no more than PLN 150 per month. Therefore, if we consider a 2-room flat, for which the rent amounts to approx. PLN 70 per month (with central heating), it becomes evident that these flats are too expensive for average-earning blue collar and white collar workers. That is why our estate has become an oasis for relatively affluent people. (…) Tenant representatives, or delegates, regardless of the difficulties should—through their help or creative criticism—persistently strive for further and necessary rent cuts, so that our cooperative does not lose its original character and become a 'blue collar workers' cooperative' only in theory' (Sieradzki 1933, p. 1).

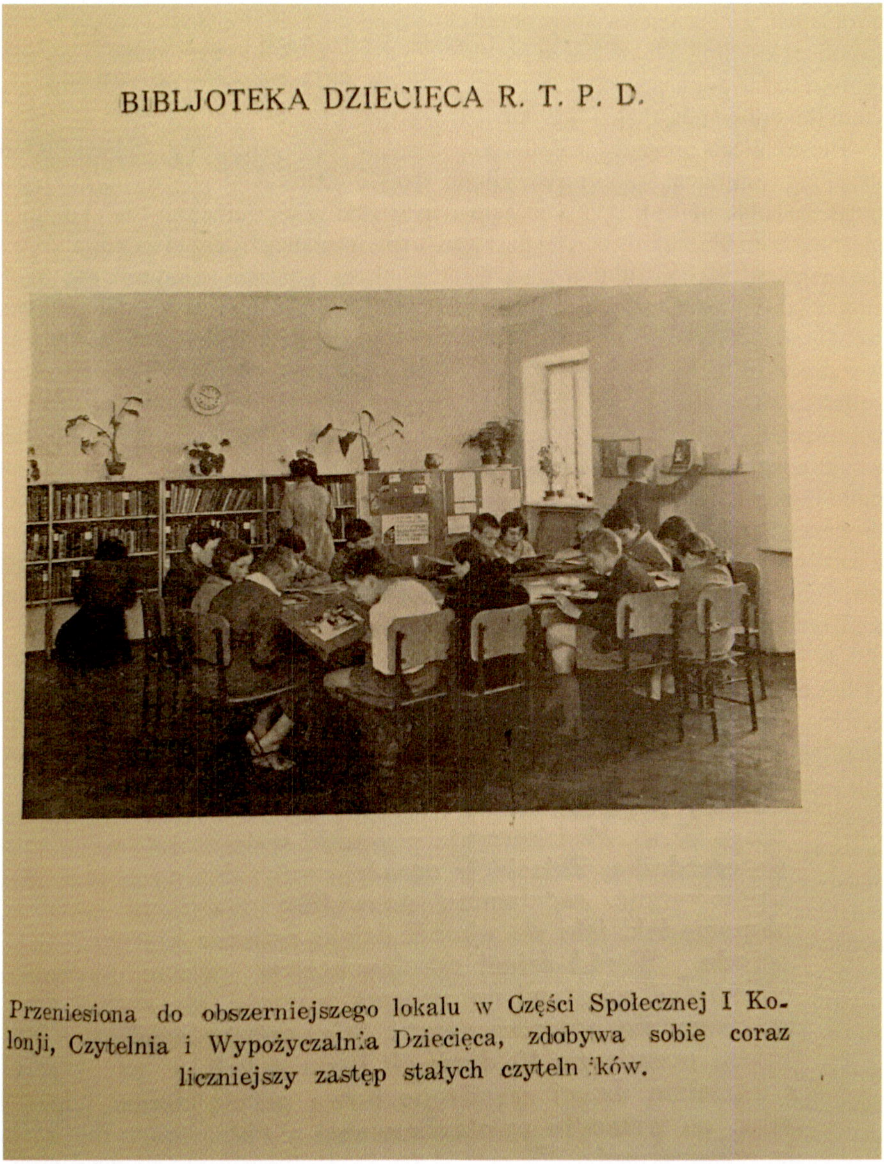

BIBLJOTEKA DZIECIĘCA R. T. P. D.

Przeniesiona do obszerniejszego lokalu w Części Społecznej I Kolonji, Czytelnia i Wypożyczalnia Dziecięca, zdobywa sobie coraz liczniejszy zastęp stałych czytelników.

Fig. 4.11 Children's Library of the Workers' Society of Children's Friends (RTPD), Report on the WHC activities for 1934. 3796, p. 79

afford coal, we burn newspapers or old crates and send children to the nearby forests to collect kindling. Or we just suffer through the cold'—said the tenants. They also refused to have a gas system installed: 'women will be tempted to cook with gas. Until the bill comes, that is' (Syrkus 1976, p. 102).

The architects prevailed over the tenants and WHC management, who pushed for installing communal toilets on each floor. The 'Toilet in each flat!' slogan, aiming to break the habits of the people who were to live in the estate, gained an almost political meaning with the architects. The flats were to perform, teach certain habits, and even the Cooperative's constitution stated that the ultimate goal was to improve the 'living culture' of its residents. It is difficult to say whether the attempt to build flats that would satisfy urgent demand, while also adhering to the principles of social housing, was successful. The apartments were of such a low standard that hygienic practices suffered and, thus, so-called minimal housing conditions were not fully met. We must not forget, however, that the experiment did not only involve designing housing that would adhere to a certain social, cultural, pedagogical and hygienic profile, but, more importantly, its main goal was to build flats on an ongoing basis, relying on the concept of housing cooperatives, engaging tenants and, in effect, bringing to life a participatory tenant model based on architectural design collaborations. The flats, therefore, were not simply a place to live, but also served as a gateway to becoming a citizen, making them, in a sense, a political act.

These acts, however, were inherently ambivalent. They did improve the quality of life of the working class, allowing workers to climb the social ladder. They did not change their class identity, however. There is no denying that members of the working class were 'on the fringes', caught somewhere between passive, impoverished pariahs and emancipated citizens (the latter earned their status mostly by submitting to WHC's logic of domination and living in a 'small, modernly appointed housing cooperative flat'). 'Residents of the estate [Rakowiec—author's note] had steady jobs and flats, both of which were the dream of all physical labourers. Their neighbours, often unemployed and living in dilapidated shacks, were envious and considered them the workers' aristocracy. In private conversations, they referred to their well-off peers as the 'red bourgeoisie' and treated them with mistrust or even hostility' (Szymański 1989, p. 90).

'Red bourgeoisie', 'PPS supporters'[19] are liminal entities. As I have emphasised before, architecture was a tool of social change, a performative challenge that encouraged the people to rise against the capitalist ideology and the Sanacja regime. On the other hand, the creators of WHC, local activists and architects had their own performative normative strategies, demonstrating their power of assigning special meaning to objects and places, but also imposing certain behaviours on individuals, both in private and in public. 'Teaching how to live' was a challenge that could be overcome using 'preserved behaviour', including a regulated set of actions, repetitions and performative instructions, which were experienced, practised and quoted countless times.

[19] A term, often contemptuously used, for describing left-wing residents associated with the Polish Socialist Party (PPS).

Thus, the Warsaw Housing Cooperative snatched the following entities from the clutches of the political and economic system: workers, their families, wives and children, and placed them on the fringes of the exploited working class. Their old social norms were ridiculed and transformed, both spatially and performatively. The working-class families had not been left lingering in the innocent state of *communitas*, however. Instead, they were reclaimed through radical practices. 'More generally, the liminal-norm operates in any situation where the valorisation of liminal transgression or resistance itself becomes normative' (McKenzie 2001, p. 50). As we can see, this type of community needs institutional violence to survive. The WHC estate resembles Krzysztof Nawratek's concept of the City, in which two types of oppression were present. 'The first is genetically pre-urban oppression—a sectarian, conformist oppression that binds people, tethers their hearts, minds and consciences. The second type of oppression comes from the city itself, representing all that the city is and how it differs from other communities. This second type of oppression is one that forces us to meet, interact and collaborate' (Nawratek 2011, p. 144). Nawratek calls it the 'empty oppression', which does not involve managing human capital in urban spaces, but rather allows citizens to independently manage their lives at the estate, community and local levels. 'The city has to use an active oppression, an active violence, to force people to interact, exchange and, finally, become dependent on each other. The new Polis, just as Aristotle's, must educate its citizens' (Nawratek 2011, p. 149).

If we adopt that perspective, Żoliborz is a self-governing City. Although I have focused mostly on the architectural and urban planning aspects, it seems that the City is more than just 'urban space'. It can be perceived as an active entity with its own will which it uses to transform the habitus of its citizens, as well as educate and invigorate them. The home, kitchen and interior decoration are no longer intimate. They take on a new social and public meaning, becoming the expression of the 'small things' political movement, class struggle and new lifestyles. They are flexible, spontaneously rising from the bottom-up, but their ideological legitimation is watched over by Żoliborz's designers (including architects and the creators of cooperative ideas). All of that makes the City a community which, on the one hand, is disciplined by the authorities (urban planning and architecture) but, on the other hand, disciplines itself (which is manifested by the aforementioned letter from a female resident written in response to the fines imposed by the authorities as a demonstration of their powers).

References

Abramow-Newerly J (2000) Lwy mojego podwórka [The lions of my courtyard]. Wydawnictwo Twój Styl, Warszawa

Austin JL (1962) How to do things with words. Oxford University Press, Oxford. https://ia801306.us.archive.org/23/items/HowToDoThingsWithWordsAU-STIN/AUSTIN%20J.L%20-%20How%20To%20Do%20Things%20With%20Words.pdf. Accessed 7 Sept 2018

Brukalska B (1929) Kuchnia współczesna [The contemporary kitchen]. Dom, Osiedle, Mieszkanie [House, estate, flat], March, pp 8–11
Brukalska B (1948) Zasady społeczne projektowania osiedli mieszkaniowych [Social rules for the design of housing estates]. Developed at the Institute of Social Economy. Wydawnictwo Ministerstwa Odbudowy, vol 22, Warszawa
Cegielski J (1968) Stosunki mieszkaniowe w Warszawie w latach 1864–1964 [Housing relations in Warsaw in the years 1864–1964]. Arkady, Warszawa
Cieślewski syn [Tadeusz?] (1930) Sztuka w pokoju [Art inside the room]. Dom, Osiedle, Mieszkanie 5:16–20
Corbusier L, Jeanneret P (1965) Oeuvre complète 1910–1929, 8th edn. Les Éditions d'architecture, Paris
Certeau M de (1984) The practice of everyday life (trans: Rendall S). University of California Press, Berkeley. https://chisineu.files.wordpress.com/2012/10/certeau-michel-de-the-practice-of-everyday-life.pdf. Accessed 7 Sept 2018
Dom, Osiedle, Mieszkanie (1929a) Dom, Osiedle, Mieszkanie 10–12:33
Dom, Osiedle, Mieszkanie (1929b) Dom, Osiedle, Mieszkanie 12:13–16
Dom, Osiedle, Mieszkanie (1932) Dom, Osiedle, Mieszkanie 3:34
Dom, Osiedle, Mieszkanie (1934a) Dom, Osiedle, Mieszkanie 2:6–11
Dom, Osiedle, Mieszkanie (1934b) Jak wybrać mieszkanie: Kuchnia według Związku Pań domu (Z.P.D.) [How to choose a flat. The kitchen according to the housewife's association]. Dom, Osiedle, Mieszkanie 1:2–14
Dom, Osiedle, Mieszkanie (1936) Dom, Osiedle, Mieszkanie 10–11:17
Heyman Ł (1976) Nowy Żoliborz 1918–1939. Architektura – urbanistyka [New Żoliborz 1918–1939. Architecture—urban planning]. Zakład Narodowy im. Ossolińskich, Wrocław
Jarzęcki J (2010) Wspomnienia Janusza Jarzęckiego. In: Pałka J, Szamruchiewicz A, Szymańska-Szwąder M (eds) Warszawa międzywojenna w pamięci jej mieszkańców [Interwar Warsaw in memories of its residents]. Dom Spotkań z Historią: Ośrodek Karta, Warszawa, pp 56–62
Joas H (1996) The creativity of action. The University of Chicago Press, Chicago
Kobieta w Świecie i w Domu (1929) Dom Tayloryzowany [A Taylorised House]. Kobieta w Świecie i w Domu 1:15–16
Kownacka M (1933) Kryzys a zielona kanapa [The Crisis and a Green Couch]. Życie WSM, February, pp 6–7
Kuzańska-Obrączkowa M (1966) Koncepcje wychowawcze Robotniczego Towarzystwa Przyjaciół Dzieci 1919–1939. Analiza założeń i funkcjonowania placówek [Educational concepts of the Workers' Friends of Children Association 1919–1939. An analysis of school principles and functioning]. Zakład Narodowy im. Ossolińskich, Wrocław, Warszawa, and Kraków
Leśniakowska M (2004) Modernistka w kuchni: Barbara Brukalska, Grete Schütte-Lihotzky i 'polityka kuchenna' [A Modernist in the Kitchen. Barbara Brukalska, Grete Schütte-Lihotzky and the Kitchen Policy]. Konteksty: Polska Sztuka Ludowa 58(1–2):189–202
Matysek-Imielińska M (2016a) Educational civil experiment: a story of a particular radical housing estate. J Crit Educ Policy Stud 14(3):251–279
Matysek-Imielińska M (2016b) Mieszkajmy razem!: pytania o odpowiedzialność Wrocławskich Kooperatyw Mieszkaniowych [Let's Live Together!: Questions About the Responsibilities of Wrocław Housing Cooperatives]. Kultura Współczesna ["Contemporary Culture' quarterly] 92:162–177
Mazur E (1993) Warszawska Spółdzielnia Mieszkaniowa 1921–1939. Materialne warunki bytu robotników i inteligencji [Warsaw housing cooperative 1921–1939. The living conditions of workers and intelligentsia]. IAiE PAN, Warszawa
McKenzie J (2001) Perform or else: from discipline to performance. Routledge, London and New York. https://monoskop.org/images/8/8a/McKenzie_Jon_Perform_or_Else_From_Discipline_to_Performance_2001.pdf. Accessed 9 Sept 2018
Mencwel A (1990) Etos lewicy. Esej o narodzinach kulturalizmu polskiego [The ethos of the left. Essay on the birth of Polish culturalism]. Państwowy Instytut Wydawniczy, Warszawa

Mencwel A (1998) Przedwiośnie czy potop. Studium postaw polskich w XX wieku [Early Spring or the Deluge. Study of Polish attitudes in the twentieth century]. Czytelnik, Warszawa
Meyer E (1928) Der Neue Haushalt. Frankh, Stuttgart
Nawratek K (2011) City as a political idea (trans: Pyzik A). University of Plymouth Press, Plymouth
Ossowski S (1967) Organizacja przestrzeni i życie społeczne w przyszłych osiedlach [Spatial organisation and social life in future housing estates]. In: Ossowski S, Dzieła [Collected works], vol 3 [Issues of social psychology]. Państwowe Wydawnictwo Naukowe, Warszawa, pp 351–370
Ossowski S (1970) Kultura robotnicza [Workers' culture]. In: Ossowski S, Dzieła [Collected works], vol 6. Państwowe Wydawnictwo Naukowe, Warszawa, pp 139–143
Oud JJ (1926) Wychowanie przez architekturę [Education through architecture]. Praesens: Kwartalnik modernistów [Praesens. Modernists' quarterly] 1:4–5
Pogorzelska J (1929) Sekcja Gospodarstwa Domowego na Naukowym Kongresie Organizacji Pracy [The household section at the scientific congress of labour organisation]. Kobieta w Świecie i w Domu 1:2–3
Pronaszko A (1929) Dialog o wnętrzu [Dialogue about the interior]. Dom, Osiedle, Mieszkanie 10–12:4–7
Reicher I (1938–1939) Sprawozdanie z ankiety o stanie mieszkań robotniczych w W.S.M. pod względem ładu, estetyki, i nowoczesnego umeblowania [Report on the survey regarding the condition of workers' flats in the WHC in terms of order, aesthetics, and modern furnishings], Warszawa 1938–1939, Izba Historii Warszawskiej Spółdzielni Mieszkaniowej [typescript, Warsaw Housing Cooperative History Chamber]
Servant' (1933) List służącej [Servant's letter] (1933) Życie WSM, March, p 3
Sieradzki J (1933) O wysokość komornego w naszych domach [On the level of rent in our houses]. Życie WSM, March, p 1
Święcicka J (1933) Pod wpływem demagogicznych wystąpień przeciwko pracownikom [Under the influence of demagogic speeches against workers]. Życie WSM, October, p 11
Syrkus S (1926) Preliminarz architektury [The architectural estimate]. Praesens: Kwartalnik modernistów [Praesens. Modernists' quarterly'] 1:6–16
Syrkus H (1976) Ku idei osiedla społecznego 1925–1975 [Towards the idea of the social housing estate 1925–1975]. Państwowe Wydawnictwo Naukowe, Warszawa
Syrkus H, Syrkus S (1937) Współdziałanie użytkowników przy opracowywaniu projektu mieszkania robotniczego [The cooperation of users in developing the design of a workers' flat]. In: Osiedla robotnicze: Studia i doświadczenia pierwszego dziesięciolecia [Workers' Housing Estates. Studies and Experiences of the First Decade of Their Existence]. Biblioteka Polskiego Towarzystwa Reformy Mieszkaniowej, Warszawa, pp 22–34
Szkutnik L (2009) Mieszkać na Żoliborzu [Living in Żoliborz]. In: Chałasińska E, Gawecka A (eds) O pięknym Brzegu. Żoliborz, ludzie, wydarzenia [About the Beautiful River Bank. Żoliborz, People and Events]. Towarzystwo Przyjaźni Polsko-Francuskiej, Fundacja 'Żoliborski Fundusz Lokalny', Warszawa, pp 199–204. Reprint of: Szkutnik L (2008) Rozmowa z anglistą, autorem podręczników i sztuk dramatycznych – Leszkiem Szkutnikiem [A conversation with an anglist, author of textbooks and dramas—Leszek Szkutnik]. Informator Żoliborza [Żoliborz's Guide], 24 April 2008
Szymański JA (1989) Warszawska Spółdzielnia Mieszkaniowa 1921–1970. Zarys dziejów [Warsaw housing cooperative 1921–1970. History outline]. Wydawnictwo Spółdzielcze, Warszawa
Wohnout W (1933) Wnętrze [The interior]. Życie WSM, January, pp 2–3
Życie WSM (1933a) W sprawie 'służących' [On 'servants']. Życie WSM, March, pp 3–4
Życie WSM (1933b) Życie WSM, December, p 6
Życie WSM (1936) Życie WSM, January, p 15
Życie WSM (1937) Życie WSM, November, p 213

Chapter 5
Sensitive Urban Planning or Critical Spatial Practice?

Abstract Both 'small architecture' and design, but also the process of planning the entire housing estate could be labelled as experimental independent critique of space (Markus Miessen), thanks to which the architects and social activists created social, cooperative, neighbourhood and tenant programmes that fostered cooperation and the formation of communities on the housing estate and socially responsible sharing of space and caring for it. Educating the tenants and animating various activities in the extensive public spaces (communal house) and common areas (courtyards, green spaces and pathways) proved to be extremely important. This education proved also emancipatory in its nature since it consisted in the production of urban entities—active citizens. In this chapter, I depict the interdisciplinary ambitions of the Żoliborz project, in which architects work together with sociologists, educators, hygienists and urban planners. As a result of these activities, the so-called living culture has been developed. I analyse in detail the concept of 'socialised individualism' developed by Barbara Brukalska, as well as the most important guidelines for the design of social housing estates included in her brochure *Social rules of housing estates design*. I confronted Brukalska's concept (which I modernised and labelled, after Miessen, as an independent practice) with Helena Syrkus' vision. Both of them are looking for the estate's central point: Brukalska calls it 'estate core', while Syrkus—'estate axis'. With Krzysztof Nawratek, I also put forward the thesis that the city must/should provide a kind of 'empty oppression'. Taking Nawratek's idea further, I analyse the housing estate's space in the context of two categories: 'spaces of interaction' and 'spaces of intimacy'. Thanks to this, I avoid using the terms of 'private' and 'public' and analyse the city itself as a political idea. The space, organised by planners and architects, is a tool that can embody this idea.

Keywords Critique of space · Socialised individualism · Barbara Brukalska · Estate core · Space of interaction · Space of intimacy · Empty oppression

WHC's urban planning and architectural features were undoubtedly avant-garde, but the groundwork for them had been laid by other European projects. The Cooperative used whatever resources it could gather and received little public aid. That did not stop the founders of WHC from pursuing their daring social, pedagogical and cultural ideas that were miles ahead of those proposed by their well-organised and

M. Matysek-Imielińska, *Warsaw Housing Cooperative*, The Urban Book Series,
https://doi.org/10.1007/978-3-030-23077-7_5

experienced counterparts from Western Europe. The construction of workers' estates in Western European countries was initiated and supervised by municipal authorities in an economic and social environment that Warsaw's housing cooperative activists could have only dreamt of. Western European initiatives, however, were mostly limited to providing housing, with only the most progressive cities introducing broader social programmes. Although the designers of Frankfurt am Mein's modernist housing estates, among them Ernst May, had initially planned to include some 'social facilities' in their designs, such as communal houses, preschools, playgrounds, rest areas, central heating boiler rooms and PA systems, most of those plans were abandoned at the end (namely the communal house, estate school and playgrounds) (May 1928).

Vienna's urban strategy of working-class housing estate development was overseen by the socialist municipal authorities, who managed to build 30 thousand flats between 1923 and 1927 (cf. Heyman 1976, p. 95). The Vienna flats were much more ground-breaking and comfortable, to the point of being perfect (cf. Vienna's Gütrel estate which inspired many of Warsaw's architects and housing cooperative activists). We can therefore safely say that, despite insufficient funding and lacking government support, WHC's activists proved to be truly pragmatic and creative. Making the best possible use of the little resources available was part and parcel of the activist-architect's daily struggle. Nevertheless, they never abandoned their ideals. Whereas their pragmatism, which paled in comparison to Le Corbusier's imagination or May's mastery, allowed Warsaw's activists to achieve their social goals. The methods that they used (including architecture and urban planning) sometimes had to be tailored to the task at hand, thus necessitating a flexible approach. In other instances, it was the social goals that defined which urban planning solutions would be applicable.

Polish avant-garde architects did not limit themselves to designing cheap houses and small flats. They wanted to introduce more general, holistic solutions to the housing issue. The Żoliborz estate had a certain atmosphere about it, a sense of caring for the social needs of its residents, which was mainly due to the humanistic nature of the architects, or more precisely, female architects: Barbara Brukalska, Nina Jankowska (Żoliborz) and Helena Syrkus (Rakowiec).

Adolf Ciborowski, Warsaw's chief architect in the 1960s, asked—with good reason, I might add—whether it was by mere coincidence that WHC became the birthplace of Polish modernist housing architecture. 'I am deeply convinced that this was a result of the social character and social principles that the Warsaw Housing Cooperative was built upon. The adopted urban planning concepts were not a result of coincidence or theoretical musings on spatial planning. For solid urban planning has the overarching goal of serving the citizens and providing them with equal and just living conditions, housing and recreation, as well as social care, to the extent that is possible using the available economic resources' (Ciborowski 1962, p. 4).

The Żoliborz colonies, however, were designed chiefly by men. They were given a free hand to realise their modernist ideas, as the land that they were building on was empty or, as some might say, untouched.

Antoni Libera provides a perfect description of their monumental urban/architectural plans in his short story entitled *Widok z góry i z dołu* [View From Above and From Below]. Uncle Oskar, one of Żoliborz's builders, tells a nine-year-old boy about the empty land stretching before them: 'And here, all around us—uncle Oskar pointed at the ground and motioned a half-circle—a workers' estate: about twelve colonies, with courtyards, a preschool and public library. The Warsaw Housing Cooperative, modernist in style. The whole thing is shaped like a trapezoid, with the base facing East. In a way, it is a retort to the Citadel, which also was built on a trapezoid plan. The city defying the fortress. A modern district rising from the ground after a hundred-year blockade. With the Wilson Square in the middle. 'Żoliborz. The Beautiful Bank' (Libera 2012, p. 30). The following excerpt takes us to a tall chimney of a Żoliborz boiler house. 'The boiler house, the heart of the estate—uncle Oskar explained with pride—that's how we build now! No more coal stoves, no more smoke, no more carbon monoxide. Central heating! Connected to over a dozen buildings over a one-kilometre radius! (…) They set off in the direction of the chimney, running down a small incline along the way, and stopped on a little square.—This is where festivals and dances are held, and in the winter it is turned into a skating rink—uncle Oskar went on.—They have thought of everything. Just as an ideal city should be' (Libera 2012, p. 28).[1]

Meanwhile, let us put aside the literary tales of great builders and focus on real architects (Fig. 5.1).

The idea of subdividing a residential estate which spans a vast piece of land and houses several thousand residents was itself interesting. The estate's subelements, i.e. colonies, were designed to organise the everyday life of residents in a way that would create neighbourly ties and a feeling of community. Colonies were therefore meant to integrate, allowing for mutual assistance between tenants to grow. As the Cooperative's basic organisational unit, colonies determined the complicated network of interdependencies between the various institutions and cross-group agreements. Such networks made it possible to set up a decentralised administrative system (which will be discussed later). The founders of WHC often stressed that their aim was to introduce a 'democratic management system of social facilities'. The Tenant Council was elected by all interested parties (tenants and shop owners) and was held accountable by them. 'As the number of residents grows, the connection between them and even the most democratic of councils will break down. As a council loses touch with its constituency, sooner or later it is bound to morph into a bureaucracy'. Thus, the idea to set up colonies of several hundred tenants each was born. 'The most needed social facilities are organised within the colonies. All of the colonies combined form an estate numbering several thousand residents' (Brukalska 1948, p. 26).

[1]Żoliborz was Poland's first residential estate that had a central heating system with a boiler room (Malicki 1957).

Fig. 5.1 View of the Boiler House from Suzina Street, Report on the WHC activities for 1934. Photo description: 'WHC estate in Żoliborz. Only WHC houses are located at Suzina Street.' 3777, p. 85

Under the photo, data set: WHC houses in the housing estate in Żoliborz

Colony	Number of buildings	Cubic capacity in m^3	Useful floor area in m^2	
			Total	Of dwellings
I	4	37735	7029	5955
II	1	27942	5224	4827
III	2	44479	8297	8297
IV	3	59624	11345	11215
V	2	43317	7887	7887
VI	3	10836	1621	–
VII	2	37968	6773	6773
VIII	1	32938	6660	6660
In total	18	294839	54836	51614

Report on the WHC activities for 1934

Richard Sennett states that cooperation may be treated as a means of achieving political or economic ends. In such circumstances, all interactions and exchanges should be ordered and disciplined. As we will soon discover, that was often the case when the estate was faced with ideological and political conflicts, mainly between socialists and communists (*jaworowszczyzna*).[2] If cooperation is a goal in itself, however, then all 'games over power and dominance' must be abandoned. Cooperation is built on free participation. That does not mean that there is no need for rules or a formal leader. 'To enable participation, the organiser may establish tacit ground rules, the conventions and rituals for exchange (...), but must then leave people free to interact' (Sennett 2012, p. 53).

Did the estate, and the movement options and routes within it, facilitate the development of such cooperation, or did it simply impose the rules of estate life on the tenants?

How should sociologists, urban planners and architects shape social ties? What should those ties be like? Do rules and regulations governing the use of various social facilities run contrary to the idea of estate collaboration?

5.1 Collaborative Space and Independent Criticism

By analysing the interrelations between the social dimension of residential estates and respect for individualism, Brukalska attempts to determine the limits of spatial organisation. She is well aware that going too far may infringe upon the freedom of individuals. Therefore, she decides to rely on informal rules of coexistence, or 'tacit rules', rooted in everyday rituals. 'It seems that the only limitations that do not strike at the very heart of freedom are voluntary agreements between the majority of counterparties and some customary norms, which also derive from tacit agreements. It is possible to organise the tenant community based on a hierarchical order of individuals (...) only on the basis of a voluntary agreement (had it not been voluntary, it would not be deliberate)' (Brukalska 1948, p. 28).

Brukalska examines the relationships in the estate with an almost sociological precision, which she owes to the conversations with Stanisław Ossowski and their collaboration in the underground Architecture and Urban Planning Studio (Pracownia Architektoniczna i Urbanistyczna, PAU). She looked into the feelings of belonging, familiarity, identity, but also 'designed' external relationships, the estate's relationship with the city.

---

[2] *Jaworowszczyzna* is a term used in the interwar period to describe the split that took place within the Polish Socialist Party due to the actions of Rajmund Jaworowski. Jaworowski tried to antagonise the Polish Socialist Party's members by voicing his support for the 'Sanacja' government of Józef Piłsudski. Thus, Jaworowski managed to split the PPS and also the Polish Socialist Parliamentary Union (Związek Parlamentarny Polskich Socjalistów), trade union organisations and even the Towarzystwo Uniwersytetu Robotniczego (TUR, i.e. the Society of the Workers' University). Cf. Naprzód (1928a, b).

Brukalska's 'social individualism' was evident in her enormous respect for the individual, but also her care for neighbourly ties, her idea of civic attitude and her attempts to encourage 'internal activity in the community', which, using her sociological imagination, she described as 'the influence of the community on its own members consisting in the members willingly taking part in the social life of their community. When a community has a disorganised and chaotic influence on its members, it clouds objective judgement and seems to lead to two ultimate outcomes: being in constant company of other people or suffering absolute solitude. Furthermore, one does not exclude the other' (Brukalska 1948, p. 29).

When setting the hierarchy of priorities governing the life of a residential estate, Brukalska states: 'The main goal of social organisation is to provide such conditions that would allow each individual to freely grow. The secondary, albeit decisive goal, is to regulate the rights to freedom and growth among the members of the community' (Brukalska 1948, p. 70). As we can see, the benefits that individuals enjoy from living in neighbourly proximity cannot replace the benefits that come with solitude, creative growth, independent thought and a reflective approach to oneself and the world.

Brukalska attempts to locate the centre and conceptualises the concept of the 'estate core' based on her observations of pre-war Żoliborz. The core should completely separate the life of the estate from the life of the entire district, allowing residents to consider it as their closest space and to create a sense of attachment. The core should be easily accessible, remaining in contact with the district, open to the outside world.

When considering various concepts of the 'estate core', Brukalska takes numerous guidelines into account, but the residents' feelings of identity are probably the most significant among them. Obviously, she does not use the above term and instead writes that the estate should be 'consistent' in order to 'be able to separate its life from the life of the city' (Brukalska 1948, p. 116). 'The estate core is a park around which the entire life of the estate is based; it is a road leading to such facilities as the library, settlement house, clubs, but one also has to cross it to leave the estate. It is also an easily accessible place for recreation and meeting other residents' (Brukalska 1948, p. 104).

Helena Syrkus, on the other hand, devised the 'social life axis' concept during her time at Rakowiec. The social life axis is a green area that runs through several estates, connecting them and becoming the main route for pedestrian traffic, also serving as a transit street. Brukalska, however, would see such a 'social life axis' as not only incapable of providing residents with the possibility of practicing solitude (because it is a loud, transit street), and because it would not be a viable route for pedestrian traffic, as it would not allow residents to take the shortest possible route to public transport stops.

Once again, Brukalska attempts to direct the life of tenants. When designing pathways and public transport routes, she makes precise estimations of resident traffic, depending on, for example, the time of day: 'everyday, especially morning traffic (…) of people going to work'; 'everyday, constant traffic between the popular social facilities'; etc. When trying to tailor the estate's shape to the habits of residents, she writes the script of everyday activities: 'we can assume that [the stop—author's note] has the highest pedestrian traffic in the estate, therefore it would be highly viable to place certain social facilities, forming part of the estate centre, as close to the stop as possible. Enabling residents to leave a pair of shoes at the shoe repair shop, drop off dirty laundry (at the laundry office), or order medication, etc. on their way to work will make for considerable savings in time and effort' (Brukalska 1948, pp. 106–107).

As a side note, we should mention that although Brukalska had modernist urban planning inclinations, claiming that she wanted to 'ration privacy' would be an overstatement. Mariusz Czubaj nevertheless claims that she had such intentions: 'the buildings and pathways were not placed to make walking distances as short as possible, but rather to lengthen them, thus forcing residents to take longer walks to improve their health'. 'Those healthy walks'—Brukalska fails to mention for some reason—were also meant to allow residents to efficiently exercise control over one another' (Czubaj 2007, p. 141). Brukalska had a very critical approach to such urban planning ideas and stated that it would be better to 'make possible, not impose, new ways of life' (Brukalska 1948, p. 36). For was not the public transport stop placed further from the buildings to allow residents to take care of various 'daily affairs', such as dropping a pair of shoes at the shore repair shop? By taking those daily affairs into consideration, Brukalska tips the hat to women and tries to plan the estate in a way that would make their lives easier. Once again, the architect tries to direct the city life of women. Being a woman herself, Brukalska is well aware of daily responsibilities expected of women which, when coupled with her social sensitivity, allows her to design the estate with the needs of female city dwellers in mind, regardless of their social status. Brukalska does not differentiate between the intelligentsia and working class. She simply wants to enable all women to lead a conscious and comfortable life in the city. Brukalska calculated that adults were spending less and less time on 'leisurely strolls' and she did not want 'that to be a permanent trend, let alone contribute to it becoming entrenched (…). Moving, resting outside, leaving work and home affairs behind, being closer to nature are the cheapest ways, from an individual and societal standpoint, of recreation' (Brukalska 1948, p. 101). Therefore, 'making the most out of free time' was one of the guiding principles of estate design. Residents should be able to spend even the smallest amount of free time that they have available in a place that is most accessible to them (parks, sports facilities, recreation facilities and open-air restaurants).

It should be noted that it was never Brukalska's intention to send residents on 'healthy walks' and expect them to keep watch over each other. Her social individualism has been misinterpreted as an attempt at imposing collective activities and 'community work', which always bring to mind socialist realist and totalitarian practices. The advent of the latter had been Brukalska's concern during the war.

The 'estate core' is, therefore, a certain type of green area through which residents walk to the most popular places. It is also a common good, allowing members of the 'working capital' to rest in the cheapest way possible. The community centre, which could be considered the 'heart of the estate', was another social facility that architects had planned. It was a place where neighbourly ties would be maintained and deepened, where residents could openly participate in public life.

Stanisław Ossowski described the communal house in great detail by comparing its role to the Inuit (Eskimo) kashim referenced in the works of Marcel Mauss. It binds social ties together and forms the centre of communal and intellectual life. 'The kazim cannot be replaced by a dispersed network of institutions spread across individual houses. The communal house, as an architectural unit that is an integral part of the estate's structure, will, therefore, become a focal point where a group of people will be transformed into a community with common values which will then spread outside of the estate' (Ossowski 1947, p. 188).

That is why it is unacceptable to build a residential estate between two avenues, for example, because the residents would seldom meet each other and the estate would soon split into two, none of which would be a distinct entity and, thus, would not develop into a local community. Consequently, the communal house would see few visitors, as the residents would feel 'not at home'. Brukalska and Ossowski mention all of that because such mistakes had been made when building the Żoliborz estate in the 1920s. Wilson Square separated the public sphere, WHC's agora, from the rest of the estate (Fig. 3.4). Did this urban planning mistake have negative consequences for the estate? Did it make it impossible to create a homely and open atmosphere? The 1934 reconstruction of the boiler house located next to Suzin street, which with time started to be treated as the 'heart of the estate' (Figs. 5.1 and 5.2) in addition to serving its role as a public utility, allowed to mitigate the consequences of the unfortunate location of Wilson Square. The old boiler house was converted into a concert hall where tenant meetings were held. Teodeor Toeplitz emphasised that public space can 'bind the colonies together into a single, indivisible whole' (Toeplitz 1933, p. 1). In fact, 'despite lacking architectural character, the communal house had drawn people from afar. The lively discussions that took place in the 'Glass House' clubs resonated with the residents. The estate schools—*despite their deficiencies resulting from meagre funding and the obstacles piled on by public schooling officials*—became a milestone in Warsaw's schooling history. At a time when the country was ruled by Ozon and ghetto benches were the norm, the Estate made its residents feel free and at home. It was a different world' (Ossowski 1947, p. 186)[3]—recollects Ossowski.

For the estate to not become a closed-off, backwards reservation, the city had to find it attractive. Brukalska analysed the relationship between the estate and the city, and introduced the term of 'external attractiveness, i.e. the influence exerted by a given community on individuals who do not belong to that community'. For an estate to be considered attractive, people from outside have to willingly take part in the estate's communal life. The factors contributing to an estate's attractiveness do not have to be luxurious in nature. 'Creativity, initiative and effectiveness are no less attractive than wealth' (Brukalska 1948, p. 29). Thus, the goal of performative

[3]Ozon, i.e. Obóz Zjednoczenia Narodowego (The Camp of National Unity) was a political organisation established in 1937 by order of Edward Rydz-Śmigły. It was accused of fascism and anti-Semitism.

Fig. 5.2 Concert and theatre hall at Suzina Street within the WHC estate, Report on the WHC activities for 1933. 3756, p. 89

Żoliborz is to provoke, attract, excite and encourage. Indeed, it has evoked different emotions and garnered enormous interest, mainly due to the intellectual, emotional and social engagement, due to its biopolitics.

> I remember—Warsaw's chief architect recollected in 1962—when I was only a child, who did not dare yet think about being an architect, hearing talk of some peculiar undertaking on Warsaw's northern outskirts, in the Żoliborz district. At the time, it was practically impossible to get there. I heard tales of impressive construction projects—a boiler house that will heat many buildings using an elaborate system of pipes. Together, the buildings will form a separate part of the city. There was talk of a laundry being built, serving all of the district's (that is, the WHC's) residents. Hence, they would not have to do their laundry in the kitchen, helping them keep their flats in good condition. The estate, I heard, would have its own horticulture facilities and greenhouses where plants and flowers would be grown to later be replanted in the inner yard gardens of estate buildings. The estate would have its own clubs, preschool, and stores run by a cooperative. It all seemed liked a beautiful, although

incomprehensible, fairy tale. We were witnessing a very radical departure from the traditional idea of housing estates as exemplified by the projects that were being built in Mokotów, Ochota, Staszic colony, the city centre and Saska Kępa (Ciborowski 1962, p. 3).

'External attractiveness' thus defined provides the local community with a necessary sense of distinction, allowing it to define the boundaries of its identity. As a tenant and sociologist, Ossowski thought about spatial organisation and social life in great length. He remarked that even the most innocent attachment to one's place, even if it is separated from the rest of the world with a lawn, will not be equal to backward isolationism if 'the place marks (…) the starting point of the universe' (Ossowski 1967, p. 370). The local community, along with the entire estate, should become a harmonised place where contradictions come together: separation and concentration, distinction from others and the 'we' identity, intimacy and community. 'If we wish to create a vibrant social life in our estates, we should not isolate individuals with lively minds and energetic temperaments from the estate's affairs. Our estates should feel as if they are in contact with the outside world' (Ossowski 1967, p. 369). Attempts were made to strike a balance between the estate's openness and its intimate, small-town atmosphere, between individualism and the social lifestyle. The centre of science was one of the means of achieving that goal.[4] On the one hand, it allowed individuals to isolate themselves in pursuit of creative endeavours, be it educational, artistic, academic or professional. On the other hand, however, the scientific centre should foster intellectual activities to 'not only prevent isolation, but on the contrary, strengthen the cultural bond between the estate's residents and other centres of culture. Opening oneself up to the cultural influence of the outside world helps counteract localism, curbs feelings of superiority, prevents residents from thinking too highly of local values, all of which is hard to escape when there is no general awareness of outside progress' (Brukalska 1948, p. 84). Brukalska's notion of opening up to the world despite feeling a strong connection with one's own place gained Ossowski's support, who stressed the importance of instilling a sense of contact with the world in the estate's residents if the estate's social life was to flourish with the active engagement of open-minded individuals. Ossowski even provided a number of specific solutions, including: 'visiting similar organisations outside of the estate, an interregional and international exchange of ideas, allowing international associations to become active in the estate. Let us feel the pulse of the world in the estate just as the pulse of the heart may be felt in each local artery' (Ossowski 1967, p. 369).

Ossowski's collaboration with the Architecture and Urban Planning Studio was an extraordinary experience. It transformed his sociological approach into one based on humanities in action.[5] Helena Syrkus, who at the time was deeply involved in

[4]Brukalska provides a detailed description of the architectural functions of the common house and the centre of science (cf. Brukalska 1948, p. 153). Ossowski intended to create a similar centre in Żoliborz to honour the memory of Stefan Czarnowski.

[5]Sociology in action differs from sociology as a behavioural science or an applied science. Sociology in action is emancipatory in nature and has more in common with the humanities than it does with social sciences.

the estate design process, recalls that 'professor Ossowski gained a lot from the close, daily cooperation with urban planners and architects. For the first time ever, as he often mentioned, he had been actively involved in the process of transforming sociological thinking, combined with urban planning and architectural ideas, into the material shape of estate, district and constellation-like design of city-regions' (Syrkus 1976, p. 241).

The well thought-through urban and social spaces of Żoliborz meant that it is far from an isolated enclave. "It is open to others, open to the city, but still manages to retain its distinct character." Social life reaches beyond its boundaries. It cannot be denied that both architects and urban planners make decisions that impose certain patterns of behaviour on residents. But that is what we would call 'empty oppression', laying the groundwork for cooperation. Although Żoliborz was a distinct estate, a separate part of Warsaw, we may nevertheless refer to it as a City, in which conflicts are present, but the City diffuses and sublimates them, it fosters cooperation with other areas of urban space using 'a-androgynous corridors' (conceptualised and implemented 'estate cores', 'internal and external attractiveness', passageways). 'What is an a-androgynous corridor? It is the basic building material of the a-a City, it is one of the possible materialisations of the city's 'empty violence' and it is a medium for the enforcement of communication between enclaves' (Nawratek 2011, pp. 151–152). Interestingly, it was not Warsaw that wanted this connecting corridor, but rather the designers of Żoliborz. 'A-a corridors are, above all, the materialisation of "empty oppression"—this means that they contain the supra-district functions (schools, nurseries, offices, larger shops, sports grounds, etc.) as well as open up areas in a restricted manner. Open spaces are too weak in themselves because they seldom play the powerful, 'oppressive' role required to force interaction, and that is why they should be located within the enclaves—where the integrated community can give them a proper meaning' (Nawratek 2011, pp. 152–153).

The concept of an a-androgynous City, or the City as a polis, has no place for closed enclaves, individual districts separate from each other. Nawratek writes: 'I am thinking about the City rather than its parts because a disintegration of the city into autonomous communities, into districts, does not make sense—their autonomy is illusory. Only the City as a whole is strong enough to protect and liberate its inhabitants' (Nawratek 2011, p. 102). Thanks to its 'internal and external attractiveness' as well as well thought-through functions of 'a-androgynous corridors, the Żoliborz estate bestows an identity upon its residents, it transforms them into citizens, but does not separate them from the city and, thus, does not try to introduce the utopian idea of Autarky. In some ways, especially those related to economy, it can be thought of as an autonomous area.

Should the Żoliborz estate be regarded as a successful attempt at building an 'a-androgynous City', which has later been defined by Krzysztof Nawratek? 'The a-androgynous City may become a political community, may become a new Polis, but only in the case of full and free participation by all—and without exception—users of the city in its social, cultural, economic, political and other aspects' (Nawratek 2011, p. 118). A 'gentle hacking', as opposed to overthrowing or destroying, of the system is a useful means to this end. 'With this approach, the androgynous myth—or rather, a-androgynous—could become an interesting alternative to both political liberalism, which developed with the Citizen, and economic neoliberalism, which deified the Consumer. Unlike these abstracts—divorced from our existence in flesh, our individual, neutral, intellectual constructs—a-androgyny is inside all of us' (Nawratek 2011, pp. 107–108). When using the term citizen, Nawratek of course meant the liberal construct of representative democracy. The citizen's role is therefore limited to making voting decisions for which he is then held accountable, as in that lies the essence of citizenship.

Despite bearing the mark of 'empty oppression', the precisely directed social life of the estate did not ration out privacy or impose certain patterns of neighbourly relationships. There is no doubt, however, that it made such relationships possible and even encouraged them. Żoliborz residents appreciate the estate's intimate character which allows social relations to develop, but also makes people feel free. 'The estate has a kind of inimitable big-city atmosphere, a pragmatic modernity among trees, colour and green areas. Space and air, a cheerful mood, an intangible sense of nostalgia. (…) A peculiar square, or actually a park, marking the intersection of history with sound, humane modernity. You can sit on a bench and for a moment forget that you are practically right in the heart of a big city' (Szkutnik 2009, p. 199). That is how Żoliborz's mood is perceived today.

The intimacy and familiarity were planned into the estate by pre-war architects, who went to great lengths to achieve that effect. One of the difficulties they had faced was avoiding acute angles. 'The already built part of the estate had posed a challenge that was mostly overcome (with the exception of the second colony) by building concentric ring roads' (Brukalska 1948, p. 137) (Figs. 5.3, 5.4, and 5.5). All of the new architectural ideas and accomplishments were discussed in the 'Życie WSM' [The Life of the WHC] magazine. 'The arc of the VII colony ties together the two obtuse sides of the triangle, the inner curvature of the arc has a centripetal effect on viewers, it binds and surrounds them. It is as if it was blocking their way and ordering them to stay. The straight line is space, distance, a long run' (Wohnout 1933; cf. Czubaj 2007, p. 65). The above comment perfectly embodies the performative character of Żoliborz's scenery. Architecture 'has a centripetal effect on viewers', orders them to do something, makes them feel safe, surrounds and seizes them. Those 'metaphors of action' show that urban planning and architecture were a part of daily life and had as big of an impact on the lives of residents as ideology and sociology. The place, its creators and users create networks that make possible the ideological shaping of space and identity (Fig. 5.6).

Brukalska's *Zasady społeczne projektowania osiedli mieszkaniowych* [Social Rules for the Design of Housing Estates] was published in 1948, before the Stalinist totalitarian regime took root in Poland, and whose coming Brukalska had anticipated. She presented the concept of social estates with a humanistic sensibility, emphasising the primacy of the individual over the collective. The architectural forms were to serve the needs of humans. In her view, the role of the architect, who is actively involved in social life, is similar to that of the sociologist. He has to learn the art of living from the tenants, deepen his social sensitivity, while also using architecture as a means of shaping tenants behaviour. It is evident that both sociology and architecture in action posed challenges for the estate's residents. Brukalska openly

IX KOLONIA W. S. M.

ROZWÓJ WARSZAWSKIEJ SPÓŁDZIELNI MIESZKANIOWEJ

Rok	Liczba członków	Udziały	Wkłady mieszkaniowe	Rezerwy	Sumy bilansowe	Wartość nieruchomości
		w tysiącach złotych				
1924	29	3	0	0.1	2	3
1925	87	5	1	0.3	173	16
1926	109	14	10	0.5	684	601
1927	345	51	53	18.0	3.095	2.364
1928	735	131	459	15.0	4.271	3.828
1929	1184	214	784	53.0	9.247	8.551
1930	1587	295	1.218	116.0	14.082	13.275
1931	1477	302	1.685	261.0	18.023	16.901
1932	1607	334	1.954	422.0	21.242	19.429
1933	1580	331	1.971	406.0	22.367	19.990
1934	1560	349	1.957	402.0	23.111	20.781
1935	1760	391	2.003	627.0	23.587	21.533
1936	1820	401	2.007	1.266.8	21.602	18.846
1937	1982	445	2.079	1.525.5	21.477	19.687
1938	2050	476	2.147	2.288.0	22.041	20.711

Fig. 5.3 Colony IX, Report on the WHC activities for 1938. 3850, p. 93

admitted that changes in government have a substantial impact on architecture. 'I could not—she writes in the Foreword—solve the myriad of difficult problems that were plaguing housing construction—a result of the combination of liberal-capitalist ideas with the new totalitarian ideals. I felt that none of those ideological approaches would do social individualism justice, as its goal was to rationally organise society and, at the same time, ensure that each individual would enjoy unbridled liberty and the freedom to use their creative abilities. As an architect, I was aware that I could unwillingly contribute to strengthening unnecessary restrictions and curbing certain freedoms. I could not remain passive, especially intellectually, in face of such a threat' (Brukalska 1948, p. 7). Barbara Brukalska's words may seem high-flown now, but if we take a moment to think about the reality that she was up against, we will understand the exulted tone of her writing. The rulers of post-war People's Poland also

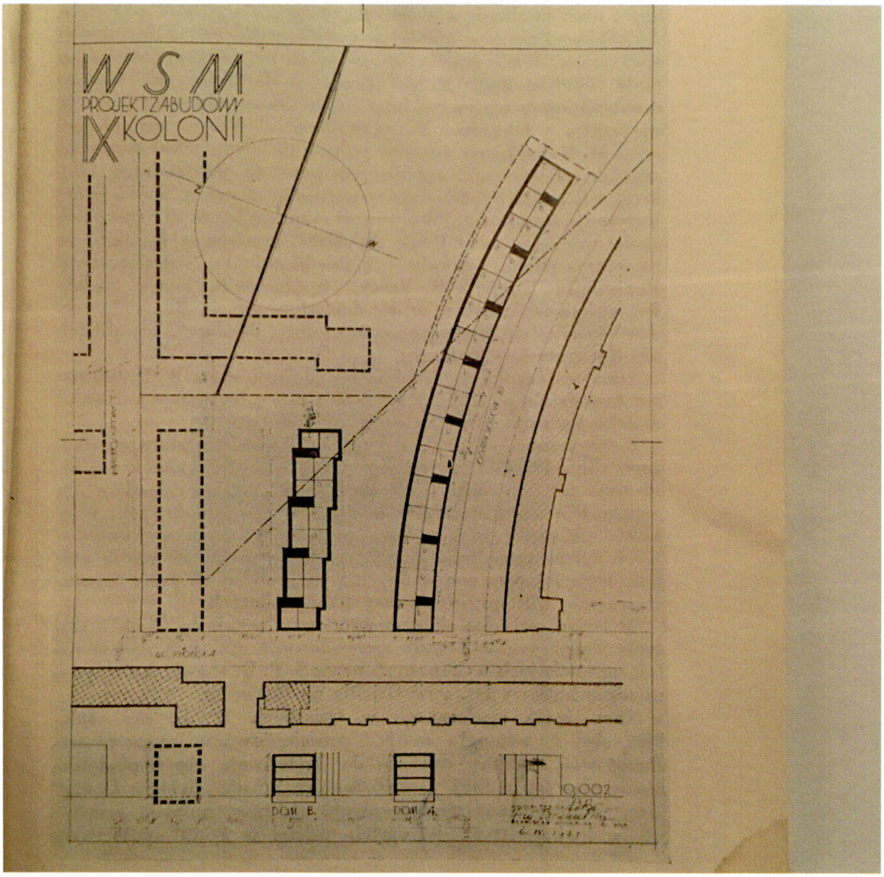

Fig. 5.4 Colony IX Development design, Report on the WHC activities for 1936. 3833, p. 93

took the time to familiarise themselves with her intentions and decided to ban further printing runs of her *Social Rules of Housing Estate Design*. Their decision sealed the book's fate. Even today, it remains completely forgotten by cultural studies scholars, and is only occasionally quoted by art historians and architects. It could, however, inspire cultural and urban activists as well as urban planners. Similarly, forgotten are Ossowski's sketches created at the underground Architecture and Urban Planning Studio (PAU). Nawratek's concept, controversial as it may be, proves that there is a pressing need to build an a-androgynous City.

Brukalska's *Zasady społeczne projektowania osiedli mieszkaniowych* [Social Rules for the Design of Housing Estates] may be considered as a very important 'study in action', an ordered collection of critical reflections on the Żoliborz experiment, which becomes new knowledge and may lay the groundwork for new ideas and concepts. 'Rather than a final "piece" of design, Critical Spatial Practice and its published by-products present inquiry, documented experiment, a discursively

Fig. 5.5 WHC Colony VII WHC, Brochure The Warsaw Cooperative Housing Society, 1938, History Chamber, p. 93

Fig. 5.6 WHC Colony III, WHC History Chamber. 2507, p. 93

argued thesis towards a spatial condition'. This condition may result in a large-scale proposal, a social event, a policy document, an analysis of spatial typologies, critical documentation of an existing situation, or a plethora of other possible formats.' (Miessen 2017, p. 37). Such activities do not necessarily have to include conducting studies, forming ideas and, finally, implementing a complete 'architectural product'. Instead, they form a speculative practice of creating scenarios to foster discussion between diverse actors. Suhail Malik's proposition, for example, uses research to define, conflict, critique and surround problems caused by earlier architectural and artistic undertakings (Malik 2014).

The practical and theoretical work of Brukalska, Ossowski and other members of the Architecture and Urban Planning Studio (PAU) may currently be viewed as an offshoot of Markus Miessen's project of which he said that 'Critical Spatial Practice should assume the role of a heterogeneous agent, one that, rather than fostering or exacerbating either/or scenarios, positions an alternative practice outside of the realm of architecture only to consider it as one of its central elements. Such collaborative, micro-political, and curatorial approaches that consists of the acknowledgement of complexity driven by an ensemble of actors prepares the ground for the realisation that the revolutionary aspect of architecture is to be found not in its *form* but in its *processes*. These processes can be called the 'construction of the democratic,' not as a romantic notion of all-inclusive modes of practice, but a means to envision and construct an agonistic space of politics' (Miessen 2017, p. 41). Both the urban planning and architectural works of Żoliborz's creators, which were discussed and analysed in the 'Życie WSM' [The Life of the WHC] magazine and the *Dom. Osiedle. Mieszkanie* [House. Estate. Flat] magazine as well as Brukalska's book may be considered as a reflective activity of a critical practice of space, who, in the words of Brukalska 'has to come to terms with the fact that before any decisions regarding the architectural form are made, studies have to be conducted until enough elements that are necessary to make the decisions are made apparent. No matter the task that the architect has to face, any solution must be preceded by expanding and deepening the foundation for creative work, by analysing the programme to not only become familiar with all of its facets, but also to be able to modify it if the need arises. This book is an attempt to formulate such principles' (Brukalska 1948, p. 13). Brukalska takes on two roles at once, thus going against the modernist tradition in which the roles were often separated. She is both an architect and a very conscious, for the time, urban planner. Furthermore, she investigates the various facets of estate life, puts them in context and thinks up alternative solutions with the aid of sociology, pedagogy and economics. By taking a collective, interdisciplinary approach, she ventures beyond the purely architectural discourse. All the time being aware that she is involved in a long-term undertaking involving complex processes of applying ideas and solutions in a space that already is filled with lively social reactions. Brukalska is aware that her applications may be rejected or modified through the process of living and other means of using space. She keeps track of the process, studies it, which allows her to use the conflicting interests of residents to her advantage. In that sense, her approach to urban planning is agonistically democratic (Figs. 5.7, 5.8, and 5.9).

The designers of Żoliborz and members of Architecture and Urban Planning Studio (PAU) did not see urban planning as a purely technocratic task, far removed from the needs of residents, but rather as an experiment in self-governance and grass-roots initiatives. The architects' modernist ambitions led to great care being put into architectural and urban planning solutions, especially those focusing on organising public space and the estate's commons. Urban planners of the time were well aware that architecture cannot exist without utopias and ideologies, as that would make it a lifeless, stone-cold, minimalistic form. When does urban planning turn into an 'event' (in a performative sense)? When it is flexible. When it is able to handle unexpected changes. When it adapts to its context and allows residents to become actively involved. For neither architects, nor the sociologists and cooperative ideologues that collaborated with them, were themselves able to create social relationships. Instead, they create networks and links that make possible the natural emergence of social relationships. They create public spaces, pathways, 'a-androgynous corridors' and social facilities. They make a general plan of how they would like to see the social relationships grow.

What was it that brought Ossowski and Brukalska closer to the independent practician approach (as defined by Miessen)? An architect or urban planner (or an activist or artist, i.e. anyone who deals in arranging space) is first and foremost a scholar who critically analyses the problems that have been conceptualised and implemented in urban space. The practice of urban planning is also a form of research, and research is a form of urban planning; therefore, the process of design and implementation creates

Fig. 5.7 WHC Colony IV—completion of construction, WHC History Chamber. 2521, p. 96

Fig. 5.8 WHC Colony IV, WHC History Chamber. 2525, p. 96

Fig. 5.9 Courtyard of the Colony IV, already inhabited, Report on the WHC activities for 1932. 3752, p. 96

new knowledge and gives birth to new ideas and concepts. Unlike the final design, the critical practice involves observing, searching, asking questions (we must not forget, however, that architecture and urban planning are scientific, artisanal, artistic fields requiring applied skills). In Brukalska's case, such practices led to the design and social undertakings on an estate-wide scale based on the principles of free access to the commons and the right to a dignified life.

There is no doubt that Żoliborz's architectural space allowed for creating informal social relationships, as it was open, flexible, dialogic and could be further corrected and expanded. It made the residents feel at home, but also had an open, (big) city atmosphere. Even today, with a widespread political and cultural awareness of 'symbolic violence' and all kinds of oppression, it seems that Żoliborz's architecture is subtle, does not discipline, does not seem striking, even though it is performative. Instead, it fosters cooperation and discreet action.

The founders of the Warsaw Housing Cooperative were guided by the idea of 'commons' which they had a very modern understanding of. Let me quote David Harvey's modern definition of the term to demonstrate that the pre-war estate was not only an urban planning and architectural challenge. Harvey states that commons are 'an unstable and malleable social relation between a particular self-defined social group and those aspects of its actually existing or yet-to-be-created social and/or physical environment deemed crucial to its life and livelihood' (Harvey 2012, p. 73).

If we were to set aside the ideology and residents for a moment and analyse the performative character of the estate, focusing (as we have done thus far) on architecture and urban planning, we could say that Żoliborz is an event in urban space. It is a result of the cultural production of space, a consequence of form and utopia joining forces, a hybrid discourse consisting of the work of architects, sociologists, ideologues and, finally, residents (the latter, as we will soon discover, often had very little freedom to shape their surroundings). Żoliborz, however, is not a specific man-made place, but a space that itself has created a viewer, resident and citizen. The architecture of the estate and the flats morphed into a separate, autonomous event, with its own motion, action, change and… emotions. We could clearly see (in laboratory-like conditions) the confrontation between space and action (especially in the sphere of interior design), conflicting visions of the organisation and location of the communal house, which is not a rigid monument to modernist form, but rather an event closely tied with living, experience and interactions, but also planned use, fixed rituals of cooperation and everyday activities. Architecture is able to serve its purpose only if it remains in touch with reality, answers the 'common' needs, and is egalitarian. 'Lower the standards of architecture? Certainly…let us lower the standards so that they fit reality, and only then will we be able to start building. (…) Building cities (…) with people in mind, thousands of ordinary people who need to have their space organised into designated living, working and resting areas. Let us give them architecture that is objective and stable enough to be considered classicist while remaining strongly connected with the present as to be considered modernist. That is the challenge faced by our generation' (Brukalska 1934, p. 6). Her views are remarkably contemporary. It is worth noting that the above quote was not a declaration of an avant-garde architect, but a manifesto of a practician with a number of achievements to her name. It should be treated as a counterpoint to the Athens Charter that had been published a year before (1993) and marked the beginning of industrialised residential housing construction on an enormous (and, contrary to the Charter's declarations, inhumane) scale.

Thanks to its revolutionary and avant-garde character, the estate served as an exhibition piece for quite some time. Starting in 1928, it received regular visits from various organised groups and special guests. The international 'Smallest Flat' exhibition was held there in 1930. In that year, the estate had its highest ever number of visitors: a total of 25,732 people. Trips to the estate were organised by, for example, the Tobacco Workers Association (Związek Robotników Tytoniowych) and various cooperative organisations as part of their Warsaw symposia. The estate was also visited by students of the University of Warsaw Sociology and Cultural History Department (1931), National Pedagogium (1936) and the University of Warsaw

Sociology Club. In 1932, 'L'Architecture d'Aujourd'hui' magazine organised an academic excursion for the members of the International Architects' Congress and on 22 August 1933, Żoliborz was visited by Henri Sellier—a distinguished French self-governance and housing activist. The estate's creators were well aware of the experimental, laboratory-like character of their work and even used those very terms to describe it.

Stanisław Ossowski saw this urban experiment as an attempt to consciously form a certain social environment: 'the WHC Żoliborz estate not only created a territorial community, but also shaped its own collective lifestyle. New cultural values were born in the estate. The community was influenced by those residents who were bound by a common ideology. The estate was no longer a local phenomenon' (Ossowski 1947, p. 186). What is curious, Żoliborz residents formed their identity by reading pieces similar to the one quoted above which were published by the estate's newspaper.[6]

Contrary to popular opinion, Żoliborz had not been planned to become a community, commune or Fourier's phalanx. WHC members were to become fully fledged citizens (not only workers, as was the case with patronage estates, not only socialists, as stipulated by the party, not only community members, which would give them a feeling of solidarity and connection, but deprive them of individuality and demand their unwavering loyalty). Żoliborz was not 'genetically pre-urban, sectarian oppression', which Nawratek fears may afflict urban enclaves. Thus, we may say that pre-war Żoliborz was a diverse group of singular entities building that which is common (i.e. a *multitude*). It was not an anarcho-syndicalist social Utopia, nor did it resemble Le Corbusier's vision of perfect society. We can risk saying that (as envisioned by Żoliborz's creators) it was mostly an experiment in citizenship, which could run as long as it was not taken over by the state.[7] Stanisław Ossowski was aware that the cooperative movement had died in post-war, Communist Poland, as it was taken over by the authorities who centralised it. Collectivist views were made mandatory. No longer were socialist believes a reflection of one's empathy. In 1947, Ossowski wrote that 'The WHC's role is less apparent today and that has led to the Cooperative losing its lustre and, with it, its status as an ideological island: it no longer opposes reality as it did before the war. (…) It was an important experiment, an attempt to create a social environment on a temporary spatial foundation.'[8] Stanisław Szwalbe voiced a similar opinion on the matter in 1970. 'It was easier to

[6]The Żoliborz Lovers Club (Klub Miłośników Żoliborza) established by the Żoliborz residents and a documentary film about the Warsaw Housing Cooperative entitled *Budujemy* [We are Building] created by Wanda Jakubowska and Józef Cekalski exemplified the self-awareness, feelings of identity and reflection upon one's place in the world. No copies of the film have survived to this day (Anonymous 1970, p. 83; Mazur 1993, p. 145).

[7]The housing cooperative idea is reminiscent of Turner's 'normative *communitas*'. If we would treat it as such, an in-depth reading of past issues of 'The Life of WSM', focusing on rites of passage and codes (such codes had been openly declared and included not only the official living regulations, but also the estate's informal 'rules of coexistence') could prove to be a feasible analytical approach. Cf. Toeplitz (1935).

[8]Ossowski (1947, p. 187). Julian Hochfeld wrote in 1945: 'Cooperatives are not an institution of the democratic state. Rather, they serve as a necessary supplement contributing to its democratic character—until the state is replaced by a cooperative, universal organisation of food producers and

organically manage the WHC when it had 2,000 or 3,000 members, to shape the ideology, foster and organise social life. That has now changed, as the WHC currently houses more than ten thousand residents. Also, the WHC was an enclave isolated from the reality created by the 'Sanacja' regime. That had brought the cooperative's members together, pushed them to act, as they were relying on each other' (Szwalbe 1970, p. 213).

The thoughts quoted above lead us to ask the question of whether Żoliborz may be considered a social laboratory, and its estate practices as experiments on both its creators and residents. I will explore the laboratory theme in more depth in Chap. 6.

In short, I would say that my analysis of the works of the 'creators of Żoliborz' lead me to believe that they should be treated as examples of Miessenian critical spatial practice. That would make it possible to interpret the urban planning and architectural writings of Brukalska and Ossowski in a more modern context and, consequently, place them in the present discourse on studies in action and critical urban studies.

5.2 Animation of a Neighbourly 'Culture of Habitation'. Emancipation? Education? Animation?

The WHC-designed flats are distinguished by their modest appointments: a small kitchen, no separate bathroom with a bathtub, no laundry or drying room. All of those shortcomings, however, were made up for by the elaborate system of social facilities. Several buildings were built which contained public baths, a swimming pool, a mechanised laundry, a cafeteria, a settlement house with a silent room, a library and a conference hall. We should keep in mind that the WHC used designs prepared in line with the principles set out by the CIAM, i.e. the smallest possible dwellings and the socially most necessary houses. Yet, living in small flats and socialising with neighbours does not transform people into citizens overnight. Tenants need time to learn how to live and not merely take up space. The cooperative administration took it upon itself to teach them how to live and established the Tenant Council (a very important achievement of civic living) in December of 1931. Each of the WHC's colonies was represented by a House Delegation (later called colony delegations) consisting of three people: the colony guardian, appointed by the Supervisory Board and two colony representatives belonging to the 'Szklane Domy' [Glass Houses] Association and the Active Cooperatists' Club.[9] The goal was to introduce a 'democratic management system of social facilities' (swimming pools, drying rooms, laundries and cafeterias). It was a self-governance system, embodying the values of direct, multi-tier and harmonious democracy, in which citizens (residents) were able to express

industrial manufacturers. (…) The cooperative movement must not be bureaucratic. (…) We are definitely not abandoning the 'Glass Houses' idea' (Hochfeld 1945, p. 260).

[9] A. Szymański provides a detailed description of the process of building and improving the representational system (Szymański 1989, pp. 186–193).

their needs and suggestions, felt responsible for the community and had a connection with the estate. House Delegations were a reflection of social norms. Bureaucratic institutions are characterised by their official rules and regulations. The delegations were to be a living social norm organism tasked with preventing the cooperative from transforming into a bureaucracy, a threat which could become reality should the direct connection between the growing, newly founded community and the management and supervisors of social facilities be broken. Stanisław Tołwiński wrote that such 'building delegates' would become 'the first universal activists of neighbourly living' (Tołwiński 1946, p. 55).[10]

It seems that the role of the building delegates was not set in stone, and indeed it had remained fluid until the end of the 'Żoliborz republic'. That is probably why it had to be rehearsed and renewed. Even in 1939, the building delegate role remained unclear. 'The report documenting the activity of the Tenant Council, containing the opinions of all colony guardians, allowed the 'Life of WHC' readers to familiarise themselves with the work of the organisational unit that receives undeserved criticism on a daily basis! Among the issues discussed and remarks made by the colony guardians, we are especially surprised by the great diversity of the activities and problems that they have to face. Mild disturbances, neighbourly conflicts and more complicated matters of community life; estate-wide activities initiated by the institutions operating in our estate, fundraising for the Teodor Toeplitz Fund, familiarising tenants with the Popular Scientific Study programme run by the 'Glass Houses' Education Department; informing tenants about the newly built Quiet Rooms in which children can do their homework; organising fun activities for children in partnership with the colony IX—such a broad spectrum of activities would make anyone who cares about the Tenant Council ask the question: what exactly are the responsibilities of Colony Guardians and what role should they play in the life of the estate? (...) The Colony Guardian's first task is to meet new tenants and help them find their way around the estate. Colony Guardians are at an advantage when meeting new tenants: they are not appointed officials; they live in the same colony, in the same housing conditions, are themselves tenants enjoying the same privileges and abiding by the same rules' (Życie WSM 1939, pp. 106–107).

The residents' perception of such guardians is interesting, as they were the polar opposite of the all-controlling Communist overlord.[11] Two caretakers in particular,

[10]We should add that a number of works were published shortly after the war as a result of the reflection on social estates and their impact on forming social ties. Brukalska, Syrkusowa, Ossowski and Tołwiński all had a similar approach. Today, we may view those publications as handbooks of social and cultural animation. Unfortunately, they have been mostly forgotten or, at best, are disregarded and treated as examples of the ideological superstructure of the Communist political regime. When the 'socialised individualism' concept is presented without appropriate context, readers who are not familiar with the principles of humanistic socialism, pre-war debates on the shape of Poland and the cooperative ideas of Adam Próchnik, Stanisław Ossowski and Jan Strzelecki among others will treat is as hollow slogans of political propaganda.

[11]Mariusz Czubaj had a similar opinion of 'building delegates', stating that they were 'in equal parts amusing and terrifying' (cf. Czubaj 2007, p. 122). His views seem to be a 'relict' of the Polish People's Republic mindset and an expression of disregard for Poland's interwar history. In truth, WHC building delegates had little in common with communist-era caretakers.

namely Marchewka and Bajurski, who were mentioned in many accounts of the estate's residents and described in Zofia Topińska's book entitled *The Promises of Żoliborz: the Formative Environment of the WHC* (Topińska 1984, pp. 13, 36, 124),[12] are perfect examples of a WHC activist, devoted to the estate and its residents, fully immersed in the cooperative's matters.[13]

Topińska's wrote about various aspects of their tasks. Because the children of Żoliborz spent most of their time outside, colony guardians had to develop a pedagogical touch in order to be able to solve the issues encountered by the youngest users of estate space. 'Children?—Bajurski said during the interview—we were raising adults, first and foremost. Otherwise, how would we be able to raise children?' (Topińska 1984, p. 13).

The memories of residents that are now being collected include ordinary, everyday spaces and the people who were somehow connected to them. The courtyard is 'a safe space, fenced off, closed for the night, guarded and cleaned by pre-war caretakers Fijołek and Wiecha (similar to other residents, they returned from exile in 1945 and worked long and hard, an attitude that is seldom encountered in this day and age, until their retirement)' (Sobiech 2009, p. 139). Even at the time, the caretakers' diligent approach to their duties was appreciated; the importance of their work, often involving educating residents, was never in doubt. The House Delegations were, of course, responsible for supervising caretakers, ensuring that the estate remained clean and orderly, fostering neighbourly ties, providing information on the material status of the residents. The disciplinary rules were enforced not only in buildings and on the streets. In some cases, flat inspections were carried out (the Collective's right to do so was stipulated in point 23 of the aforementioned housing savings books), and in 1933, all of the Żoliborz estate flats were inspected at the behest of the Tenant Council.[14] Katarzyna Gmochowska, a contemporary researcher of Żoliborz,

[12]Jakub Bajurski, a carpenter by trade who had been active in the construction workers' labour union, became involved in the WHC after he was fired from work for his membership in the PPS party. At first, he worked at the boiler house and later became the 1st Colony caretaker.

[13]In 1936, an article was published in issue 5 of 'Życie WSM' describing how Bajurski, the 1st Colony caretaker, had a box mounted on the fence outside of the building where residents could dispose of old bread that was then donated to the poor living in shacks in Marymont (Życie WSM 1936). Although the residents often called for 'ideological awareness' and 'socialism', WHC's daily life was not devoid of social inequalities and exploitation of servants. Bajurski, true to his socially aware attitude, often intervened: 'There were a few cases which Jakub Bajurski could refer directly to the disciplinary commission (…) which strongly resonated with the community. But the servants' issue was more general in scope. In order to solve it, Bajurski sought assistance from Magdalena Białkowska, representative of the Tenant Council, building delegate and member of the cooperatists' club. The club decided to act and managed to persuade the Cooperative's management to provide them with a two-room flat in Colony II where former servants who were looking for work could find temporary shelter (Topińska 1984, p. 15).

[14]Exterminators had to clear almost 40% of flats of bed bugs. The infestation was a result of the tenants neglecting to clean their homes. It was decided to carry out yearly inspections of flats and examine the belongings of all new tenants. Cases of extreme neglect were relegated to the Disciplinary Commission. The 'Penalty Rules for Breaking House Rules' entered into life on 1 January 1929. Their main goal was to curb subrenting. Tenants who felt wrongly accused could appeal to the WHC Court consisting of Tomasz Nocznicki, Juliusz Rydygier, Antoni Burkot and

admitted that: 'WHC authorities had the right to an evident intrusion into the privacy of residents in the form of flat inspections (point 23); however none of the residents who had lived in the estate mentioned that such incidents had indeed happened. One of the women whom I have interviewed summed up the often amusing concrete rules of conduct aimed at changing the habits of residents, the various obligations and bans, with a simple albeit blunt remark: 'it was a harmless regime" (Gmochowska 2008, p. 293). After some time, residents required no more prodding to renovate their flats, and the estate's daily life was taken over by grass-roots organisations as well as self-government, cooperative and self-help institutions (Warszawska Spółdzielnia Mieszkaniowa 1938; Życie WSM 1934b). There is no doubt that flat inspections were carried out and it seems that they were not rare occurrences.

> Due to the vermin infestation, the estate Administration, in accordance with paragraphs 15 and 23 of the Housing Rules, decided to approve flat inspections that will begin on 1 April of this year. The inspections will be carried out on a colony by colony basis according to the order and dates set out below. In accordance with paragraph 23 of the Housing Rules, Cooperative Tenants are required to be present during the inspection or leave the keys to their flat with the caretaker. If the tenant is not present, the inspection will be carried out in the presence of the Colony Guardian. Reports will be prepared after each inspection. The following items will be inspected: furniture, walls, floors, paintings, etc. If any traces of vermin are found, the Administration will call in an exterminator and charge the tenants with the fees incurred. Additionally, the Administration will refer each case to the Disciplinary Commission (paragraph 2 letter E of the Penalty Rules) (Życie WSM 1933, p. 7).[15]

> We receive many complaints related to the extermination of insects in the flats. The Tenant Council pushed the Cooperative to declare war on vermin. We are aware that no amount of coercion will rid our estate of the problem. Tenants need to show good will and initiative, as only regular cleaning of flats and furnishings will help us to completely eliminate vermin. The Administration should, first and foremost, provide effective assistance. We will only use coercion where there is no desire to act on the part of the tenants or the contamination is so serious that hand cleaning is impossible. Such an approach is bound to make inspectors more understanding. Alas, their duties are not easy or pleasant (Życie WSM 1934a, p. 6).

With time, such collective campaigns led to the habits of the intelligentsia, i.e. 'cultural coexistence', the tacit rules of cooperation learned at home or in social environments, also being internalised by working-class tenants, who wanted to have even the slightest influence on the estate's daily life.

Maria Belsinger. 'The Disciplinary Committee and WHC Court had an important impact on forming and entrenching respect for public property and habits of neighbourly coexistence. They ruled in less important cases that would otherwise be relegated to national courts' (Szymański 1989, pp. 188–189). For me, however, the above serves as further proof of the estate's educational and disciplinary character, not to mention its independence from the Sanacja government and state institutions.

[15]In December of 1933, 'Życie WSM' wrote that flat inspections had been carried out in 95% of flats. The remaining 5% were not inspected as the tenants were not present at the time of the inspection or refused to allow entry to inspectors. It turned out that most flats were clean (92% of inspected flats) and those residents who failed to properly clean their homes were sent memos reminding them about basic principles of cleanliness and informing them that further inspections would be carried out in the future.

The informal and silent rules, however, were often 'undermined' by the cooperative authorities, whose goal was to institute a system in which the Tenant Council cooperated with the estate's various organisations and self-help/administrative institutions. The goal was to develop a sense of comanaging public property. That is why the Tenant Council became one of the most important elements of WHC's formative system. The 'WHC Constitution' gives endless possibilities of exerting influence, fostering efficient cooperation and, in many cases, participating in the Cooperative management's decision process'—wrote Emanuel Freyd in an article published by the 'Życie WSM' magazine. He addressed all individuals (including Tenant Council delegates) 'who feel responsible for the fate of the Cooperative and Estate and take part in the life of our 'phalanstère', as far as their vigour, skills and competence allow, participating in meetings, electing delegations and sharing their critical views. (…) These frail and weak forms have to be entrenched through collective efforts, if we are to show the capitalist world around us that we do not need greed and profits, but rather solidarity and joint effort, to 'create a new way of life with no outside assistance' (Freyd 1933, p. 1).

I probably would not be at all interested in the meticulous accounts of flat inspections, the amusing House Delegations ordeal and matters relating to keeping staircases clean if it did not highlight what neighbourly living used to be like in those days. The 'cooperative republic' and 'socialist Polish average' slogans are rarely associated with the interwar living experience. They most often bring to mind the less fondly remembered communist times when the principles of cooperativism were shattered through centralisation, abolishing self-governance and ridiculing the local supervision system.[16] 'There cannot be a worse opinion of the tradition of 'subbotniks' than there is today in Poland. In fact, it is hard to imagine anyone in Eastern Europe taking the idea of subbotniks seriously. Despite this, I think it is worth looking at what we unthinkingly reject. The consumer's relationship with the neoliberal city is based on money but none of us want to be reduced to mere consumers' (Nawratek 2011, p. 148). Today, as was the case a century ago, we need citizens who are responsible for their surroundings, who feel connected and identify with the city, who shape it into a space of interaction and cooperation.

Today, we may view the activities of interwar activists as an anti-capitalistic strategy of urban participation, or even 'urban guerrilla warfare', neighbourly self-aid, cooperative activities borne out of resentment for municipal and state authorities (or, according to Freyd, capitalist reality). Those events should remain in our collective memory, as the micro-stories of life in Żoliborz are proof that we have not learned much from history,[17] especially if we look at the still relevant demands of today's

[16]Note that we should adopt a historical approach to the Polish People's Republic and recognise the different periods in its past. There is no doubt, however, that many complained of the lack of 'genuine caretakers' all through the communist era. That issue was the inspiration for the *Alternatywy 4* television series (directed by Stanisław Bareja, aired in 1983). *Wojna Domowa* [Home Front] (directed by Jerzy Gruza, aired in 1965 and 1966) was another television show that focused on the issues of housing and neighbourly relationships.

[17]In 1947, 'Kurier Codzienny' [The Daily Courier] published Andrzej Ziemęcki's fantasy novel entitled *Schron na Placu Zamkowym: opowieść o Warszawie z 1980 roku* [The Castle Square Shelter:

'Tenant Action' Association, which advocates for humane living conditions. One of its campaigns is called 'Clean Stairwells' and its purpose is to improve the cleanliness of public spaces and publicise the egregious working conditions and salaries of cleaners. The 'Tenant Action' website contains 'Poradnik dla lokatorów i lokatorek' [Tenant's Handbook] and the 'Nasze Nadodrze' [Our Nadodrze] newsletter whose editorial staff are well aware of the importance of housing assistance, as were the WHC activists before them. 'Tenant Action's' social activism is a specific form of protesting (Akcja Lokatorska 2018). It is a rallying cry for those who oppose real estate developers, for-profit investors and municipal authorities, all of whom are responsible for taking over public space and raising rents. It very well may be the case that such protests will become a tactic for creatively using urban space as a public good. A countergentrification strategy is using effective social means of preventing the seemingly unstoppable gentrification of Nadodrze and other similar districts.[18]

In Brukalska's days, Polish authorities (she referred to them as liberal-capitalist) perfected the use of common goods (both artificial, such as the effects of productivity, and natural, that is natural resources) to manipulate and create social divisions on the most fundamental plane. Thanks to the concept of cooperatives (when compared to its post-war centralised incarnation, WHC was mostly a sound cooperative), whose main goal was to reclaim common good in urban space (fresh air, light, vegetation and running water), thanks to direct democracy and the social self-control that it breeds, thanks to cultural capital and emotional engagement in social matters, gentrification may be effectively stopped. The estate's activities were not profit-driven, but instead based on 'social individualism', collective work (including physical work) for the benefit of cooperative members (nowadays we would say that value has been reattributed to its creators).

a Story of Warsaw in 1980] (Ziemięcki 2012). The novel's main protagonist, Jerzy Wasilewski, falls asleep in a bomb shelter during the Warsaw Uprising and does not wake up until the year 1980. He emerges from the shelter into a modern city, with two-level streets and a cutting-edge subway system, and sees people communicating using radio-telephones. The buildings, he finds, are powered by nuclear cells. The future political system, however, resembles pre-war socialism (based on principles of humanism, self-governance, and cooperativism—in short, nothing like the actual political reality of the Polish People's Republic). Ziemęcki orders his protagonist to fall asleep with a copy of Spengler's *The Decline of the West*, and after waking up in 1980, the protagonist starts voraciously eating up cooperativist literature. He may have deemed pre-war cooperative thought as equally modern and progressive as nuclear cells and flying taxis. Unfortunately, both the spectacular technology and cooperativist reality presented in the book were only works of the author's imagination.

[18]Such campaigns were run by Poznań's Rozbrat squat, for example. The squatters used various tactics typical for urban movements: squat defense protests, 'Food Not Bombs' activism, DIY bicycle repair, etc.

5.3 Transformation of Urban Entities: Passive Tenants or Active Citizens?

The aforementioned House Delegations may be stereotypically likened to Giorgio Agamben's notion of a camp ('Today it is not the city but rather the camp that is the fundamental biopolitical paradigm of the West.'[19]) (Agamben 1995) in which the border dividing public and private life is dissolved creating a space which Agamben called the 'bare life'. In my opinion, however, the disciplinary regime of the Żoliborz estate may be analysed in the context of the cultural productivity of urban entities, but also the productivity of the principles of cooperation and reciprocity. Hence why, I am not convinced by Gmochowska's interpretation. She claimed that 'despite its humanistic meaning, the concept of 'experiment', when used to describe WHC's avant-garde rules, implies that the modernist designs were lacking in sensitivity when judged by the standards of modern-day discourse (especially if it is further implied that the experiments had been carried out on humans). Zygmunt Bauman wrote about the precarious ambivalence resulting from creating social order and its destructive implications: the selection and classification of people. Bauman claims that in its desire to introduce order into the world, modernity creates a rational, hygienic reality, but also contributes to generating waste, which together form the order—chaos continuum' (Gmochowska 2008, p. 290).[20] The cooperative's policies, however oppressive they might have been, were not based on the principle of isolating and controlling socially undesirable elements to the benefit of the rest of the population. The cooperative was not a divisive tool of social disposal (as the camp metaphor would lead us to believe). It was not akin to Agamben's definition of the carnival. The intermingling of law and lawlessness (today we would consider violations of one's home to be an example of anomie), of that which is private with that which is public, may lead to the emergence of new urban qualities, spaces of interaction, spaces in which new ways of living are created, in which political acts of reclaiming dignity and one's rights to the city take place. Żoliborz activists instilled urban living habits in the residents; they campaigned for the right to cleanliness (flat inspections) and intimacy ('a separate bed for every child'). These goals were achieved through

[19] I am aware that spatial and urban issues are not Agamben's focus. Nevertheless, the Italian philosopher is very popular among today's urban activists (cf. Diken 2004; Perera 2002). House Delegations had nothing in common with camps. To consider them as such is a misinterpretation at best or ethical abuse of camp survivors at worst.

[20] I find using Bauman's idea to expose the modernist ambitions of the Żoliborz estate to be unfortunate. Bauman himself used a food producer's cooperative metaphor to describe active creator–user strategies. As an aside, it should be noted that the Polish 'Społem' cooperative had been the initial model for the metaphor. In the opinion of Poles, however, 'Społem' was a highly bureaucratic institution that disregarded food producers. Eventually, Bauman decided to use Rochdale's Society Equitable Pioneers cooperative (established in 1844) as a model for his metaphor. That is all the more interesting if we take into consideration that Bauman was very familiar with Żoliborz's intellectual and social environment, as his mentors, the Ossowskis and Hochfeld, lived there. It was them who taught him, as he used to say, what sociology is and just how much responsibility is involved in the work of a sociologist (Bauman 1997, pp. 127–140). It should be mentioned that Bauman's texts included in the Polish edition were devoted to Ossowski and Hochfeld.

educational efforts and rhetorical tricks. The activists could not have relied solely on 'tacit rules' if they wanted to create a system of neighbourly cooperation in the estate. By abiding by the cooperative's rules, the tenant community may become a space of transformation and achieve social change. As a space of cooperation, the city requires its residents to adopt certain prosocial attitudes, meaning that they are always obliged to submit to the social rules of the game and, in that sense, become victims of 'empty oppression'. From today's perspective we can clearly see that the Żoliborz estate was a temporary space in which a specific kind of order and 'bare life' were established (not allowing laundry in flats and hanging clothes out to dry from windows, organising public swimming pools, asking residents to let fresh air into their flats, maintain cleanliness and submit to other everyday norms), however it also led to the production of new (self-determining, emancipated, conscious) types of identity and formed new lifestyles: such as shaping new habits of rest and recreation at home, using pathways as a space for recreational walks, becoming conscious of the intimacy of bedrooms, having separate rooms for children. Therefore, Michel Hardt's and Antonio Negri's idea of common goods may be used as a better vantage point from which to view the 'Żoliborz republic'. What is curious, Hardt's and Negri's analysis starts with a set of assumptions that are similar to Agamben's: the heterogeneity, boundlessness and timelessness of metropolises governed by biopower and their lack of demarcation between public and private spheres. Those assumptions, however, lead Hardt and Negri to a different set of conclusions, mainly due to their adopting a different definition of biopolitics. According to Hardt and Negri, the *multitude* created in cities leads to democratic shifts which, in turn, create counterspace and heterotopies, with their inherent liminality and state of community established through resistance against the dominant order. Żoliborz's multitude created space in the non-space of the ruling Sanacja government, in times of economic crisis and stark social inequality. It was, as Hardt and Negri would put it, a rite of passage from a factory to an urban factory of multitude. The multitude, which was an entity separate from the working class, was not a people or a crowd but, due to its self-governance, production of self-reliance (economic and, to a point, legal autonomy), spatial organisation and even strict disciplinary rules, morphed into a conglomerate of singularity. Żoliborz is a joint work of its residents who reclaimed their rights to the city. A history of exploitation and poverty, solidarity in standing up against the reality that they were faced with and collective interest allowed for building a network of multitude with the ultimate goal of reclaiming the natural common goods, but also leading to the creation of artificial commons. Hence, 'social work' should be treated as biopolitical work based on skill, engagement and trust. Freyd's words that I have quoted earlier in this chapter come to mind: 'As far as our strengths, skills and competence allow—'Life of WHC' readers convinced themselves—(…) we are able to rely solely on collective work and efforts, and not the desire to profit, to 'create a new life for ourselves' (Freyd 1933, p. 1). The Żoliborz city—estate, therefore, manages its own capital, which was mostly human in nature in the most literal meaning of the term. And shaping the estate involves managing the time, bodies and emotions of its residents, but more importantly, their social responsibility. It also entails hard biopolitical work on the common spaces, using the law as an institution making possible strategic takeovers.

Common interests, but not necessarily common ideologies, reclaiming the right to fresh air, cleanliness, light, intimacy, caring for the well-being of residents and raising children became the foundation upon which the longevity and vitality of the network of multitude were built. Although we may say that the Żoliborz estate was totalitarian in nature, Hardt's and Negri's perspective would lead us to believe that it was an all-encompassing undertaking, consisting of producing new living conditions.

Initially, the estate was built with workers in mind, aiming to morph them into fully fledged citizens, but with time it became more geared towards the needs of the intelligentsia. Today, we may view it as an autonomous struggle that crossed lines of social hierarchy and class structure. By replacing the contradictions of identity and differences by the complementary values of commonality, action and singularity, the estate was a source of new life. We can safely say that it was an attempt to politicise architectural modernism which led to bringing two theretofore foreign classes together, providing opportunities for 'metropolitan meetings'.

> In those social and customary conditions, the class makeup of the estate had to be backed by strong arguments, given that the community labelled itself as socialist and pro-working-class. (...) [Andrzej Wróblewski[21]—author's note] claimed that the community was very diverse and colourful. In the flat above him lived a tram driver by the name of Zalewski/whose first name he could not recall/, most likely a communist. Below him lived a police officer. During the occupation, he proved his allegiance to the WHC community by helping out other residents using his contacts in the police force. He also had a neighbour, name of Franciszek Kornacki, a cattle sales agent for the Adamczewski Company. He joined WHC because his family has traditionally held leftist views—his father was a member of PPS and took part in the 1905 revolution. (...) Tomasz Nocznicki, an old man with a long, silver beard, one of the leaders of the radical wing of the people's movement, lived in 1st Colony. (...) In his accounts of WHC, Andrzej Wróblewski emphasised that he had finally felt safe. He knew that he had found a place where he felt at home, where the social and political attitudes of those around him were humanitarian in essence, and that his socialist moral sensibility was not being challenged. The feeling of safety accompanied his wife and him since they moved to 1st Colony to live alongside workers, artisans, clerks, professors, etc. Even though they came from Vilnius, where they had mingled in progressive circles, they were surprised by the direct relationships between individuals of unequal social standing. The lack of an inferiority complex among so-called simple people in relation to the 'state' was striking. It was something that they had never seen before, something entirely enticing (Topińska 1984, pp. 12–13).

Thus, the Żoliborz estate became a *multitude*, Warsaw's heterotopy, a counter-space, a new 'metropolis', a new a-androgynous City.

How should the metropolis be defined? 'The metropolis might considered first the skeleton and spinal cord of the multitude, that is, the built environment that supports its activity, and the social environment that constitutes a repository and skill set of affects, social relations, habits, desires, knowledges, and cultural circuits. The metropolis not only inscribes and reactivates the multitude's past (...)' (Hardt

[21] Andrzej Wróblewski left Vilnius for Warsaw in 1937 following the trial of the editorial board of 'Po Prostu' [Simply Put], a magazine published in Vilnius by 'Front' Leftist Students Association.

and Negri 2009, p. 249). In order for the metropolis to be for the multitude what the factory was for the industrial working class, it must be a site not only of encounter but also of organisation and politics. This could be a definition of the Greek concept of *polis*: the place where encounters among singularities are organised politically'.[22] Such encounters, aiming to produce common goods, shape a new social body that is able to voice its interests, pursue its desires, undertake living work and create an intersubjective space of exchange.

If we were to examine Hardt's and Negri's concept of joyful encounters, however, we would see that it is the weakest link in their theory, especially when applied to the analysis of metropolises. At their heart, 'joyful encounters' are not political but affective. They form immature, ephemeral neo-tribal communities (Maffesoli). They become evident in *jacqueries* which are usually built on outrage. Their activity is not directed inwards, aiming to improve living conditions in the socially created entity, but outwards, against the Empire. Yet, their power is not enough to weaken the Empire. David Harvey asks: 'Are all those screaming right-wingers interrupting the health-care reformers in the United States an instance of singularities in motion as a jacquerie? They are certainly erupting in a seemingly infinite rage against the capitalist state's attempt to impose a new form of biopower on their world.' (Harvey 2009). Is this a way for singularities to achieve objective and collective political presence? After all, anachronising the city does not recognise it as an entity, but rather destroys and dismantles it. The way it does it, however, is more seductive or, in fact, 'more gentle'. Joyous encounters are last-ditch efforts. When considering public strategies of urban struggles, Andy Merrifield abandons the right the city discourse in favour of the encounter discourse (Merrifield 2013). 'Insurgent encounters', having no constant, material form, but rather exploding like a rebellion, like pent-up social emotions, are necessary when citizens engage in a revolutionary struggle. The rights discourse is ineffective in such conditions and may even hamper the process of resolving social issues. It all depends on who is defining what the 'right to the city' really means. Thus, reclaiming the right to the city may be just a political manoeuvre.

---

[22] Hardt and Negri (2009, p. 254). As an aside, we should also quote the following excerpt from David Harvey's review of *Commonwealth*: 'For many years now, I and others have been arguing that the exclusive focus in Marxian political theory on the working classes in the factories made no sense. It was theoretically wrong because it ignored the production of urbanisation, the production of space and all the workers employed in such activities. It was historically inaccurate, given how many of the revolutionary movements in the history of capitalism have been focused as much on urban discontentment with the quality of daily life as on factory-based grievances (the Paris Commune, the Seattle general strike, the Tucumán uprising of 1969, the Shanghai Commune, and so on), and even when there were key movements in the factories (e.g., the United Auto Workers strike in Flint, Michigan, in the 1930s and the Turin factory councils of the 1920s), it always turned out that organised support in the neighbourhoods (the women's support groups in Flint and the communal 'houses of the people' in Turin) played a critical but uncelebrated role in the political action. The emphasis on the factory was also programmatically inept because struggles over what Henri Lefebvre dubbed 'the right to the city' could have provided a far broader basis for a revolutionary conjoining of urban social movements and work-based politics' (Harvey 2009).

'[Q]uestions of rights are, first and foremost, questions of *social power*, about who *wins*. The struggle for rights isn't something 'recognised' by some higher, neutral arbiter. Instead, for those people who have no rights, rights to the *cité* must be taken' (Merrifield 2014, p. 137).

The Żoliborz estate, however, used a different set of tactics. In order to transform the city into an entity, citizens were allowed to participate in WHC's decision-making processes by building an interface, a space of cooperation and relationships, allowing the Cooperative to prove that it had the backing of progressive citizens in its dealings with state authorities (manifesting its *social power*) and set out on the 'long march through the institutions'. Therefore, we may agree with Merrifield when he writes that urban spaces of the *cité* are public spaces not in the formal sense (collectively owned, managed by the state), but because they foster activity and social demonstrations of identity. The goal is not to build space, but to build a public entity. The struggle is not between public and private, but between passive and active.

Unlike Hardt and Negri, Merrifield does not claim that 'joyous encounters' are ephemeral and neo-tribalist in nature; however, he does believe that they create 'sovereign spaces'. The encounter category better reflects the relationship between politics and the corporeal/affective sphere—it makes apparent the interactive nature of the entities and their activities based on communication, cooperation and exchange. Whereas the right to the city is more abstract, socially objectivised and, in a sense, incorporeal.

Although Żoliborz's strategy was not devoid of 'joyous encounters', which will be discussed at more length further in this work, the notion of building alternative spaces on the outskirts of the capitalist system, a set of counterspaces subject to a different set of rules, is far more convincing. Such spaces are used to test (based on biopolitical work, rituals of cooperation and estate customs, skills and convictions) social capital, the accumulation of which does not generate profits in the economic sense but rather allows regaining dignity due to the collective use of space that has been unlawfully appropriated. It does not constitute a rebellion against Authority, or the System, but rather 'hacks' it to the benefit of the collective.

Despite focusing solely on the architecture of Żoliborz from an art historian's perspective, Łukasz Heyman's analysis includes and highlights the turning points in WHC's history. 'Work on the discussed complex [building complex designed by Bruno Zborowski—author's note] was set to start in 1930. Yet, during a programme conference called by BGK, the Ministry of Treasury delegate objected to one of the facilities (the nursery) being financed with public funds. Today, we cannot be sure what motivated his decision, but it nevertheless exemplifies the authorities' approach to WHC, an approach consisting in placing obstacles in the way of a Cooperative whose aim was to work towards 'the housing needs of the working class" (Heyman 1976, pp. 108–109). Such were the problems of Żoliborz's cooperative activists who wanted to see WHC as an 'institution' that would help the municipal authorities by taking over the issues of housing for the poor and providing them with cooperative aid that would replace the municipal welfare system. WHC took on the duties of the City (as defined by Krzysztof Nawratek) by running its own library, children's clinic, soup kitchen, reading room, training courses, school and kindergarten. In other words, it created its own 'a-androgynous corridors'. WHC could, therefore, perceive itself as a 'partner' in the city's economy thanks to its modern infrastructure increasing the city's income (gas, electricity, water and sewage) among other factors. Marian Nowicki lists the endless difficulties that the city piled on the Cooperative: 'demanding that it cover the cost of connecting it to the city's water and sewage systems; the Cooperative had to level the ground to lay electrical and gas lines; the municipality demanded that rent be paid for green areas' (Nowicki 1934, p. 2). Nowicki's list of complaints goes on and on. Żoliborz activists campaigned for common good, which forms the basis for citizens' rights: 'As the capital's citizens, who are better than others in that we make efforts that without any doubt improve the city's utilities and take over many of the municipality's responsibilities, we have the right to demand that no more obstacles be put in our path and assistance, even material, be granted to the WHC. WHC's impact on the life of the capital can no longer be disregarded!' (Nowicki 1934, p. 2).

Żoliborz activists, therefore, purposefully took over official institutions and infrastructure to ensure wider access to basic services and goods. Such a strategy is a far cry from 'joyous encounters'. As such, it has certain pitfalls. If the authorities respect the right to self-determination and acknowledge that a given community has the right to the city, then they will stop providing public services to such communities, thus making them responsible for setting up their own networks of services. Such a 'participation nightmare', consisting in shifting responsibilities on urban activists and expecting them to show initiative, is a means of covering up evident 'self-exploitation' and self-funding.

Was Żoliborz an embodiment of state-enforced participation or did it resemble Castells' 'urban question', which is a form of struggle for dignified life and 'collective consumption'? Urban planners and architects proved to be the most influential actors who, we can safely say, took the side of the citizens. The estate's design and the movement paths within it fostered organisation and cooperation, but also enforced certain disciplinary rules of estate life. They created an architectural and social environment in which residents could be active, new collaboratory skills could form, new lifestyles, patterns of intimacy, recreation, cleanliness could emerge, new affects and desires could be created. The city has to manifest 'empty oppression', influence its residents, and help create connections, if it is to function. The city cannot be a solely emotional community based on neighbourly ties and intimate relationships. Rather, it is a network of institutions, self-governance and direct action with or against the authorities, formal and informal rules that discipline and create a sense of identity.

References

Agamben G (1995) Homo sacer: sovereign power and bare life, p 117. http://www.opa-a2a.org/dissensus/wp-content/uploads/2008/03/agamben_giorgio_homo_sacer.pdf. Accessed 7 Sept 2018

Akcja Lokatorska (2018) Akcja Lokatorska [Facebook]. https://pl-pl.facebook.com/Akcja.Lokatorska/. Accessed 7 Sept 2018

Anonymous (ed) (1970) Żoliborz wczoraj, dziś, jutro [The past, present and future of Żoliborz]. Książka i Wiedza, Warszawa

Bauman Z (1997) Culture as consumer cooperative. In: Bauman Z (ed) Postmodernity and its discontents. New York University Press, New York, pp 127–140

Brukalska B (1934) Architektura pospolita [Common architecture]. Pion 2(3):6

Brukalska B (1948) Zasady społeczne projektowania osiedli mieszkaniowych [Social rules for the design of housing estates]. Developed at the institute of social economy. Wydawnictwo Ministerstwa Odbudowy, vol 22, Warszawa

Ciborowski A (1962) Rola WSM w kształtowaniu stolicy [WHC's impact on the Poland's capital] Życie osiedli WSM [The life of WHC estates] 1:3–4

Czubaj M (2007) W stronę miejskiej utopii. Szkice o wyobraźni społecznej [Towards the Urban Utopia. Essays on social imagination]. Wydawnictwo SWPS, Academica, Warszawa

Diken B (2004) From refugee camps to gated communities: biopolitics and the end of the city. Citizsh Stud 8(1):83–106. https://doi.org/10.1080/1362102042000178373

Freyd E (1933) Życie WSM, Sept:1

Gmochowska K (2008) 'Szklane domy'. Żoliborska WSM. Miejsce uwikłane w historię ['Glass houses'. Żoliborz WHC. A place Entangled in history]. Konteksty: Polska Sztuka Ludowa 62(3–4):286–294

Hardt M, Negri A (eds) (2009) Commonwealth. Harvard University Press, Cambridge, Massachusetts. http://korotonomedya.s3.amazonaws.com/Michael_Hardt_%26_Antonio_Negri_-_Commonwealth.pdf. Accessed 7 Sept 2018

Harvey D (2009) Commonwealth: an exchange, analysis. Artforum 48(3): 210–221. http://www.korotonomedya.net/kor/index.php?id=27,316,0,0,1,0. Accessed 5 Apr 2018

Harvey D (2012) Rebel cities: from the right to the city to the urban revolution. Verso, London, New York. http://abahlali.org/files/Harvey_Rebel_cities.pdf. Accessed 9 Sept 2017

Heyman Ł (1976) Nowy Żoliborz 1918–1939. Architektura—urbanistyka [New Żoliborz 1918–1939. Architecture—Urban Planning]. Zakład Narodowy im. Ossolińskich, Wrocław

Hochfeld J (1945) O istotną treść spółdzielczości [On the fundamental contents of the cooperative movement]. Przegląd Socjalistyczny [Socialist review] 1(XI); reprint in: Hochfeld J (1946) My socjaliści [We, socialists]. Wydawnictwo Spółdzielnia Wydawnicza 'Wiedza', Warszawa, pp 259–260

Libera A (2012) Widok z góry z dołu [View from above and from below]. In: Falicki B (ed) Żoliborz dzielnica marzeń: wybór tekstów [Żoliborz. The estate of your dreams. A reader]. Stowarzyszenie Żoliborzan, Warszawa, pp 23–57

Malicki Z (1957) Pasja postępu. Z doświadczeń budownictwa WSM w ciągu 35 lat [The passion for progress. 35 Years of construction in the WHC]. Życie Osiedli WSM 2:8

Malik S (2014) The value of everything. Texte Zur Kunst 93, Summer:66–79

May E (1928) Das soziale Moment in der neuen Baukunst. Das Neue Frankfurt 2(5):77–83

Mazur E (1993) Warszawska Spółdzielnia Mieszkaniowa 1921–1939. Materialne warunki bytu robotników i inteligencji [Warsaw Housing Cooperative 1921–1939. The living conditions of workers and intelligentsia]. IAiE PAN, Warszawa

Merrifield A (2013) The politics of the encounter. Urban theory and protest under planetary urbanisation. University of Georgia Press, Athens, GA

Merrifield A (2014) The new urban question. Pluto Press, London. https://pl.scribd.com/read/314636395/The-New-Urban-Question. Accessed 7 Sept 2018

Miessen M (2017) Crossbenching: towards a proactive mode of participation as a critical spatial practice. Centre for Research Architecture, Goldsmiths College/University of London, London. http://research.gold.ac.uk/20980/1/VIS_thesis_MiessenM_2017.pdf. Accessed 7 Sept 2018

Naprzód (1928a) Cała Polska robotnicza potępia robotę rozbijaczy warszawskich. Znamienne uchwały organizacji robotniczych w całej Małopolsce i na Górnym Śląsku [The whole of Poland's working class condemns the work of Warsaw's disruptors/troublemakers. Significant resolutions of workers' organisations in Lesser Poland and Upper Silesia]. Naprzód [Forward] 248, 27 October

Naprzód (1928b) Jaworowszczyzna. Naprzód [Forward] 249, 28 October

Nawratek K (2011) City as a political idea (trans: Pyzik A). University of Plymouth Press, Plymouth

Nowicki M (1934) W.S.M a władze miejskie [WHC and the municipal authorities]. Życie WSM, October:1–2
Ossowski S (1947) Zagadnienie domu społecznego w osiedlu [The issue of the communal house in the estate]. Życie WSM, July–Aug; reprint in: Ossowski S (1970) Dzieła [Collected works], vol 6. Państwowe Wydawnictwo Naukowe, Warszawa, pp 184–188
Ossowski S (1967) Organizacja przestrzeni i życie społeczne w przyszłych osiedlach [Spatial organisation and social life in future housing estates]. In: Ossowski S, Dzieła [Collected works], vol 3 [Issues of social psychology]. Państwowe Wydawnictwo Naukowe, Warszawa, pp 351–370
Perera S (2002) What is Camp? Borderlands. Electron J 1(1). http://www.borderlands.net.au/vol1no1_2002/perera_camp.html. Accessed 7 Sept 2018
Sennett R (2012) Together. The rituals, pleasures and politics of cooperation. Yale University Press, New Haven
Sobiech B (2009) Żoliborz... Moja mała ojczyzna [Żoliborz... My Little Homeland]. In: Chałasińska E, Gawecka A (eds) O pięknym Brzegu. Żoliborz, ludzie, wydarzenia [About the Beautiful River Bank. Żoliborz, People and Events]. Towarzystwo Przyjaźni Polsko-Francuskiej, Fundacja 'Żoliborski Fundusz Lokalny', Warszawa, pp 136–146
Syrkus H (1976) Ku idei osiedla społecznego 1925–1975 [Towards the idea of the social housing estate 1925–1975]. Państwowe Wydawnictwo Naukowe, Warszawa
Szkutnik L (2009) Mieszkać na Żoliborzu [Living in Żoliborz]. In: Chałasińska E, Gawecka A (eds) O pięknym Brzegu. Żoliborz, ludzie, wydarzenia [About the Beautiful River Bank. Żoliborz, People and Events]. Towarzystwo Przyjaźni Polsko-Francuskiej, Fundacja 'Żoliborski Fundusz Lokalny', Warszawa, pp 199–204. Reprint of: Szkutnik L (2008) Rozmowa z anglistą, autorem podręczników i sztuk dramatycznych—Leszkiem Szkutnikiem [A conversation with an Anglist, author of textbooks and dramas—Leszek Szkutnik]. Informator Żoliborza [Żoliborz's guide], 24 April
Szwalbe S (1970) Jeszcze o WSM [More on the topic of the Warsaw Housing Cooperative]. In: Anonymous (ed) Żoliborz wczoraj, dziś, jutro [The past, present and future of Żoliborz]. Książka i Wiedza, Warszawa, pp 211–214
Szymański JA (1989) Warszawska Spółdzielnia Mieszkaniowa 1921–1970. Zarys dziejów [Warsaw Housing Cooperative 1921–1970. History outline]. Wydawnictwo Spółdzielcze, Warszawa
Toeplitz T (1933) Serce WSM. Przemówienie inauguracyjne ob. Teodora Toeplitza wygłoszone na uroczystości oddania nowej sali do użytku mieszkańców Osiedla [The heart of WHC. Teodor Toeplitz's speech at the opening of a new hall in the WHC]. Życie WSM April:1
Toeplitz T (1935) 12.XII 1925–12.XII. 1935. Na dziesięciolecie WSM [12 Dec. 1925–12 Dec. 1935. On the occasion of the tenth anniversary of the Warsaw Housing Cooperative]. Życie WSM, 12:1
Tołwiński S (1946) Zagadnienia samorządu w świetle doświadczeń demokracji ludowej [The issues of self-governance in the light of the history of people's democracy]. Czytelnik, Warszawa
Topińska Z (1984) Żoliborskie obietnice: o wychowawczym środowisku WSM [The promises of Żoliborz: the formative environment of the WHC]. Wydawnictwo Spółdzielcze, Warszawa
Warszawska Spółdzielnia Mieszkaniowa [Warsaw Housing Cooperative] (1938) Sprawozdanie [1938 WHC report], Warszawa
Wohnout W (1933) Wnętrze [The interior]. Życie WSM, Jan:2–3
Ziemięcki A (2012) Schron na Placu Zamkowym: opowieść o Warszawie z 1980 roku [The castle square shelter: a story of Warsaw in 1980]. In: Haska A, Stachowicz J (eds) Śniąc o potędze. Narodowe Centrum Kultury, Warszawa, pp 289–294
Życie WSM (1933) Odpluskwianie mieszkań [Home exterminations]. Życie WSM, March:6–7
Życie WSM (1934a) Odnawianie mieszkań [Flat renovations]. Życie WSM, Sept:6–7
Życie WSM (1934b) Życie WSM, Sept:6
Życie WSM (1936) Życie WSM, May: 86
Życie WSM (1939) Rola opiekuna kolonijnego [The role of the colony guardian]. Życie WSM, April:106–107

Chapter 6
An Old or New Urban Issue?

> Enough of relationalities and immaterialities! How about concrete proposals, actual political organisation, and real actions?
>
> (Harvey 2009, pp. 210–221)

Abstract The collaborative forms of urban creation and consumption direct my analysis towards the 'urban question'. Following Manuel Castells, I ask how, by organising their everyday life on the housing estate outside the capitalist economy, self-sufficiently, and from the bottom-up, the residents created cooperative forms of everyday supply, consumption and even organisation of work. Collective consumption was created in a manner resembling modern autonomous zones (Chris Carlsson or Hakim Bey) in accordance with the principles of social economy. Following the everyday life of Żoliborz residents, I also consider questions posed by Merrifield, for whom Castells' 'urban question' is becoming outdated by today's standards. Merrifield leaves the 'right to the city' logic behind in favour of the 'logic of joyful encounters', spontaneous Occupy movements and something akin to the 'jacqueries' analysed by Michael Hardt and Antonio Negri. In the midst of Żoliborz's public activity forms, I search for manifestations of what Merrifield called the 'new urban question'. Cooperative trade, gardening and plant cultivation as well as cafeteria and health care, developed at the cooperative estate level, became not only cooperative, self-help economic institutions, but above all, ideologised anti-capitalist strategies. Cooperative thinkers provided a broad theoretical reflection on the above strategies, skilfully combining cooperativism with socialism. Activists and reformers, on the other hand, trying to improve the functioning of the cooperative, taught residents rational shopping and saving habits.

Keywords Social economy · Autonomous zone · Anti-capitalist strategies · Cooperative trade · Rational management

'We are not—Próchnik claimed—a fortuitous gathering of people connected by the concern of having a roof over their heads. Neither are we connected [...] by belonging to one professional group. We are an ideological group [...] we are bound by the

M. Matysek-Imielińska, *Warsaw Housing Cooperative*, The Urban Book Series,
https://doi.org/10.1007/978-3-030-23077-7_6

same social strata. We all, each and every one of us, belong to the world of labour. None of us classifies as a member of the class of owners, and nobody here lives off the exploitation of other people's labour. We all make our living by the sweat of our necks and minds. What is more, there are no employees with very high salaries, or so-called dignitaries, among us, who are neither materially nor mentally different from the class of owners. We all must work hard to maintain ourselves. And there is nothing that brings people closer than **common social conditions**. Furthermore, we are bound by common **cooperative beliefs, faith in solidary, unionised work and a clearly negative attitude towards the capitalist world**. Finally, we share a common striving for a system of social justice. Bonded together by such deeply rooted factors, we form a tight-knit fraternal community which strongly separates itself from the surrounding world' (Próchnik 1933, pp. 1–2).

The manifesto published in the WHC Żoliborz estate bulletin can be treated as a performative autonomist manifesto expressing the desire to build on the outskirts of a dominant capitalist system a kind of autonomous zone generating the conditions of its own existence through biopolitical work. Żoliborz residents can thus practice 'self-development of living labour' and 'the ability to cooperate in creating social relations'.

This manifesto can also be treated as a demand for the 'right to the city', the right to urban life. Henri Lefebvre's 'right to the city' was quite broad and, as such, it could contain all sorts of meanings. Many urban movements, defining themselves as active subjects of urban policy, are standing up today for the 'right to the city'. The 'right to the city' is thus subjected to multi-lateral articulations, transforming them into various ideological camps which justify the struggle for this space. David Harvey had a straightforward approach to Henri Lefebvre's definition of the concept: 'The right to the city is, as was noted at the outset, an empty signifier full of immanent but not transcendent possibilities. This does not mean it is irrelevant or politically impotent; everything depends on who gets to fill the signifier with revolutionary as opposed to reformist immanent meaning' (Harvey 2012, p. 136).

It should be recalled that Lefebvre derived the 'right to the city' from Engels' analysis of the processes of relocating marginalised social classes. The co-author of *The Communist Manifesto* was convinced that it is possible to regain the right to the city solely through overcoming capitalism, and the struggle for this right takes on a revolutionary form by transforming the relations of production and, hence, property relations (Engels 2010).

Following his line of reasoning, Lefebvre insisted on trying to dream up a city unlike any other, a city that is not governed by capitalist principles. It was the working class who was to be the subject of this urban revolution, because its resourcefulness was objectified and commodified as part of the modernist model of the factory city. Henri Lefebvre's idea, however, was much broader and more open, because he saw the city as an 'imaginary realm', a space of uncommodified exchange, a space for the unfettered expression and meetings that compound into 'urban life' (Lefebvre 2010). The entire human experience is concentrated in urbanity. As a side note, it is worth pointing out that Lefebvre built his vision of the city in opposition to functionalism and, more broadly, modernism, associating both with hyperrationality and with technocracy. Yet, he drew attention to the unlearned lesson of interwar modernism, unlearned because of the rapid acceleration of urbanisation. This process, dating back to the 1920s, meant that 'a large portion of workers and middle classes was quartered in tolerable conditions, but in housing estates built without any urban or architectural invention. (…) Urban development has been dominated by economic, social and cultural segregation'.[1] Lefebvre, similar to other avant-garde architects, overestimated the ambitions of the signatories of the Athens Charter and underestimated the social programmes of the 1920s. He was correct, however, when he spoke out against the segregationist policy in mass housing construction projects.

This was perhaps because pre-war Western modernists implemented a programme of socially indispensable housing construction, but did it only as architects. Meanwhile, in Poland, these objectives were pursued by 'urban activists', as the founders and visionaries of Żoliborz would have been dubbed today. The branching roles of modernist architecture are a consequence of the diverging paths taken by avant-garde architects, which have been described in Chap. 1.

The other important feature of the Żoliborz's housing estate is its heterogeneous structure. Even though the estate was meant for the working class, workers constituted only 30–35% of its inhabitants and ultimately the estate was associated with the intelligentsia.[2] Therefore, what could have been assessed as a failure of the Żoliborz project ultimately proved to be an effective safeguard against social and cultural segregation.

Lefebvre himself was in favour of the social production of space. When asked, in this context, about the role of urban planners and architects, Lefebvre saw it as interventionist rather than technocratic and engineering: somewhere between macro-architecture and micro-urbanism. The space that interested him was to be calibrated: 'It is no longer a village, but not yet a big city. The research of several of the most eminent architects of the era focused on this issue, including Constant's research in Amsterdam, Ricardo Bofill's in Spain and the studies of several sociologists, such as Mario Gaviria' (Lefebvre 1979). The idea of the 'right to the city' is not, as I have emphasised before, a matter of urban organisation of space, but rather of 'the collective ownership and space management'. Lefebvre, however, did not find satisfactory solutions for such space management once private property would be abolished. 'Nationalisation has catastrophic consequences because it transfers the absolute ownership rights to the State. Municipalisation makes apparent drawbacks and limitations. There is, of course, socialisation, i.e. a situation in which the people

[1]Lefebvre (1979, p. 90). Translation based on the Polish text. Although Lefebvre's *Le Droit à la ville* appeared in English as *Right to the City*, *La bourgeoisie et l'espace* was omitted in the English edition which is why the translator based her translation on the Polish edition.

[2]In December 1931, the share of workers in the total number of cooperative members dropped to 35.9%, 34.9% at the end of 1933, 32.4% in 1934, and finally in 1935 they constituted only 29.5% of all cooperative members. This share of workers started climbing again in 1936, as a result of the construction of the Rakowiec estate thanks to cheap credits of the Workers' Estates Association (Towarzystwo Osiedli Robotniczych), and later of Colony IX in Żoliborz. Consequently, the percentage of workers among the cooperative members rose to 38.3 as of 31 December 1938 (Szymański 1989, p. 82).

occupy and appropriate social space by overcoming property relations' (Lefebvre 1979). This, however, immediately raises suspicion about the utopianism of such a project. 'I agree with the position of those who believe that all social movements of space users are conducive to the takeover of power by the people. But to what ends should the power be used? (…) "Quality of life"? A different lifestyle? "Changing one's life"? Yes, certainly yes, but this change is unthinkable without considering the space of the entire planet. This does not preclude, however, the creation here and there of adequate spaces whose assimilation, freeing from ownership, can serve as an example' (Lefebvre 1979).

Lefebvre's idea was 'intercepted' by the counterculture movements of the 1970s, becoming a political demand which has undergone various transformations over time. Manuel Castells summarised it perfectly in 1975 by bringing up the following images: riots in the streets of Barcelona's industrial suburbs, where workers fought for the right of a district of more than one hundred thousand inhabitants to a hospital care, street revolts in Milan or protests in Bogota, where the city dwellers demanded that politicians fulfil their election promises of clean and green districts. Castells described protesters who gathered in a square in Stockholm to oppose cutting down trees, French demonstrators who stood in defence of Les Halles de Paris which the city authorities intended to demolish and American activists who organised a symbolic funeral of a car engine in a gesture of protest against environmental pollution. It is obvious that for Manuel Castells, who was strongly impressed by these events and Lefebvre's concept of 'the right to the city', these 'urban actions' were in fact a series of demands for 'a series of basic rights (housing, amenities, health, culture etc.) which have developed over time and have been wrenched from the bourgeoisie and the state. The social administration of these has become more and more collective and interdependent. (…) Public consumption, that is to say housing, public amenities, transport, etc. thus becomes simultaneously an indispensable element for the functioning of the system, a permanent objective of workers' demands and a deficit sector of the capitalist economy' (Castells 1977, p. 43). The Castellsian demand for the 'right to life', that is urban life, is defined in terms of 'public consumption', understood as all public goods and services, good-quality housing, parks, schools and clinics. There is a conflict going on between urban planners and social movement activists in the realm of reproduction. City planning and management policies are implemented mainly in the interest of the capitalist class which strives to ensure that reproduction consumes as few resources as possible. Social movements, on the other hand, which were established by the working class and various other entities, try to gain as much ground as possible in terms of life comfort and dignity, thus proving that the unequal and unfair distribution of goods and social facilities is rooted in class divisions.

Thus, the Spanish sociologist asks for an egalitarian access to these goods, but his demands are not limited to the sphere of economy. Demanding the 'right to life' is more about experimenting with forms of ownership and a universal access to urban goods rather than creating a functional and attractive city.[3] And the city's coercive measures must be on the side of grass-roots social movements. 'Public consumption' gives rise to a contradiction within the capitalist city between: 'the individual way of appropriating higher living standards ('doing your own thing') and the objectively collective way of managing this process. To the extent that urban organisation forms a whole (…)' (Castells 1977, p. 43). The above contradiction pushes the state to actively engage in solving urban problems and managing urban space. The state—Castells writes—fulfils 'its role as investor in the economic sphere[4] and as administrator in the technical and political spheres. These dual roles enable the state to act as the real planner of the daily lives of the masses and, under the guise of 'organising space', it is really concerned with predetermining how everyone should spend their time' (Castells 1977, pp. 43–44). Thus, the city is planned from the top-down and all of its problems are solved using 'neutral', 'technical' and 'rational' means. Urbanity is a concentrated way of life, and its contradictions can be overcome by urban planning and clerical decisions. It is therefore an instrument of domination and power.

Castells does not believe in participative models in which 'good citizens' discuss among themselves their visions of space and the ideas for their implementation. He does not believe in it mainly because he knows that such conversations are subject to the principles of 'rationality' (presented and evaluated by technocratic officials), 'the superiority of the technical imperatives presented to them as unquestionable'. That is why, it is 'these urban protest movements and not the planning institutions which are the real instigators of change and innovation. It remains to be seen if there is any possibility of urban change without general social change; in other words, total political change' (Castells 1977, p. 45).

This is an extremely important question, because it concerns the degree of radicality of actions taken by urban movements. They can lead open anti-capitalist subversive campaigns to break and disrupt the system. Such acts would be considered revolutionary. They can also be active on the outskirts, in the crevices and on the brinks of the capitalist system. Then, their actions would be considered reformist and pragmatic, involving waging small wars within the system, setting out on 'a long march through institutions'.

Castells' ambitions are strictly revolutionary and he cannot imagine, just like the Żoliborz activists, changing social structure without changing the ruling class and bringing on a revolution in the area of political power. Despite these radical aspirations, he seems to be pragmatic. He knows perfectly well that '[t]here can be, (…) within the general process of change (…), phases and skirmishes which will alter, in an unstable and partial way, the general logic of urban organisation' (Castells 1977, p. 45).

---

[3] I have doubts whether the category of experimenting is adequate to the description of demands for access to housing in the case of Castells' deliberations. Perhaps terms from the field of materialistic analysis would be more adequate.

[4] Today, of course, the investor in the field of economics is embodied by networked, global capital (no longer private, in the sense that one cannot identify a specific owner and hold them liable). Castells tries to rectify those reflections, which originated in the 1970s, in his subsequent works. This is also one of the reasons why Andy Merrifield regards Castells' urban question as outdated and obsolete by today's standards (see Merrifield 2014).

This is a very important moment, because it shows the evolution of the idea of the 'right to the city' (the right to urban life): from Lefebvre's anti-capitalist revolution inspired by Engels to the gradual, partial and unstable, yet revolutionary in intention, appropriation of the city by activist movements. Finally, David Harvey argues that reformist and pragmatic urban actions also prove to be an effective weapon in the fight for urban spaces. Harvey appreciates, for example, the actions of urban activists in Brazil based on legalistic procedures which produced clauses in the Brazilian Constitution that guarantee the 'right to the city' (Harvey 2012, p. xii).

It is conspicuous that although Castells and Harvey demand the 'right to the city' by starting with an anti-capitalist rebellion and deeming it necessary to redefine the sanctified institution of property rights, their visions focus not so much on the economic base of capitalism, but rather on the willingness to rectify the ideological superstructure of the existing regime present in the city. Hence, it can be argued that their positions are in fact rather social and culturalist, and not only materialistic (as Castells is often perceived). They are directed primarily at shaping a culture of cooperation and initiating a collective transformation of urban lifestyles. For these authors, the potential of freedom is rooted in urban relations, mainly those located in 'public consumption'. They are also universalist when the 'right to the city' becomes an obvious human right to them.

At the very beginning of *Rebel Cities*, Harvey declares: 'The right to the city is, therefore, far more than a right of individual or group access to the resources that the city embodies: it is a right to change and reinvent the city more after our hearts' desire. It is, moreover, a collective rather than an individual right, since reinventing the city inevitably depends upon the exercise of a collective power over the processes of urbanisation. The freedom to make and remake ourselves and our cities is, I want to argue, one of the most precious yet most neglected of our human rights' (Harvey 2012, p. 4). In order to be able to shape themselves, to collectively create spaces for common life, culturalistic claims have to involve much more than providing the right to the commons: access to sunlight, green space, fresh air, silence and even the right to 'collective consumption'. All of those things are made possible thanks to the concrete rituals of intersubjective exchange and cooperation, establishing silent rules, as well as building institutions based on legalism. Therefore, the findings of Antonio Negri and Michael Hardt will be of extreme importance for our considerations here. Although Negri and Hardt foresaw the emergence of an exclusively biopolitical city, their thoughts on how the urban commons are produced are convincing and should be discussed at more length. They recognise that the commons are largely produced by what economists call *externality effects* (i.e. the effects that are not taken into account by the market) and which can have both negative (e.g. pollution and traffic congestion) and positive consequences (e.g. successful social meetings). However, in broader terms, there is no doubt that people, through their daily activities, create the social world of the city (*cité*). In doing so, they create something common that others can use.

This lengthy introduction to contemporary concepts, which are extremely popular among today's urban scholars, was needed to show how 'the right to the city'—both empty and meaningful—acquires a new meaning. The term has a performative character insofar as it stimulates and provokes in urban activists' diverse articulations, thanks to which various entities define themselves in the context of urban politics.

Próchnik's manifesto contains the characteristics of a collective entity which formulates its right to urban life. Such an entity constitutes a *multitude* whose self-identification is shaped by the category of 'living off the very work of their hands and brains' and the lack of capital. The multitude places itself outside of the lifestyle determined by property relations.[5] We will look at how it determines its demands and with what content it fills the 'right to the city'. Is the Żoliborz culturalist experiment an instrument of revolutionary action aimed at dismantling the capitalist system, or is it rather an attempt to build—within the capitalist system—cooperative enclaves, autonomous zones, a reformist and defensive action rather than a revolutionary and offensive one?

6.1 Anti-capitalist Tactics and Economic Strategies

In the 1960s, Stanisław Tołwiński argued that the Żoliborz estate founders were reluctant to embrace the revolutionary ambitions of rebuilding the political system. Obviously, one should take into consideration yet another political context in which that claim was made; but still, it is worth quoting if only for the sake of answering the question that has been raised. As a long-time president of the WHC, Stanisław Tołwiński realised the insignificance of the workers' cooperative movement which—based on cooperation with trade unions—'could play only an auxiliary role in a capitalist market economy, consisting mainly in educating and training personnel in areas that would be important for the future economic system. It could contribute to raising awareness of the working classes, to awakening trust in their own creative powers and in the great importance of mutual aid. It could also facilitate political work and help to expose the bourgeois state. (…) The Warsaw Housing Cooperative has never embraced the utopia of resolving the housing issue in Poland by developing cooperative housing' (Tołwiński 1961, p. 5). Similarly, Szwalbe knew that the face of the regime would not be changed by cooperatives, but by the attitudes of specific people, i.e. cooperative members who put cooperative ideas into practice. Let us also quote the opinion of Poland's leading cooperative ideologist and WHC activist, Adam Próchnik. Próchnik's brochure titled *Ideologia spółdzielczości robotniczej* [The Ideology of the Workers' Cooperative Movement] was published as a supplement to the May issue of the 'Życie WSM' magazine in 1937. Therefore, it would have easily reached the residents of the Żoliborz housing estate. In his text, Próchnik analyses the development of the cooperative movement and its relations with the capitalist system. Even though Próchnik envisions a political revolution, which will hand over power to the working class, and a social upheaval, which will destroy the economic foundations of capitalism, he knows perfectly well that the

[5]Such multitudes are open and only those who, having the capital, 'live from the exploitation of someone else's work' may be excluded.

cooperative movement itself will fail to do so. 'The faith, therefore, that the social system can be revolutionised only through the cooperative movement is a utopia that is as harmful as any utopia can be. Cooperativism ceases to be utopian only when it focuses on its proper role, i.e. forming one of the branches of the working-class liberation movement. The task of changing the system may be achieved through the combined, harmonised activity of the entire labour movement and not just a single type of worker organisation' (Próchnik 1937, p. 4). This does not mean, however, that cooperativism was deprived of this emancipatory potential. Próchnik had very ambitious plans for the cooperativists. And the events that took place within the Warsaw Housing Cooperative in Żoliborz in the interwar period were an attempt to make those ambitious dreams reality. The opposite may also be equally true: by publishing *Ideologia spółdzielczości robotniczej* [The Ideology of the Workers' Cooperative Movement] in 1937, Próchnik chronicled the achievements of Żoliborz residents. Cooperativism encompasses three aspects responsible for the three most important areas of social transformation. First of all, it is about protecting the economic interests of the working class and the world of labour. Secondly, cooperativism can also create economic forms which reject the capitalist principles of for-profit activity based mainly on competition and rivalry, and introduce socialist principles of cooperation to meet people's needs. By creating new forms of socialist economy, cooperativism tests them and 'subjects them to an empirical baptism of fire'. And thirdly, all of the above means that cooperativism must educate future generations of people to be socially sensitive and open to cooperation, and 'whose goal would be to take over the mechanism of a socialised economy' (Próchnik 1937, p. 4).

Thus, the Warsaw Housing Cooperative was of a dual nature: economic—it was a cooperative enterprise—and culturalist (pursuing educational and emancipatory ambitions). 'Combining those two realms was very unusual in Poland at the time: as a non-ownership cooperative, it was a new, unknown form of entrepreneurship of clearly socialist character, and thus it was a ground-breaking development in the housing economy; as an association, it gave socialist ideas a safe space to grow in. That must have radiated outwards, gaining both sympathisers and enemies' (Głodowska 1962, p. 24).[6]

The WHC is an extensive complex of surrounding organisations cooperating with each other so efficiently and closely that often the residents themselves did not know which institution assisted them, and which formal entities initiated the everyday life of the housing estate. This peculiar network of distribution of new urban lifestyles offered by the cooperative permeated the everyday life of the estate residents via various invisible but also openly manifested 'channels'.

[6]The author of these memoirs, written in 1962, appropriates the basic concepts of bottom-up and non-state cooperativism for the benefit of the state ideology, but also appropriates the ways of acting, important for party members, attributing them to the members of the Communist Party. The author points out that the cooperative concept rejected political neutrality from the very beginning and was formed around the struggle in the name of Communism. That, of course, was not true and indicates that the author uses the strategy to intercept bottom-up practices and attempt to 'nationalise' them. The interpretations of this type of reflection should be very careful and take into account the obvious historical and political context of Poland's post-war reality.

The scope of both social and economic activity was of course determined by the necessities of everyday life. The estate was situated far from Warsaw's city centre and lacked convenient public transportation connections. So the objective was to try to provide as many conveniences as possible within the estate and create conditions for the estate's efficient functioning, including: grocery shopping, all kinds of services and facilities for raising children, i.e. preschools, medical clinics, a school, playgrounds, a place to rest for adults, courtyards and green areas. Although it may seem that economic activity had to be conducted within the estate out of necessity, there was an ideological justification behind it. Próchnik convinced WHC members to show solidarity with its institutions. 'We must do what we can to ensure that our community, united within the WHC, ceases to be divided into two categories: those who only live here, and those who take an active part in various domains of our lives. The Warsaw Housing Cooperative, the 'Szklane Domy' [*Glass Houses*] Association, the Cooperative Inn, the Workers' Friends of Children Association [RTPD, Robotnicze Towarzystwo Przyjaciół Dzieci—author's note], the 'Marymont' Club, etc.—all of them may seem to be separate institutions, but only from a legal standpoint and only apparently. For us, they actually constitute an integral and organically interconnected whole. This is not an ordinary tenement house in which several merchants rent out space to do business in. We actually form a community which organises and arranges our entire lives. Just like a citizen of a state who cannot limit himself to merely living within its territory but should also engage in the entire life of the state, attend its schools, use its courts and institutions, rely upon its economic, political and cultural foundations, likewise members of our community should be its active participants. We can achieve this primarily through propaganda and educational work' (Próchnik 1934b, p. 4). So, if a housing estate was conceived as a cooperative, and the principle of cooperativism was the basic rule underlying its economic life, then ideological considerations, and not pragmatic aspects, quickly led to the development of a complex system of economic, social, educational and cultural cooperative institutions.

Doing the laundry or even bathing was often impossible in mall flats. What was needed therefore was a bathhouse, a laundry room and a drying room. Residents required assistance in renovating their flats at a time when the estate was still being built. Shops and cafeterias were badly needed. In trying to meet all these needs, the estate authorities organised 'collective consumption' by introducing a new urban lifestyle project aimed at eliminating inequalities in access to various lifestyles and commons that were previously treated as a privilege reachable only by the bourgeoisie. By that I do not mean collective bathhouses or laundries, both of which were merely tools, but rather providing access to hygiene, calm and intimacy, availability of ready-made meals, access to work (women's emancipation), etc.

It bears repeating that access to natural commons ('the beauty and charm of nature'), as well as a specific housing *habitus* (the need for isolation, relaxation, intimacy, a separate room for a child, access to a place and time for resting and social gatherings) ceased to be treated as a luxury, and people started to fight for their egalitarian distribution. Thus, the estate's residents implemented important projects aimed at the emancipation of the working class and women. It is worth noting that implementing such projects was not a widespread practice of the then city authorities. The housing estate's residents owed everything to its founders' initiative and to numerous formal and informal organisations bringing together mostly cooperative members and estate residents. Hence, the answer to the question of whether we are dealing with 'a cooperative or cooperativism' seems obvious. Likewise, the cooperative's independence from the city authorities also seems clear. When Teodor Toeplitz wrote about European housing practices, he pointed out that providing housing was an obligation of public institutions in most countries. However, municipalities that built working-class estates, both during their construction and after their completion, often managed them financially, administratively and technically. In cooperativism, however, the idea was to make all these duties 'rest not in the hands of people appointed by the City Board or the City Hall, but in the hands of the cooperative members themselves—shareholders who feel much more at home in the same house than in a house owned by the city or the City Hall' (Toeplitz 1925). The ideology of real housing cooperativism was based on the involvement of the cooperative members and their bottom-up self-organisation. The cooperative 'may and must be a socialist ideology—all its characteristic features logically stem from it—its democratic management board, the allocation of apartments based not on the privilege of capital, but taking into account housing, family and social determinants of the cooperative's members—and eliminating the possibility of speculation and subletting. And furthermore—socially providing all necessities that may be provided in such a fashion: nurseries, preschools, reading rooms, meeting rooms, etc. (in the field of education and culture)' (Toeplitz 1928, p. 299). Toeplitz saw cooperativism as an autonomous sphere, which—when managed from the bottom upwards by the estate's residents, i.e. cooperative members—effectively blocks the city's attempts at claiming ownership. The cooperative's constitution did not allow private ownership of the flats and stipulated that even with the WHC being liquidated, its assets would not be distributed among its members.

6.2 Everyday Life—The Epitome of Non-material Work

The story about Żoliborz women is extremely interesting because it demonstrates how this project of emancipation was implemented and whether it succeeded or not. For those reasons alone, it deserves a separate study.[7] Here, however, it is necessary to outline one of the numerous women's activities in the Żoliborz district, namely the Active Cooperatists' Club [*Koło Czynnych Kooperatystek*].

One of the biggest enthusiasts of the cooperative movement at the time was Dr. Maria Orsetti.[8] No women's cooperative organisation has ever developed in Poland, and yet the idea of cooperativism has matured in a remarkably creative manner

[7] I wrote about the role of Żoliborz women in organising the estate in: Matysek-Imielińska (2017).

[8] Apart from her cooperative activity, Maria Orsetti also cooperated with the People's University (Uniwersytet Ludowy) founded by the PPS Left Wing Party, and in the years 1919–1920, together with Jan Hempel, ran the 'Książka' [*Book*] Bookstore Cooperative.

here. Maria Orsetti became a delegate to international conventions of cooperatists organised every three years in European cities (including Basel, Vienna, Stockholm, Paris and Zurich), while, on a local scale, Orsetti coorganised a women's cooperative movement as part of the Association of Consumer Cooperative Societies. Thanks to her efforts, women's cooperative societies were established all over the country, although the model society was founded in Warsaw's Żoliborz district. It was at the invitation of Orsetti among others that Honora Enfield, the first secretary of the International Women's Co-operative Guild, came to Poland and visited the Warsaw Housing Cooperative. A meeting with her was organised at Stanisław Tołwiński's flat. Despite the fact that only 30 women came to the meeting, the decision was made to establish the Active Cooperatists' Club [*Koło Czynnych Kooperatystek*]. The Club was founded at the end of 1929, and not without obstacles, requiring much effort and numerous conversations with the housing estate residents. 'Dr. Orsetti pushed through the name of 'Active Cooperatists', because she thought that each of the female cooperatists should have a function for which she would be responsible' (Święcicka 1963, p. 54). The first cooperative women's organisation in Poland focused on the idea of cooperation and self-help.

The Active Cooperatists' Club [*Koło Czynnych Kooperatystek*] aimed to 'raise the level of knowledge and social education among women by removing all obstacles which inhibit the active participation of women in the work of national and international cooperative organisations'. That is all that was written on the matter in the WHC activity report from 1930 (Warszawska Spółdzielnia Mieszkaniowa 1930, pp. 138–139). Thanks to the active female cooperatists, the legal awareness among women has spread, and legal bills were discussed: matters concerning birth control, matrimonial law, and social welfare. The members of the Active Cooperatists' Club included, among others, Maria Orsetti, Janina Dłuska, Janina Święcicka and Zofia Żarnecka.

Women wanted to work in various areas of public life: economic, social, and cultural. It was believed, however, that what they knew best was everyday chores, housekeeping, shopping, doing the laundry and keeping the home clean. And that is why women initially invested their time and energy only in spheres which we would call 'reproductive'. They were well organised, resourceful, proactive and effective enough to support the neighbourhood cooperativism to such an extent that they proved that the ultimate core of biopolitical work is not the production of objects for subjects, but subjectivity itself. They abandoned reproduction in favour of biopolitical work, which focused on affective labour and building social relations.

Wishing to provide the housing estate with a comprehensive infrastructure, the Żoliborz Cooperative established cooperation with Warsaw's Consumer Cooperative (one of Poland's oldest and most numerous cooperatives). Warsaw's Consumer Cooperative ran stores in Żoliborz. A 'Workers' kitchen' was also created for the builders of new workers' colonies where they could listen to the music from a radio provided by the 'Szklane Domy' [*Glass Houses*] Association (Szymański 1989, p. 156).

The housing estate's residents often complained about Warsaw's Consumer Cooperative shops, mainly due to supply shortages, high prices and impolite staff. In addition to the inconveniences reported by the estate's residents, there was also a political conflict because the Warsaw's Consumer Cooperative was dominated by supporters of Rajmund Jaworowski, who was involved in a dispute with the Żoliborz cooperative. Therefore, the Warsaw Housing Cooperative management founded their own cooperative, the 'Cooperative Inn', in 1931, which took over the Warsaw's Consumer Cooperative stores and also ran a cafeteria, a fuel warehouse, and expanded the network of grocery stores (Fig. 6.1). The 'Cooperative Inn' thus became the only entity which traded within the housing estate premises and brought together the WHC members as shareholders. The ideas of the 'Cooperative Inn', inspired by active women cooperatists, were to make shopping as convenient as possible. However, these ideas turned out to be more than just consumer actions: the estate's residents would hang out a shopping bag on the door handles in the evening, into which fresh bread, butter and milk would be put the following morning. Such a form of shopping, born out of a sense of comfort and proximity to the consumer cooperative in the estate's residents' opinion, turned out to be an effective consumer strategy based on mutual trust (Mazur, 1993, p. 129). The cafeteria run by the 'Cooperative Inn' cooperated with stores and estate residents could still buy the most needed products there (including tobacco!) even after the stores had closed. At first, people were rather reluctant to visit the cafeteria, deterred by quite high prices. Also, they had not yet warmed up to the idea of eating out. So, the offer had to be diversified: portioned and low-calorie meals and home delivery were introduced. All WHC residents were able to buy breakfast, lunch and supper in the cafeteria, while the stores stocked groceries, sweets and candies, as well as industrial and paper products, and even fuel. Thus, residents became relatively self-sufficient and could do their daily shopping within the housing estate premises. However, that made them cease to be city dwellers, and their urban lifestyle became closed or limited to the Żoliborz district (ironically, the freedom and comfort they could enjoy within their housing estate was the cause). A 'Cooperative Inn' was also opened in Rakowiec in 1935.

The 'Cooperative Inn' cafeteria was a contentious issue and the subject of constant discussions that continued throughout the period of the housing estate institutions' development. Still, in the first half of 1939, due to the fact that the 'Cooperative Inn' cafeteria generated losses, its liquidation was debated. The WHC management was well aware of the cafeteria's poor financial condition; still, they wanted to keep it going. 'Our cafeteria is at the level of an average nineteenth-century soup kitchen. The meals are prepared according to the patriarchal methods of the kitchen handicraft, based on the traditional patterns of self-taught housekeepers. We do not take advantage of the achievements and progress that the 20th century and modern knowledge have brought to the kitchen. Our kitchen does not use modern hygiene innovations nor the modern theory of rational nutrition and cooking to arrange menus. We never do anything by ourselves nor conduct propaganda activities to promote rational nutrition among the users of our cafeteria. When it comes to the aesthetics of the premises and dishes, much can be done even under the present conditions. (...) As a result of general discussions led and statements made by the Cooperative's management, it should be stated that everyone unanimously deemed the current state of cafeteria as highly unsatisfactory and everyone, except for Mr. Wrzos, decided that the reorganisation of the cafeteria should be the main task of the 'Cooperative Inn' for the coming year' (Życie WSM 1934, p. 112).

Fig. 6.1 The 'Cooperative Inn' Shop in the WHC Colony II, WHC History Chamber. 2498, p. 125

The goal of the Active Cooperatists' Club was, of course, to promote the idea of cooperativism and women's help in the organisation and development of local cooperative institutions. The most important practical task that women faced was cooperating with the 'Cooperative Inn'. Women prepared a questionnaire which was aimed to diagnose the reasons for the members' dissatisfaction with the activities of the 'Cooperative Inn'. The results of the questionnaire were analysed and a recovery plan was prepared on their basis. Shop committees were established which included housewives who were also Active Cooperatists' Club's members. The committees were an important link between the shops' and consumers' interests. They monitored the efficiency of service, cleanliness and aesthetics of window exhibitions, and made note of supply shortages. The cooperatists were also performers who effectively and efficiently influenced the consumers' attitudes. 'Education consisted in teaching housewives how to shop in a rational manner. They achieved this by lecturing at the meetings of the Active Cooperatists' Club, during which they explained how to make daily, weekly and monthly shopping plans and convinced women that they should refrain from going to the store several times a day, or sending their children instead, because they find at the last minute that they are missing certain ingredients that they might need. Research has been carried out to find out at which hours the most frequently visited shops were the most crowded; there have been attempts at regulating these peak hours by encouraging housewives not to visit shops at certain hours and to inform them when the shops are the least crowded' (Święcicka 1963, p. 55).

Do remember that this economic system was not founded on profit, nor the cult of the consumer. The plan was to build cooperativism on the involvement, equal participation and input of both sellers and buyers. This form of economy was to be based on mutual courtesy and respect between sellers and buyers, their care for the comfort of shop assistants and the buyer's satisfaction.[9] The social economy made the shop committees run awareness-raising campaigns aimed to introduce a two-hour lunch break for female shop assistants. 'After a few weeks of such propaganda, the turnover during these peak hours decreased so much that the stores could be closed. The break, however, did not affect the stores' total turnover. Thanks to this painstaking procedure, and taking into account that the stores had to compete with private trade, we have prevailed and our overworked staff were given two-hour breaks', Janina Święcicka recalls with pride (Święcicka 1963, p. 56).

Jan Andrzej Szymański explains in fairly great detail that 'the store clerks worked 14 h a day, and that was particularly striking in the consumer cooperatives within the Żoliborz estate. (…) The 'Cooperative Inn' was the first consumer cooperative outpost in which a lunch break was introduced. However, this break was not introduced in the stores of the second colony, in order to allow the residents to shop there throughout the day and not lose customers to privately held grocery stores' (Szymański 1989, p. 160).

Yet another interesting activity conducted by the 'Cooperative Inn' was the introduction of retail vouchers for the employees of the WHC and Social Construction Firm [*Społeczne Przedsiębiorstwo Budowlane*]. These retail vouchers could only be redeemed in 'Cooperative Inn' stores. However, this attempt to go beyond the 'monetary economy' raised objections among some WHC employees. This dissatisfaction led to an open conflict (which had political undertones). The 'Życie WSM' [*Life of the WHC*] magazine from October 1933, being the voice of the WHC management, reported: 'In early October, a number of leaflets were distributed at the construction site of colony VII B to protest against salaries being partially substituted with 'Cooperative Inn' retail vouchers. The 'anti-voucher' leaflets expressed hostility towards the Central Construction Workers' Union [*Centralny Związek Robotników Budowlanych*], Warsaw Housing Cooperative, Social Construction Firm [*Społeczne Przedsiębiorstwo Budowlane*], and 'Cooperative Inn'. In a word, they lay blame on everything and everyone. This tone reveals, to a large extent, the distinctly aggressive character of this performance and thus frees us from the obligation to engage in discussion with its authors. Their purpose is to break down the professional organisation and undermine confidence in social institutions that were built and are run with tremendous effort of the working class' (Życie WSM 1933, p. 8). Such performative speeches and attempts to achieve domination by the so-called rebels (probably a group of Jaworowski's supporters) and representatives of the Cooperative management may be seen as a public spectacle in which the order of meanings and the

[9]It was necessary, however, to encourage this kind of attitude. Wanda Wasilewska, for example, formulated various kinds of authoritative speech acts. In her performative speech act: 'Please, show a little bit of courtesy in our stores', she called for culture and gave multiple examples of incidents proving the lack thereof in the 'Cooperative Inn' stores (Wasilewska 1936).

forms of action are constantly being established or confirmed. The leaflet campaign became a manifestation of the lack of understanding for cooperativism and social economy and was interpreted as 'rabble-rousing', while its authors were presented as troublemakers.

Meanwhile, let us set that topic aside and return to Żoliborz's active cooperatists who also held customer loyalty competitions on behalf of cooperative stores, teaching how to systematically keep accounts and finances under control in order to avoid 'living on credit'. It was the cooperatists that came up with the idea of morning home delivery of groceries and set up a household appliance rental service ['we started with renting out vacuum cleaners and never got further than that' (Święcicka 1963, p. 61)]. Therefore, everyday shopping was not only a consumer matter, or an ordinary activity falling within the domain of economy. Although the above mechanisms can now remind us of modern and rational principles of Western capitalism, they were supposed to instil in the estate's residents a sense of inclusion, cooperation and interdependence, efficiency and good organisation of work. 'Gradually, women began to truly comprehend the political and educational importance of the cooperative movement. (…) They became attached to their cooperative stores in which they bought essential everyday products. This economic activity acquired relevance and social character in their eyes. It was important for them that their voices regarding the matters related to running a cooperative and adapting cooperative trade to their needs had been heard' (Święcicka 1963, p. 56). Although the activities of the active Żoliborz women (not only those associated with the Active Cooperatists' Club, but also other female residents and domestic servants) initially had economic causes, they ultimately became a matter of women's activity in the public sphere and an element of their emancipation. Performative actions, incitements and educational performances proved to be effective to the extent that women were becoming aware of their ability to make a change. It can therefore be said that the aim was to 'liberate living labour from the capitalist regime in order to open a possibility of creative and independent activity' (Editorial team of Praktyka teoretyczna 2012, pp. 31–32). 'The hardest task is involving female residents of the WHC in social life—and all of them probably know the 'Cooperative Inn' inside out: only some of the children and young people from the housing estate attend the school and the settlement house, but all of them have been to the Cooperative shop and have seen with their very own eyes the social form of a cooperatively owned store. The 'Cooperative Inn' is faring well, its turnover is growing each year, and the number of its members will soon exceed a thousand. Hundreds of people visit cooperative stores every day. The 'Cooperative Inn' has turned into a club where you can meet all of your neighbours. It goes without saying that we should have a special and cordial attitude towards it and it should become something more than merely a shop which supplies us with good produce at low prices' (Haubold 1937, p. 204). Could the readers of the 'Życie WSM' [*Life of the WHC*] magazine, living within this housing estate and doing the shopping at the 'Cooperative Inn', disagree with such an image? 'More than merely a shop', 'publicly-owned property', 'a club where you can meet all of your neighbours'—even if those statements were not really true, the estate's residents were certainly trained to imagine 'a city that they do not yet know'. The training included both monthly and daily exercises. The former included reading articles (published in the 'Życie WSM' [*Life of the WHC*] magazine), whereas the latter consisted in simply doing the shopping.

The 'Cooperative Inn' was supported by all institutions operating within the WHC premises. And its activity was often used during housing estate events, especially on the International Day of Cooperatives, or the Labour Day (1 May).

The network of connections between various institutions and informal activities was very dense in Żoliborz, and it is difficult to determine which of them initiated various undertakings. Szymański reports that the 'Cooperative Inn' also coorganised many meetings, lectures and courses (including the Course for Social Workers and the General Socialisation Course, among others). Święcicka, on the other hand, mentions that the active cooperatists organised various courses and help for domestic servants, raising awareness of their rights to a dignified life and work. It was painful to see, for example, 'homeless people seeking a job or even an overnight accommodation who were wandering about in this socialist housing estate. These were domestic servants looking for any job. The members of the Active Cooperatists' Club decided to organise a small hotel for these homeless women where they could find temporary lodgings and also train themselves in running a household. 'We wanted to help women to acquire qualifications and, consequently, a better remuneration'—explained Święcicka. We were also planning to run a Club for domestic servants, where they could spend their Sunday afternoons free from work' (Święcicka 1963, p. 60). The plan succeeded, and its activities went far beyond this modestly outlined assumption. In 1933, the Active Cooperatists established the Warsaw Branch of the Trade Union of House Caretakers and Domestic Servants [*Związek Zawodowy Dozorców i Służby Domowej*]. Over time, however, it fell under the influence of the communists (Szymański 1977).

The cooperatists also ran a work agency, thus becoming coordinators of the internal labour market within the housing estate consisting of various plants, shops, cafeterias, the Social Construction Firm, the Horticultural Centre and other institutions. As you may remember, the cooperatists were also part of the home delegation, often reviewing tenants' requests to reduce or postpone the repayment of rent and financial aid applications. Therefore, a loans and savings 'Mutual Assistance' bank was established in Żoliborz, which constituted one of many forms of support for tenants.[10]

[10] J. A. Szymański reveals its name: Spółdzielnia Oszczędnościowo-Pożyczkowa 'Pomoc Wzajemna' ['Mutual Assistance' Loans and Savings Cooperative]. It was created on the initiative of the 'Glass Houses' Association, and anyone who lived off of their own labour could become a member. The Cooperative's aim was to support WHC tenants financially and prevent usury. WHC deposited over PLN 5,000 into a 6% investment account with the 'Mutual Assistance' Loans and Savings Cooperative, with the proviso that the Cooperative would grant loans to WHC members against their housing savings books. The activity of the 'Mutual Assistance' Loans and Savings Cooperative was suspended soon after, because—as Szymański wrote—'the environment of the WHC estate turned out to be too narrow a base for its activity' (Szymański 1989, p. 175). The 'Szklane Domy' [Glass Houses] WHC Members' Assistance Association often provided help for the unemployed. In 1933, 163 unemployed members of the Association were given jobs; the Association also provided help for people threatened with eviction, and people with no health insurance were provided with medical care. Szymański described this form of activity of the 'Glass Houses' Association in

In fact, it does not matter which institution (the Active Cooperatists' Club or the 'Cooperative Inn') started those initiatives. What matters is that they created a dense network of cooperation and mutual aid, thus blurring institutional boundaries and creating space for organised undertakings, producing their own living conditions.

Although it is worth noting here that in male narratives, the 'Cooperative Inn' was the initiator of economic life, while in the memories of Żoliborz women, it was rather the Active Cooperatists' Club which organised and took care of this sphere of life.

In 1930, Maria Orsetti established the first cooperative laundry room in Żoliborz, which was used by the cooperative members working there and user-members, i.e. residents of the housing estate (Fig. 6.2). The laundry was invaluable in relieving the housewives from the cumbersome and exhausting task of washing clothes. Residents, however, did not know how to use the public laundry, leading to frequent quarrels and disputes (they often brought in linens that were too dirty and damaged). It also turned out that its services were too expensive for the working-class residents of the housing estate. The WHC management did try to reduce costs and encouraged residents to develop hygiene habits and banned residents from doing the washing in their flats. The laundry room nevertheless continued generating losses. Along with the Cooperative's changing social structure and the influx of intelligentsia, the frequency of using the laundry room by the domestic servants and professional laundresses increased. Janina Święcicka, however, mentions that the cooperative laundry room was working efficiently and managed to break even (Święcicka 1963, p. 59).

Maria Orsetti's concept of the cooperative laundry room was inspired by her visits to the UK, where she saw how cooperative and communal laundries operate. She admired the high standard of the equipment and specialised machines, thanks to which working women were not physically exhausted and had perfect conditions for rest: cafeterias, relaxation rooms and hygienic working conditions. 'The cooperative laundries made everyone benefit from the human-machine cooperation: customers, i.e. cooperative members who benefit from extremely low prices for doing the laundry, and the employees who benefit from the best possible working conditions' (Orsetti 1934, p. 6). It seems that Maria Orsetti was mostly concerned about the lack of hygiene among the Żoliborz residents. 'It must be impartially admitted that the British laundries have an easier task than ours, thanks to the habit of daily bathing that is deeply rooted in all the layers of British society' (Orsetti 1934). Orsetti tried to pass on her educational intentions, social awareness and sensitivity to employee issues to the WHC management, hoping that it would significantly improve women's work comfort within the housing estate. However, the Żoliborz's laundry turned out to be a far cry from the British ideals, and its atmosphere provoked many quarrels. The guardian of colony IV, Zofia Hryniewicz (a woman), called it the 'women's hell' with 'a horrific atmosphere that only lessons in mutual goodwill and kindness could remedy' (Życie WSM 1939, p. 81). It is worth emphasising, however, that the

---

detail (Szymański 1989, pp. 122–126). The scope of this aid and its diversity were very vast. The Cooperative management tried at all costs to keep the WHC members within the Cooperative.

Fig. 6.2 Opening Ceremony of the Central Laundry Room on 20 June 1931, Report on the activities of the WHC for 1931. 3520, p. 131

laundry provided employment for the wives of the unemployed residents of the housing estate, thus supporting seven WHC families. Cooperativism supported by women established new forms of employment. Many household chores were performed by female employees who were encouraged to join the worker cooperative. This meant that pay started to be demanded for work that had been socially unappreciated before, treating it as a form of reproduction rather than production. On the other hand, social insurance was introduced and working time was normalised.

6.3 Experiencing the Field—How to Physically Learn Social Relations?

The *modus operandi* of the Independent Horticultural Farm (Samodzielne Gospodarstwo Ogrodnicze),[11] which operated in Żoliborz since 1932 and was designed to care for the development of courtyards and green areas, was also interesting (Fig. 6.3). There was a garden for children and a petting zoo kept by the Workers' Friends of Children Association (RTPD, Robotnicze Towarzystwo Przyjaciół Dzieci). The school garden served educational and pedagogical purposes, and vegetables were grown there to feed the animals at the petting zoo. The estate's residents could buy flowers, fruits and vegetables, and obtain advice on horticulture there. In 1936, the so-called guest house for sick plants was established, where plants and flowers requiring care were accepted for storage during the holiday season (Fig. 6.4). Planting flowers on balconies, terraces, courtyards and squares was very popular.

In accordance with the WHC organisational principles, the Horticultural Farm had its own departments (school garden, courtyard maintenance, and a food and commerce department) and was supervised by: representatives of the WHC management, representative of the Workers' Friends of Children Association and the manager of the Horticultural Centre.

The organisation of the school garden is a perfect example of the ideological and social character of the housing estate's economic model that was instilled in Żoliborz children. The garden's organisational committee was composed of an interdisciplinary group of enthusiasts: physicians, architects, naturalists, agronomists, as well as amateur gardeners, i.e. WHC residents and a school teacher. In the interwar period, the role of natural sciences was dominant, and natural sciences occupied a special place in the school curriculum within the WHC secular environment, considered not only as life sciences but also as social sciences. The school garden was a kind of team workshop where participants could learn systematic and responsible work, rules of coexistence, cooperation and respect for physical work.[12] The garden was divided into three parts:

- 360 m^2 was allotted to individual beds (called 'patches') cultivated by older children and common plots which were cultivated collectively by teams of children from preschools, the settlement house and first and second grade pupils.
- the second part of the land contained a school garden where grain was grown, mainly for the animals at the petting zoo, and experimental plots for biological teams of children (an alpine garden, sand dune plants, pools for plants and aquatic animals, and a herbal garden) were also set up.

---

[11] I wrote about this issue in: *Polityka ogrodnicza Warszawskiej Spółdzielni Mieszkaniowej: ogródek i prawo do miasta* [The Horticultural Policy of the Warsaw Housing Cooperative: a Garden and the Right to the City], Matysek-Imielińska (in press).

[12] In addition to plant cultivation, children also made gardening tools themselves during the workshops (e.g. garden frames, boxes for the animals at the petting zoo).

OŚRODEK OGRODNICZY W. S. M.

Nasz Ośrodek Ogrodniczy zaspokaja potrzeby ogrodnicze nie tylko W. S. M., ale także znacznej części spółdzielczego Żoliborza.

Fig. 6.3 WHC Horticultural Centre, Report on the WHC activities for 1938. Photo description: 'Our Horticultural Center satisfies the gardening needs not only of the Warsaw Housing Cooperative residents, but also of a significant part of the residents of the cooperative Żoliborz. 3854, p. 132

Crops obtained from collectively cultivated fields were also donated to the school cafeteria. In addition, children took part in selling vegetables, learning management and economy mechanisms along the way. With time, a school zoo and a 'geography' yard were also created.

- the third part of the plot was allotted to the growing of vegetables, flowers and ornamental plants for school decoration, seedlings for flowerbeds, shrubs and fruit trees. The work was organised so that each child could take part in various activities (preparing land for cultivation, propagating plants from seedlings, sowing and planting flower bulbs, cultivating perennials and setting up garden frames).

Fig. 6.4 Interior of the greenhouse and plant clinic, Report on the WHC activities for 1932. 3749, p. 132, p. 132

> The garden-workshop was used by children from the primary school and *gymnasium* as well as children from neighbouring schools, based on special agreements. Gardening (2 h weekly) was included in the curriculum.

Children could also sign up for the school garden programme individually. Children from age 10 received their own patches (from 3 to 5 m^2). Young gardeners signed a written agreement specifying their rights and obligations, and providing tips on how to use tools and obtain professional advice. The agreements defined the rules of coexistence, cooperation and obliged children to work as a team in a collective garden. The children received garden logs where work history and plant observation were recorded. At times, all gardeners would gather to debate the requests and remarks regarding the organisation and life of the garden. New solutions were introduced, tools were upgraded, and the crop improvement concept was developed.

The work in the school garden was organised by a nature teacher, Stanisław Żemis, who after the war, in 1958, summarised his overall interwar pedagogical experience thusly: 'Instilling gradual independence in children through horticultural work, moving from the collective gardens of preschool children to the individually cultivated patches, was adapted to the development and experience acquired by children, and cultivating a collective garden socialised children and taught them the rules of coexistence and cooperation. The physical effort, regularity and precision required by horticultural work taught children respect for all human work without resorting to moral lessons and finger-wagging speeches' (Żemis 1958, p. 10). This educational and, in a sense, economic system based on horticultural aspects dismantles both the property right and the rivalry characteristic of capitalism, directing the young Żoliborz residents towards cooperation. But it is also important that such an organised system be founded on regular, collective work, self-discipline, good organisation and craftsmanship. There is no place for short-lived, emotional impulses. Principles such as responsibility, respect for nature, physical work and collective neighbourly actions are *de rigueur*. 'On the site of the future Suzina Street, where a preschool, a boiler house and a cinema as well as WHC colonies would be built later on—small patches (*Zagonki*) had been set up at the turn of the 1920s and 1930s. The land was cultivated and an old shed was repurposed through social work. Each of the WHC children could—under the guidance of a nice, elderly lady, Julia Zubelewicz, and a young gardener, Mr. Pawełek—work in the fields: sow and plant various domesticated and ornamental plants. We even had greenhouses where we planted houseplants and seedlings of various vegetables: tomatoes, cabbage, and kohlrabi. We transplanted delicate plantlets with tiny tongs and stakes carved out of wood. Mr. Paweł was rigorous and strict:—'Those tongs are more suited for trapping crocodiles than transplanting semperflorens begonias'—he would often say and ordered the children to carve more precise tweezers out of a wooden splint. We also had our own petting zoo in the fields. There were dogs, chickens, geese, ducks, Guinea fowl, rabbits, guinea pigs, and probably a goat. For many of our colleagues who have never been to the countryside, 'Zagonki' were the first opportunity to learn more about plants and animals, to experience the joy of being close to nature and taking care of plants and animals. Gardeners observed plants germinate, sprout and grow, blossom and mature, from seedlings planted with their own hands and cultivated by them. Could there be something more beautiful than the first radish or pea pod that you have planted and cultivated on your own? And what a joy it was to bring my mother the first flower from my own flowerbed!' (Nowicka 2009, p. 14).[13]

Large families received allotment gardens, and of course, the management of the allotment gardens was established, headed by Edward Osóbka.

Residents took good care of the estate's green areas, thus forming strong ties with it. In one of the WHC reports from 1930, we can read: 'the fact that last year, at the

[13] As a side note, it is worth mentioning—and this fact is quite significant—that these memoirs were published in 2009 in the contemporary 'Życie WSM' monthly, in the section 'Z kart historii' [Pages from History]. The editors of the monthly made sure to instil the historical identity and the models of living in an estate having such a long and noble tradition into the minds of contemporary WHC inhabitants.

time of lilac blooming, the tenants spontaneously decided to stand sentinel during the night to protect flowers from being damaged by pests proves just how much they care for flowers (sic!)' (Warszawska Spółdzielnia Mieszkaniowa 1930, pp. 100–101). Residents of the housing estate worked on a voluntary basis in order to beautify courtyards; children's playgrounds and sandpits were arranged, and even a small swimming pool with a shower. It is worth mentioning that sandpits and playgrounds for children were a complete novelty at the time (Czałczyńska-Podolska 2010).

Adam Próchnik, who recognised that green spaces are one of the most basic common goods and an indispensable element of modernist architecture, formulated the basic principles of the WHC's 'horticultural policy', emphasising its environmental, anti-capitalistic and social character. 'We managed to leave behind narrow streets and the drab reality of tenement houses, we abandoned both the suburb hovels and the sad, high-rise city buildings, to look for a wide open space where we could build our houses—we were not only trying to get sunlight and air for the working class, but we also wanted to surround people with green lawns and the scent of flowers. We told ourselves: 'the beauty and the charm of this world must no longer belong only to the rich and the mighty, to the lucky chosen ones; it must become available to the working class as well. We told ourselves: gardens and green spaces must no longer be a privilege' (Próchnik 1934a, p. 1).

Próchnik's ideological zeal led him to use anti-capitalist rhetoric and mount resistance against the existing economic divisions. 'A private capitalist would squeeze as much profit from rent as possible to maximise his earnings; he would build up his property as to the absolute limits. But the worker cooperative would be nothing like that. The cooperative also has to abide by the principles of percentages because it exists within the capitalist system. In the case of the estate, however, the percentages are not dictated by its desire to turn a profit but rather are extorted by capital. But despite that, the cooperative (…) is guided by its social policy. Here we have vivid proof of the difference in creative activity depending on whether the starting point is the pursuit of profit or the desire to satisfy social needs' (Próchnik 1934a, p. 1). On the other hand, the cooperative idea was both educational (knowledge of nature and the surrounding world, respect for physical work) and aesthetic (beauty that brings solace to our eyes, ensuring internal harmony and calm).

6.4 Żoliborz—An Autonomous Zone?

All the Żoliborz colonies were built by the Social Construction Firm (*Społeczne Przedsiębiorstwo Budowlane*), which was set up for that purpose, and managed by architects-activists, which guaranteed an efficient and economic organisation of work geared towards quality, not profit. The Renovation & Painting Worker Cooperative (*Spółdzielnia Pracy Remontowo-Malarska*) was established in 1933 and carried out minor repairs, flat renovations, dealt with key duplication and reglazing broken windows. In 1937, the Warsaw Bookstore Cooperative (*Warszawska Spółdzielnia Księgarska*) began to organise its activities and managed to carry out many successful campaigns up until 1939: it imported school textbooks for Żoliborz children and organised book fairs. It also published a book by Adam Próchnik titled *Piętnastolecie Polski* [The Fifteenth Anniversary of the Polish State].

The housing estate had the necessary economic infrastructure to satisfy the residents' everyday needs, ensure decent resting conditions or, according to Marx's theory, decent conditions for the reproduction of productive forces for which 'collective consumption' was a basic matter. Particular attention was paid to ensuring that the entities operating within the housing estate were of a cooperative nature. Thus, the municipal branch of the 'Społem' Cooperative Bank was opened in the 1st colony in October 1933, and in 1936, it was transferred under the management of the 'Cooperative Inn'. The Warsaw Housing Cooperative also rented its commercial premises to a pharmacy and postal and telecommunications offices.

The economic dimension of the housing estate life presented here so far should be supplemented of course with the remaining social, cultural and educational spheres (in those times, health care was considered to be an inseparable part of education). Each area of activity was formally organised, institutionalised, but also subject to constant adjustments, improvements and regulations.

'The work of the cooperative and all of the institutions and social organisations operating within the housing estate was supervised by the WHC management. Members of the WHC management were included in the management and supervisory boards of various economic institutions established on their initiative and in cooperation with them; they ensured that the ideological and pedagogical work of these institutions complied with the cooperative's current programme' (Szymański 1989, p. 177). Hence, it can be said that the WHC was a total institution, shaping—in the ideological sense—all spheres of public life and often interfering with the private sphere as well. On the other hand, this consistent picture painted by Jan Szymański can be easily undermined. The authors of articles, the ideologues of this project, construct it dynamically, on an ongoing basis. And the social order was constantly being renegotiated and re-established, and not fixed once and for all. Although the cooperative ideology sets general aspirations, the goals of specific initiatives are initially undefined. They become more pronounced and specific only when decisions are made about the use of certain measures. Only when cooperative members discover that some tools are available to them and some innovations can be implemented do they discover objectives that were not apparent before. This is not a revolutionary strategy, but rather a reformist one. Yet, it efficiently combines theory and practice, while consistent action plans are used to structure and legalise the entire undertaking. Cognition and action are inextricably linked with each other in everyday life.

The Warsaw Housing Cooperative may be seen as a well-functioning organisational performance: efficient, effective and at the same time socially sensitive, as a perfectly organised institutional *communitas* with a transparent separation of powers and prevailing hierarchies. A certain ambivalence, however, can be felt: the WHC wanted to teach people how to live, wanted the tenants to become city users, citizens of public spaces, consciously and responsibly using the municipal common goods which were, in the interwar period, often distributed unevenly and unfairly, being a privilege of the bourgeois class. On the other hand, the activists were only starting to learn how to organise social housing estates, i.e. the pioneering functioning of the local community. Colin McFarlane (2011) associated this type of knowledge transfer between the estate residents and the initiators of urban policies with translocally 'learning the city', a process rooted in an unusual 'expertise'—localised in a specific place and largely autonomous—which has an 'amateur' character and is, after all, a production of knowledge coming from activists and urban movements which have a holistic perception of the city. In order to be able to teach such a large number of residents how to live in such different areas of public life, it was necessary to create a complex institutional system which would provide a framework for these educational ambitions. Yet, the search for the principles of urban life did not take place in university auditoria, nor did it draw from academic solutions. Rather, it was based on imagination and a practical critique of capitalism; social emotions and intellect were merged, and political organisation was combined with the spontaneity of everyday life. All those who had their vision of the housing estate (Toeplitz, Tołwiński, Szwalbe, Próchnik and many others) can be described, as has been repeatedly stressed, as Gramsci's organic intellectuals. They did not forget about what was concrete, practical, mundane and quotidian. They treated the reality of the estate as an area of decommodification, and non-capitalist, political subversion, confirming the importance of imagination as a meaningful productive force and a political power that is responsible for creating an alternative to the dominant state of affairs. They stood by the housing estate's residents.

But how to encourage different people to cooperate and actively participate? What forms should these 'amateur expertise' take on? The local activists' response was clear: social relations are founded on ordinary everyday experiences, such as the possibility of doing convenient and cheap shopping, efficient transport, clean dwelling, bringing up children in a healthy environment. Therefore, concrete progress in everyday matters would prove the effectiveness of collective action. If these experiences are to be shaped by cooperation, should they be based on voluntary exchange and informal relations or institutionalised forms and top-down (but not state-inspired) actions?

Richard Sennett wrote: 'The community organiser had, and has, to engage poor people who feel paralysed, whether as foreigners or simply losers in the capitalist game. To rouse people from passivity, the organiser has to focus on immediate experience, rather than dramatising, say, the evils of capitalism; that big picture is likely to root even more deeply someone's sense that it is hopeless to get involved. To enable participation, the organiser may establish tacit ground rules, the conventions and rituals for exchange (…), but must then leave people free to interact' (Sennett 2012, p. 53). On the other hand, however, informality is associated with the risk of disorganisation and does not necessarily have to work in all areas of the WHC estate. Therefore, it is certainly not a rule by which lifestyles should be organised, nor is it advisable to use it to teach how to lead an urban life. The suggestions put forward by Żoliborz residents were closer to Castells' concept, for whom creating community ties must lead to something concrete and does not equal free and informal coexistence (the coexistence is pleasant and safe, however). It requires rules, rituals and principles. Action needs structures and standards to bring long-lasting and important results (Castells 1985).

As it seems, the WHC activists decided that building numerous self-government institutions, interconnected by a cooperation and exchange network, would be the solution. Not a single, top-down authority, but extensive and branched connections that would make it possible to continuously redefine cooperation. Such mobile solidarity, often based on experimentation and flexible cooperation, allowed tenants to use various, often combined, forms of assistance and employees to change the forms of action and see the problems of the entire housing estate from a wider perspective.

The cooperative management members managed the work of the WHC Social and Educational Fund Board (*Zarząd Funduszu Społeczno-Wychowawczego WSM*), the Bureau of the Administrative Commission of the Housing Estate (*Prezydium Komisji Administracyjnej Osiedla*) in Żoliborz and in Rakowiec, the Main Ad Hoc Financial Assistance Committee for the Housing Estate Tenants (*Główna Komisja Doraźnej Pomocy Mieszkańcom Osiedla*). They also managed the activities of various other committees, i.e. economic, audit and regulatory. The latter drew up regulations governing the working principles of all institutions and organisations operating within the Żoliborz housing estate. Propaganda and disciplinary committees were also established. The cooperative management took part in the cultural and educational activities of the 'Szklane Domy' [*Glass Houses*] Tenants' Mutual Assistance Association and even in the works of the Żoliborz branch of the Workers' Friends of Children Association (RTPD, Robotnicze Towarzystwo Przyjaciół Dzieci), which was an autonomous unit independent from the WHC management, or rather subordinate to the Main RTPD Board.

Years later, Szymański admitted that: 'the general activity of the Żoliborz branch of the Workers' Friends of Children Association was supervised by Dr. Aleksander Landy whose views matched the WHC's approach to education. The institution received subsidies from the Cooperative and implemented the socialist educational system. Therefore, it was more dependent on the Cooperative management, which provided the necessary resources for it to function, than the RTPD superior authorities headed by Tomasz Arciszewski' (Szymański 1989, p. 180).

The WHC management also established a Social and Educational Fund (Fundusz Społeczno-Wychowawczy) which financed social, educational and cultural activities. The financial resources necessary for its activity were obtained by charging fees each time a member was inducted into and removed from the cooperative register and when they applied for the allocation or transfer of flats. This fund was used by various institutions operating within the estate, including: 'Szklane Domy' [*Glass Houses*] Tenants' Mutual Assistance Association or 'Marymont' Workers' Sports Club. As various initiatives proliferated within the estate and the activities of organisations and institutions expanded rapidly, an Intercompany Consensus Commission (*Międzystowarzyszeniowa Komisja Porozumiewawcza*) was established in order to elaborate rules for using the communal area of the 1st Colony, so that a unified work plan and timetable could be established. A database of the tenants' finances was also set up. The Tenant Council was established, represented by members of House Delegations (I wrote about these organisations in Chap. 3).

> We cannot decide what to admire most in our WHC cooperativist township: its powerful and continuous development, or the flourishing social and economic institutions which are the main feature distinguishing WHC from other housing cooperatives (Kantorska 1934, p. 5).

Let us try to think about such a housing estate for a moment, for example, on the outskirts of modern Barcelona or London—it could be interpreted as an 'autonomous zone', created from the bottom-up as a result of various urban policies being formed by both founders and its tenants. Conscious of the futility of the struggle for the commons and collective consumption against the representatives of the city's official policy and state authorities (I mentioned the mistrust towards such 'negotiations' in the introduction to this chapter by recalling Castells), they organised an 'interception' strategy of the official institutions to widen access to public goods and services. Yet, even though they did not accept private property rights, which are central to a capitalist economy (no resident of the housing estate owned their flats, no matter how much time or money they had invested in it regardless of the standard they maintained their flats in), they did apply to the city for sectioning off undeveloped areas and for obtaining low-interest loans. On the other hand, however, they weakened in a sense (on a narrow area) the legitimacy of state authority institutions, such as courts, by introducing their own disciplinary committees, calling the Cooperative's constitution the Constitution, organising internal trade, and even an internal labour market, and finally the Cooperative Bank which was run by the 'Cooperative Inn'.

I do realise, of course, that pre-war capitalism differed from both the capitalism described by Castells and the one we know today. My goal was to draw attention to the human, collective experience of capitalism, to the autonomous organisation of the work process (autonomous in the sense of its independence from capital) and the reversal of the relationship between capital and the forces shaping the logic of the development of production relations. Even if in this chapter, I have analysed only Castells' 'public consumption', I have also pointed out (after Harvey, Negri and Hardt) that the entire social life became Żoliborz's production field. Therefore, my research interests include all of human activity and experience. The category of 'biopolitical work' used by the authors of *Commonwealth* refers to an activity that 'generates itself its cooperation conditions', becomes more autonomous and learns the ability to resist, by using network forms of cooperation. Owing to them, a sustainable economic base has been created, and the social rules which have been developed become entrenched and provide support for future activities. Tenants can refer to their own institutional order that is already working beyond their control. They do not have to mobilise their social emotions every time in order to manifest resistance to their 'joyful encounters'. It requires, however, an involvement in continuous work for the benefit of the housing estate as an institution. If Hardt's and Negri's analyses of common goods and the concept of biopolitical work prove useful, the transition from the discourse of rights to the discourse of 'joyful encounters' should prove to fail in the aspect described here before. This does not mean, however, that Żoliborz residents did not engage in such jacqueries. However, what bonded the residents together, thus allowing them to form an autonomous area of collective cooperation, was their cooperative and educational work. 'This educational work will be fulfilled by no other form of the labour movement, because only in the workshop and in contact with an active and positive economic activity, only in contact with specific tasks and by overcoming difficult situations can the values necessary to make full use of our productive capacities be created' (Próchnik 1937, p. 4). Próchnik was sceptical about the ideas of the founders of utopian socialism, who imagined that a parallel socialist system could be created on the sidelines of the capitalist system. 'Socialism does not mean an escape of the chosen ones to the isle of eternal socialist bliss; rather, it is a struggle to transform the entire world. We take part in this struggle and do so of our own accord' (Próchnik 1937, p. 4). Próchnik, on the other hand, realised that under the conditions of the capitalist system the cooperative movement does not have comprehensive development opportunities; it will not be able to break the power of capitalism. Forming a political movement is the only way that they could prevail in this struggle. In addition to the economic and political front, Próchnik saw a 'pedagogical front which brings together the cultural, educational and sports movement of the working class and aims to influence the feelings, reasoning and character of the broad masses to bring up fighters who would fight against the old world and build a new tomorrow. (…). So, if cooperativism cannot claim to be a factor that will handle the task of social reconstruction by itself, it still has an important task to fulfil' (Próchnik 1937, p. 4).

Contemporary researcher Patrick Cuninghame emphasises that autonomism is not an attitude of escape from capitalism as it requires living within the system: 'So autonomy is not *in*dependence, rather it is the *inter*dependence of the various sectors of the multitude inside, against and beyond capital' (Cuninghame 2010, p. 454).

It is important, therefore, to treat Żoliborz as a social movement which does not seek to take over or abolish power, nor does it fight against the capitalist system. It is, however, a local autonomous enclave, 'replacing' in a sense the extensive and non-functioning systems of state representative democracy and majority rule with local principles of direct democracy and self-management in all aspects of life. This does not mean, however, that the metaphor of 'struggle' which has been used here should be considered useless. Autonomy must have been actively gained. And Żoliborz was not an island of happiness isolated from the rest of the political and economic world. For the time being, I leave the following question open—did the housing estate founders and its management unintentionally create new and excessive power structures and were they interested in taking active part in these already existing state authorities?

'Collective consumption' is thus shared in Żoliborz, in order to fill it with quite specific political and grass-roots content, subjectifying the previously marginalised milieus. And the commons created there embody experimental and prefigurative demonstrations of self-management.[14] The principle of self-management and

[14]Andre Pusey writes about creating 'new commons' and about 'prefigurative demonstrations of self-management' in autonomous centres. For describing bottom-up ideological and political practices based on experimentation and the *do it yourself* principle, the terms 'prefigurative politics and anticipatory institutions' are often used. Cf. The first term was introduced in the 1970s by Carl Boggs: 'By 'prefigurative', I mean the embodiment, within the ongoing political practice of a movement, of those forms of social relations, decision-making, culture, and human experience that

autonomy manifests itself in many areas of everyday life, extending to experimenting with new forms of collective life organisation, including economy and management principles, based on the equality of social and economic goals and the idea of cooperativism.

According to Chris Carlsson, the *do it yourself* principle developed in grassroots production and service practices, not solely to make profit, but to restore the meaning of work and to build community ties, creates a different way of life, new lifestyles and new milieus (Carlsson 2008, p. 52). Such a model is implemented today by a number of various urban movements whose members implement autonomous strategies, for example in the area of urban agriculture, organise workshops in which broken appliances are repaired, or create and develop IT systems and Internet tools.[15] *Nowtopia* proposed by Chris Carlsson is a kind of 'city-centric anti-capitalism'. It is an autonomous zone based on socially useful work which is carried out outside the market economy and the logic of profit-making. The work is done in leisure time in order to implement the ideals of community life. Carlsson sees this type of work as a tool to combat the commodification of the basic forms of everyday activities. According to him, the creation of *nowtopia* is not a rebellion of the working class, but rather an expression of opposition to an alienating and meaningless work. The author's ambitions, however, are neither revolutionary nor anti-systemic in a confrontational manner. *Nowtopias* are more akin to temporary autonomous zones, or rather local and temporary activities that open up space for decent living conditions, based on direct democracy and the principles of autonomy. By developing both well-thought-out strategies and temporary small tactics of opposition in this manner, they create networks of various 'social centres'. 'These 'autonomous zones', through the process of occupying and opening up space that would otherwise be private and closed, facilitate the creation of life 'held in common'. Employing the practice of self-management and principles of autonomy, participants aim to create an example of an alternative to contemporary capitalist society (…). Social centres make private space 'common' and are run on non-profit values. They act as both an ideological and material form of opposition to capitalist logic and its enclosures' (Pusey 2010, p. 178).

It is worth adding that the creation of self-management and sovereign work zones takes place not only in urban space. It is also a path outlined by modern peasants and small farmers who have declared 'food war' on the capitalist agrarian industry, advocating for 'food sovereignty'. Therefore, the Via *Conpensia* International Organisation aims to spread communal and collective forms of ownership that will enhance environmental responsibility and equal distribution of agricultural products. The path to social and economic reorganisation leads through self-management, diversity and cooperation. 'Indeed, one finds movements of 're-peasantisation,' as entrepreneurial

---

are the ultimate goal' (Boggs 1977; cf. Pusey 2010). The second category is used by the Argentine historian, publicist and left-wing political activist Adamovsky (2011).

[15]The main inspiration for this trend is the movement promoting manufacturing and repairing of appliances and equipment, utilising abandoned objects and spaces, creating gardens, etc., which has existed in the USA since the 1960s. The 'Whole Earth Catalog' magazine can be perceived as a forum of this movement. Cf. Carlsson (2008, pp. 47–48).

farmers abandon capitalist farming and an increasing numbers of urbanites take up small-scale agriculture. 'One might even consider the possibility that, as van der Ploeg puts it, 'the emergence of urban agriculture in many parts of the world signals the emergence of new numbers of (part-time) peasants and a simultaneous spatial shift of the peasantry from the countryside towards the big metropolises of the world' (Bello 2009, p. 146; cf. also Ploeg 2008, p. 276). So the following question may be asked: could the WHC, with its idea of Horticultural Farm and estate gardens, become an inspiration and an example of a 'base' organisation exemplifying good practices in this area?

The cooperative was an 'enclave of socialism', as Zofia Topińska described it. Its residents 'wanted to live in an environment which joins together to make its livelihood and satisfy its cultural needs, organises the education of children in accordance with the ideals of social justice, shunning the obscurantism, intolerance, anti-Semitism, and raging nationalism of those days' (Topińska 1984, p. 10). In this sense too, the cooperative, sectioned off on the city's outskirts, was a space where everyone could count on the help of the estate's social organisations or neighbours. They would receive support when faced with the threat of unemployment. The entire cooperative economic system, however, could only work with the active participation of the estate's residents. The residents, in turn, were probably incited by the WHC management in their activities. Analyses which focused on how this power structure works for the benefit of the subjects and by using those subjects allow us to see the potential for the production of alternative subjectivities (Foucault 1982). Performative speech acts on the pages of 'Życie WSM' [*The Life of the WHC*] magazine created a kind of obligation, conceived as conventional quotations of a previously performed statement. And thanks to that they really had an impact. Tenants were obliged to act not through manifesting some real political body, but through the continuously renewed game of conventions and intentions. It was a perfect example of subjectification of power (Foucault). Printed messages and public calls became a space for staging basic conflicts which were apparent at various levels: within the management, between the estate institutions and in their relations with the tenants. Thus, they became the main operational tactic of all involved parties (domestic servants, tenants and the housing estate officials). Hence, the shift towards preparing affect displays, performances, information, pictures and festive celebrations, etc., becomes visible. They aimed to direct the production of subjectivity, sometimes replacing the state power institutions, the official economy or the church.

Finally, we can see this economic enclave as a laboratory in which the urban lifestyle was tested along with the accompanying social economy based on the principles of self-management and cooperation. As a result of this collective experiment, the entire 'city' became part of a set of experiences in some sense prepared by the estate founders, activists and Żoliborz residents. Urbanity is not a stable structure here, but a living network of inner energy, creating loose, flexible and portable boundaries, allowing people to block the influence of the outer city that led to the peripheralisation of people.

Żoliborz, considered today as an empirical example of intentional actions and struggles conducted 'against the republic of property', precisely because of its material dimension, exposes what seems to be the weakest point of the *Commonwealth,* which was pointed out by David Harvey in his criticism of Hardt and Negri: it was not specified how the revolutionary transformation should affect the material foundations of daily life. That is why, the British geographer estimates that: '[...] The suspicion lurks, and there is a lot of evidence in *Commonwealth* to support the point, that it is precisely because Spinoza did not have to be concerned with such mundane things that his formulations are so attractive. They permit Hardt and Negri to bypass consideration of the material basis of revolutionary endeavours in favour of abstract and, at the end of the day, somewhat idealist formulations (Harvey 2009, p. 116). That is why, taking up Harvey's challenge and having full awareness of the historical distance between Żoliborz and modern experiences, I tried to describe the materiality and 'collective consumption' that 'produce' the city dweller who, in turn, transforms and reorganises this housing estate base and space into an urban subject. 'How about concrete proposals, actual political organisations, and real actions?' In the next chapter, I will attempt to tackle this challenge by describing what specific actions have been taken in the sphere of 'institutionalised culture'.

The urban, which I described using Castells' categories in order to highlight its materiality and concreteness, contrary to Merrifield's view does not seem archaic nor obsolete. Merrifield interprets *The Urban Question* 'as engaging with the state as much as with capital, as subverting and re-appropriating state power, as demanding collective public goods, as rallying for relatively autonomous self-management of these public goods' (Merrifield 2014, p. 54). The new urban question is much wider in scope. I will leave aside the most obvious political and economic change resulting in an almost complete privatisation of publicly owned property which was diagnosed by Merrifield. It is an area of more extensive intervention and resistance against the endless reproduction of capital. According to Merrifield, the urban is a real and normative battlefield. 'In the past, in the 'old' urban question, scholars like Castells looked toward the urban to resolve the problem of building a social movement. Now, we need to build a social movement to resolve the problem of the urban' (Merrifield 2014, p. 15). So for Castells, the main problem was the question of how the state manages, orchestrates, plans and funds collective consumption, thus keeping its own political and ideological legitimacy. Urban movements, acting in the spatial unit of collective consumption, i.e. in the city, became a new political entity that assumed a different role than trade unions or political parties. The urban (a spatial unit of social reproduction) was passive and stable. It was a question to which urban social movements struggling for their right to the city should find an answer. Merrifield's new vision of the urban defined it as a space in which 'capital *productively* plunders: capital now actively dispossesses collective consumption budgets' (Merrifield 2014, p. 13). The difference is not only that Castells assumes that urban movements will confront the state, which is not present in the city today, but because the city is ruled by capital. What differentiates these urban questions and makes one of them 'old' is the way of perceiving the urban. When viewed from today's perspective, the urban is a performative space that produces and actively dispossesses. '*[L]and, labour, and capital* aren't conceived as mere things, but are, respectively, *landlords, workers, and capitalists,* actual living bearers of processes' (Merrifield 2014, p. 48). Merrifield sees the urban as the incarnation of that which is collective. 'Outside of human woof and weft the urban creates nothing, is nothing. The urban serves no purpose and has no reality outside of human reality, outside of exchange and union, outside of human proximity and concentration, outside of human encounter and intensity. Nodes of intensity that resonate, that connect with other nodes of intensity, that fuse together and create energy and electricity, incandescent light' (Merrifield 2014, p. 30).

No one should underestimate, however, Castells' findings. The confrontation of the old and new urban questions allows us to highlight various woofs of struggles taking place within the city. The spaces of collective consumption, which had been hacked and then recovered by the housing estate's residents, did not remain passive after all. Merrifield shows us how to conceive them today, treating the housing estate as a carrier of actually existing processes. All these economic projects, as tangible and concrete as they could have been, worked effectively, going far beyond their economic goals. And the 'urban' produced there manifested itself actively in the forms of direct democracy, in effective grass-roots actions, and autonomy, conceived as the condensation of urban culture and material tissue.

References

Adamovsky E (2011) Anty-capitalism (trans: Trigona M). Seven-Stories Press, New York

Bello W (2009) The food wars. Verso, London

Boggs C (1977) Marxism, prefigurative communism, and the problem of workers' control. Radical America 11:99–122. https://libcom.org/library/marxism-prefigurative-communism-problem-workers-control-carl-boggs. Accessed 7 Sept 2018

Carlsson C (2008) Nowtopia: how pirate programmers, outlaw bicyclists, and vacant-lot gardeners are inventing the future today! AK Press, Edinburgh, Oakland, West Viriginia

Castells M (1977) The class struggle and urban contradictions: the emergence of urban protest movements in advanced industrial societies. In: Cowley et al (eds) Community or Class Struggle? Stage 1, London

Castells M (1985) The city and the grassroots. University of California Press, Berkeley

Cuninghame P (2010) Autonomism as a global social movement. Working USA: J Labor Soc 13(4):451–464. https://doi.org/10.1111/j.1743-4580.2010.00305.x

Czałczyńska-Podolska M (2010) Ewolucja Placu zabaw: Koncepcja dla dzieci w Europie i Stanach Zjednoczonych = Evolution of the Playground. A Concept of Children Play-Space in Europe and the United States. Przestrzeń i forma 13:73–88 http://www.pif.zut.edu.pl/pif-13_pdf/B-01_Czalczynska-Podolska.pdf Accessed 1 Feb 2017

Editorial team of Praktyka teoretyczna (2012) Wprowadzenie. Rzecz-pospolita i aktualność komunizmu. Genealogia i krytyki polityczno-filozoficzne projektu dobra wspólnego [Introduction to the Polish edition of Commonwealth – Commonwealth and currency of communism: genealogy and political-philosophical critiques of the common]. In: Hardt M, Negri A (eds) Rzecz-pospolita. Poza własność prywatną i dobro publiczne (trans: Editorial team of Praktyka Teoretyczna). Korporacja ha!art, Kraków, pp 7–71

Engels F (2010) The housing questions. In: Marx K, Engels F, Collected Works, vol 23. Lawrence & Wishart Electric Book, [s.l.]. http://www.hekmatist.com/Marx%20Engles/Marx%20&%20Engels%20Collected%20Works%20Volume%2023_%20M%20-%20Karl%20Marx.pdf. Accessed 7 Sept 2018

Foucault M (1982) The Subject and Power. Critical Inquiry 8(4):777–795

Głodowska W (1962) Wspomnienia [Memoirs]. Życie Osiedli WSM January:24
Harvey D (2009) Commonwealth: an exchange, analysis. Artforum 48(3):210–221. http://korotonomedya.net/kor/index.php?id=27,316,0,0,1,0. Accessed 5 Apr 2018
Harvey D (2012) Rebel cities: from the right to the city to the urban revolution. Verso, London, New York. http://abahlali.org/files/Harvey_Rebel_cities.pdf. Accessed 9 Sept 2017
Haubold O (1937) Czym jest dla nas 'Gospoda' [What does the 'Cooperative Inn' represent to us?]. Życie WSM, Nov:203–204
Kantorska A (1934) Nowy ład w pralni [New order in the laundry]. Życie WSM, April:5
Lefebvre H (1979) La bourgeoisie et l'espace [The bourgeoisie and space] (trans: Jałowiecki B, Jałowiecki P). In: Jałowiecki B, Nurek S (eds) Krytyka społeczeństwa kapitalistycznego w pracach socjologów zachodnich [The critique of capitalist society in the works of western sociologists]. Uniwersytet Śląski, Katowice, pp. 88–102.
Lefebvre H (2010) The right to the city. In: Lefebvre H (ed) Writings on cities (trans: Kofman E, Lebas E). Wiley-Blackwell. https://monoskop.org/images/4/40/Lefebvre_Henri_Writings_on_Cities.pdf. Accessed 7 Sept 2018
Matysek-Imielińska M (2017) Projektantki i performerki przedwojennego osiedla Żoliborz [Designers and performers of the pre-war Żoliborz housing estate]. Przegląd Kulturoznawczy 34:521–536. http://dx.doi.org/10.4467/20843860PK.17.035.8202
Mazur E (1993) Warszawska Spółdzielnia Mieszkaniowa 1921–1939. Materialne warunki bytu robotników i inteligencji [Warsaw Housing Cooperative 1921–1939. The living conditions of workers and intelligentsia]. IAiE PAN, Warszawa
McFarlane C (2011) Learning the city: knowledge and translocal assamblage. Wiley-Blackwell, Malden, MA, Oxford
Merrifield A (2014) The new urban question. Pluto Press, London. https://pl.scribd.com/read/314636395/The-New-Urban-Question. Accessed 7 Sept 2018
Nowicka M (2009) WSM – RTPD – Szklane Domy [WHC (Warsaw Housing Cooperative) – RTPD, Robotnicze Towarzystwo Przyjaciół Dzieci (Workers' Friends of Children Association) – 'Szklane Domy' (Glass Houses) Association]. Życie WSM, September:14–15
Orsetti M (1934) O angielskich pralniach społecznych [On British social laundries]. Życie WSM, October:5–7
Van der Ploeg J (2008) The new peasantries. Earthscan, London
Próchnik A (1933) Współżycie ideowe czy walka polityczna [Ideological coexistence or political struggle]. Życie WSM [Life of the WHC], October:1–2
Próchnik A (1934a) Polityka ogrodnicza WSM [The WHC horticultural policy]. Życie WSM [Life of the WHC], July:1–2
Próchnik A (1934b) Spółdzielnia czy spółdzielczość [A cooperative or cooperativism]. Życie WSM [Life of the WHC], Sept:2–4
Próchnik A (1937) Ideologia spółdzielczości robotniczej [The ideology of labour cooperativism]. Życie WSM, May:1–4
Pusey A (2010) Social centres and the new cooperativism of the common. Affinities: A J Radic Theory, Cult, Action 4(1):176–198. http://ojs.library.queensu.ca/index.php/affinities/article/view/6155. Accessed 20 Nov 2016
Sennett R (2012) Together: the rituals, pleasures and politics of cooperation. Yale University Press, New Haven
Szymański JA (1977) Problem zatrudnienia służby domowej na terenie WSM w latach 1927–1939 [The issue of employing domestic servants within the WHC estate in the years 1927–1939]. Życie Osiedli WSM (addition to the 'Domy Spółdzielcze' monthly) 1–2
Szymański JA (1989) Warszawska Spółdzielnia Mieszkaniowa 1921–1970. Zarys dziejów [Warsaw Housing Cooperative 1921–1970. History outline]. Wydawnictwo Spółdzielcze, Warszawa
Święcicka J (1963) O pracy spółdzielczej wśród kobiet w latach 1928–1939 [Women's Cooperative Work in the years 1928–1939]. In: Wspomnienia działaczy spółdzielczych [Memories of cooperative activists], vol 1. Zakład Wydawniczy CRS, Warszawa, 1963, pp 51–69

Toeplitz T (1925) Istota spółdzielczości mieszkaniowej [The essence of housing cooperativism] (editorial). Robotnik 245: 7 September. Reprint of: Szymański J.A. (1989), pp. 292–296
Toeplitz T (1928) Robotnicza spółdzielczość mieszkaniowa [Workers' Housing Cooperatives]. Robotnik 10: 10 January
Tołwiński S (1961) 40-lecie WSM [The 40th anniversary of WHC]. Domy Spółdzielcze 12:5
Topińska Z (1984) Żoliborskie obietnice: o wychowawczym środowisku WSM [The promises of Żoliborz: the formative environment of the WHC]. Wydawnictwo Spółdzielcze, Warszawa
Warszawska Spółdzielnia Mieszkaniowa [Warsaw Housing Cooperative] (1930) Sprawozdanie [1930 WHC report]. Warszawa
Wasilewska W (1936) Życie WSM, March:50
Żemis S (1958) Ogród szkolny RTPD na Żoliborzu jako koncepcja metodyczno-wychowawcza [RTPD School garden in Żoliborz as a methodological and educational concept]. Życie Szkoły 12
Życie WSM (1933) Życie WSM, October:8
Życie WSM (1934) Walne Zgromadzenie 'Gospody Spółdzielczej' [The 'Cooperative Inn' general assembly]. Życie WSM, April:112
Życie WSM (1939) Życie WSM, March:81

Chapter 7
Workshop? Settlement House? Laboratory?

Abstract Everyday life in Żoliborz and the lifestyle shaped there direct my attention to the category of the common good understood not only traditionally, as the *common wealth* of the material world—the air, the water, the fruits of the soil, and all nature's bounty (the natural commons) but also, as defined by Hardt and Negri, 'those results of social production that are necessary for social interaction and further production, such as knowledges, languages, codes, information, affects, and so forth'. Hence, moving away from economic issues and living conditions, but not abandoning them altogether, I examine the 'results of social production'—namely, all these forms of cooperation, neighbourly assistance, new forms of sociability and spending free time, the emerging institutions bringing together the members of the cooperative and designating, to some extent, their *modus vivendi*. Therefore, I analyse the models of power structures and the urban culture produced on a housing estate, starting with a description of a very dense institutional network: a common room, a school, a nursery, vegetable gardens and green areas, playgrounds and reading rooms, and special interest clubs—spaces of self-education. To do this, I use the performative tools proposed by Richard Sennett and explore various forms of cooperation and being together. I describe some institutions using the metaphor of a workshop, others—the metaphor of the common room. These models, devised by Sennett, proved useful for the analysis of various forms of cooperation. Being together is, at times, the objective, but sometimes also a tool for achieving other political, economic or social goals. The category of rituals and the production of social habits turned out to be extremely important here, allowing me to describe the techniques associated with the forming of and experiencing collective identity.

Keywords Common good · Rituals of cooperation · Participation · Laboratory · Workshop · Common room/settlement house

7.1 Institutionalisation of the Cooperation Ritual

The city uses a wide range of various institutions, techniques and urbanised spaces in order to 'mould' its residents, city users. These institutions and techniques, however, are properly distributed and dispersed. To what extent such decentralisation reduces

M. Matysek-Imielińska, *Warsaw Housing Cooperative*, The Urban Book Series,
https://doi.org/10.1007/978-3-030-23077-7_7

their efficiency and, at the same time, the efficiency of the city's operations is, of course, a matter that needs to be analysed every time. David Harvey assumes that decentralisation is always a worthwhile objective, but it entails the installation in the office of a certain power structure. 'It is simply naive to believe that polycentrism or any other form of decentralisation can work without strong hierarchical constraints and active enforcement'—David Harvey claims (Harvey 2012, p. 84).

The city creates its citizens mainly thanks to two spheres: the material sphere related to the physical, architectural or urban-planning character, and the immaterial sphere of symbols, rituals, regulations and institutions. The areas of this impact are obviously not separate, but rather overlap. It is difficult to imagine an architectural form without a symbolic dimension. It is also difficult to find social institutions, for example, a 'communal house'—as described by Ossowski using the Inuit metaphor of 'kashim'—without urban-planning assumptions, leaving it on the sidelines of the local community.

Thus, the city is an area based on coercion (spatial planning, institutional power of norms and regulations), on a certain hierarchy of power which ensures the setting of demarcation lines between order and chaos, and on the structural organisation of urban activities. The entities which create an 'urban machine' which places citizens in specific spaces, assigning them specific roles, are the institutions, regulations, provisions of law and tacit agreements. However, in order to assume these roles, citizens often experiment and test different ways of living, developing social relations and coping in this culturally, semiotically and urbanistically dense space.

The intangible sphere of the city assumes certain conventions and rituals which constitute one of many methods of the symbolic disciplining of city dwellers. They can act through orders and prohibitions, setting sanctions and causing fear. They can also create a sense of community and foster neighbourly ties. They operate on the basis of simple mechanisms, e.g. bringing people together on the occasion of ordinary daily activities which lead to fostering neighbourly relations, or using a bathhouse or a laundry room which bridges the boundary between the private and the public, thus giving rise to the need to compromise when it comes to the organisation of everyday life. The functional diversity of institutions and the 'good space' favours building an integrated community, while avoiding at the same time its homogenisation.

Undoubtedly, cities, as Harvey argues, are founded on power and violence. And they cannot function without either of them. The Żoliborz estate (city within the city) is no different. Here, however, one should ask by whom, for what reason, and by what means is this 'urban machine' regulated. What makes it capable of self-management and not break up into pieces, not be subject to fragmentation, and be an entity capable of managing itself?

Is this relatively small Żoliborz community similar to the hippie *communitas* analysed by Turner which, after all, were not entirely based on direct democracy and community management? They often fell under the seductive power of strong individuals who, in an open or masked manner, subjugated its residents by manipulating their behaviours.

What made the Żoliborz republic avoid elitism and a strong leadership role was the power of institutions and mechanisms that transcended the community (even though they developed within it). Self-management and at the same time self-awareness of the order of norms and permanent ideological framework were possible mainly due to the continuous processes of reminding, quoting, repeating, discussing and 'celebrating' it. These tactics of self-awareness made the adopted behaviours and conventions be constantly refreshed and negotiated, thanks to which they became neither a natural nor a common *modus operandi*. The initiatives such as the Żoliborz Lovers Club (Klub Miłośników Żoliborza) or a documentary film about the WHC titled 'Budujemy' [We are Building] were therefore not surprising. On the pages of the 'Życie WSM' [*The Life of the WHC*] magazine, the cooperative's assumptions were constantly revised and confronted with everyday activities, while its management board proposed and properly enforced the ideals outlined in the Cooperative's constitution. Żoliborz owes its self-awareness and subjectivity to the power of that which is inside. The cooperation rules are set up, after all, in groups, not outside them. This is where the initiatives, plans and activities stemmed from. Żoliborz is not a passive district, an urbanised space whose development strategies are based on 'external investments', but represents rather an emancipatory potential which stimulates its residents to engage in different activities. It did not stem from an explosive emotional community and was far from *jacqueries* or joyful encounters. The activity of the Żoliborz district manifested itself thanks to an organised structure of institutions and formal ventures initiated from the top down, but also of informal meetings. It was supported by the conviction that every resident is worthy, can be responsible for something[1] and has a lot to offer. Each one of us lacks something and turns this lack into community actions, defining thus community interests. This is where smouldering tensions between those who have something to offer and those who expect support flare up. Here, the residents' lack of self-sufficiency manifests itself. In Żoliborz, everyone is 'connected' to the estate institutions which inspire its residents to commitment and communication, thus making the estate alive. Here, the boundary between the private and the public becomes blurred. Instead, there is a division of space into interaction/collaboration and intimate space (Nawratek 2012, p. 24).

Let us not be misled by the vision of an undivided, homogeneous community. It is not true that the only boundary generated there was the division into that which is within the estate and that which is outside of it. There were also boundaries, conflicts, divisions and ideological skirmishes within the estate. However, this internal differentiation did not constitute an obstacle for the Żoliborz residents. On the contrary, the conflict is an immanent feature of both the city and social life in general; it prompts the need for transferring skills and ideas and leads to the compulsion to mediate, to resolve disputes and at the same time to continue self-determination. The boundaries between people with different social and cultural capital are not impassable: they become the space for connections and interactions. Moreover, in Żoliborz,

---

[1]Maria Orsetti used this argument to justify the name of the Active Cooperatists' Club (Koło Czynnych Kooperatystek), see Chap. 5.

these boundaries were institutionalised in the form of meetings, clubs, readings and special interest clubs—while initiated from the top down, eventually taken over and developed by residents themselves. Becoming a sphere of mediation, they turned out to be something beyond the residents' control, a separate entity in which its symbolic and ritual dimension manifested itself. This institutionalised boundary has been porous and permeable, and connected representatives of different classes, statuses, professional positions and ideological standpoints; it has also connected Żoliborz's internal world with that which is outside of Warsaw, thus complementing the capital's 'cultural offer'.

The housing estate is not only a world of (architectural and institutional) resources, but a self-governing subject of politics, culture and economy, an interface through which residents act together and communicate with each other through dispersed and hierarchical subsystems managed by a visible, undisguised power structure which disciplines these spaces of interaction and cooperation, thus disciplining what is shared and common. Does it leave some space for intimacy away from its influence?

It is, of course, easy to have an impression that the estate authorities in Żoliborz are omnipresent and can find their way into residents' personal and home spaces at all times. It is like that because modest housing equipment requires the transfer of most activities to the sphere of interaction. A library and a quiet room provide residents with space for silence and retreat, while a bathhouse provides them with hygienic solutions. And while staying in such a space, we must agree on interactions which are regulated and adjusted to the institutions' objectives. The house stretches into the housing estate's public space thanks to public squares or a cafeteria, and there is no place here for refusal of social relations. That is why the estate founders, Żoliborz activists and also the estate residents are constantly negotiating norms, examining and commenting on the 'principles of cultural co-existence', and on the sharing of space with others and by others. This, in turn, prompts the need for creating regulations, constitutions, committees and subcommittees. Does this mean that residents are not allowed to interact freely? Are the residents allowed for innovation, spontaneity, experimentation? Settlement house or workshop? Which of these metaphors describe better the institutional dimension of the housing estate? (cf. Rewers 2014, pp. 22–25, 56–65). I derived the inspiration for these questions from the metaphor of the settlement house, laboratory and workshop proposed by Richard Sennett to describe complicated social relations based on solidarity and cooperation. These metaphors (beside library and *translatorium*) are also used in the analytical area of cultural urban studies, so it is worth using them to answer the question what shape the Żoliborz housing estate took. The questions asked here put the spotlight on the city's one additional figure, namely the laboratory, also because social estate designers themselves (Ossowski, Hochfeld, and Szwalbe) also had such associations. In order to try to answer the above questions, we need to look at the institutional context developed within the area of Żoliborz culture.

7.2 Tenants' Mutual Assistance Association

On 30 December 1926, as part of the implementation of the Cooperative's statutory goals, the 'Szklane Domy' [*Glass Houses*] Tenants' Mutual Assistance Association was established (Fig. 7.1). Clearly, the Association's name refers directly to Stefan Żeromski's novel *Przedwiośnie* [Early Spring] and reflects the general intellectual atmosphere of the cooperative founders.[2] The inspiration for the 'Glass Houses' Association was derived from French organisational models. Teodor Toeplitz, an extremely enlightened expert on housing, was well-versed in the European solutions of these issues and often derived the knowledge about cooperative solutions based on tenants' cooperation from abroad. This time, he benefited from the pioneer experiences of the mayor of Suresnes, a factory town in the suburbs of Paris. Henri Sellier was both president of the Office for Affordable Social Housing (Urząd Budowy Tanich Mieszkań) and organised tenants' mutual aid societies. In the 1930s, Sellier visited Żoliborz, which he viewed as a perfectly delivered objective of a housing estate. Thus, using the French organisational experience, an association was established whose ideological background was already based on Polish models, derived this time from Edward Abramowski's concepts. *Związki przyjaźni* [The Friendship Bonds] assumed that this mutual help carried out in groups of people who knew each other personally would concern all social matters, but above all is to be implemented on the basis of self-help, and not as a charitable organisation.[3]

The constitution defined the objectives and the scope of the association's activities, its sources of financing (membership fees, monthly contributions, subsidies granted by the cooperative or municipality[4]) and the management's organisational structure and their forms of operations. It also included 'a ban on conducting political or religious discussions at meetings'. The Warsaw Housing Cooperative included members and sympathisers of both the Polish Socialist Party and the Polish Communist Party. An exchange of views on political issues could lead to the deepening of the already existing ideological divisions, which was not difficult under the political conditions of the time. 'This ban has also removed the possibility of suspicion by the Sanacja regime of conducting political activity incompatible with the spirit of the bourgeois state' (Szymański 1989, p. 92). The cooperative members were usually people ideologically involved and, although it was difficult to remain neutral, tried to restrain the 'political struggle', which was in line with one of Robert Owen's Rochdale Principles.

---

[2]The attitude of 'Żeromszczyzna' [*sensitivity to social injustice and the suffering of the Polish nation*—translator's note] as an intellectual posture was broadly described by Czubaj (2007, pp. 42–54). I elaborated on the topic of 'Żeromszczyzna' in Chap. 3.

[3]It is worth noting that the Żoliborz cooperative architects were after all quite critical of these ideals. Ossowski doubted friendship (see Chap. 3). Maria Orsetti, in turn, opposed Robert Owen's version of socialist utopia, treating it as an anti-cooperative, non-participatory vision, forcing people to the assumed state of happiness (cf. Orsetti 1926).

[4]Detailed data and amounts of contributions are provided by Szymański (1989, p. 92).

Fig. 7.1 House of the 1st Colony—the « Glass Houses » Association's seat as of the end of 1932. A library, a magazine reading room, mind game rooms and a dark photo club are located on the ground floor. On the first floor: The « Glass Houses » Association's office

However, during the meetings at the Discussion Club (established on 13 October 1932 as part of the 'Szklane Domy' [*Glass Houses*] Association), vehement interlocutors unleashed emotional political quarrels. 'Discussion evenings were interesting, and political topics caused heated debates. The speakers' extremely radical and sometimes even communist approach to the topic did not suit the 'Szklane Domy' [*Glass Houses*] Association Management Board. The Association had to clearly manifest the observance of neutrality principles, for example, in 1933 revoking the membership of four people for 'seeking to spread confusion and to engage in political struggle within the estate' (Filipczak 1963, p. 156). When Adam Próchnik, after Stanisław Szwalbe's resignation, became the president of the 'Szklane Domy' [*Glass Houses*] Association, he immediately decided to dissolve the Discussion Club. Here again, authoritative speech acts appeared as a voice which was supposed to reach the housing estate residents through the 'Życie WSM' government mouthpiece. In October 1933, Próchnik published an article titled *Współżycie ideowe czy walka polityczna?* [Ideological Coexistence or Political Struggle?]. His voice is a performative call to 'implement the slogan of a working class common front in the fight for economic liberation and political rights' (Próchnik 1933, p. 2). However, it did not end with the authoritative speech acts. Spectacular behaviours resulting from political involvement which also emotionally engaged the estate residents were the domain of Żoliborz life within the estate organisations. The rebellious members of the dissolved club set up the 'Committee of Sixteen' which tried to change the existing management board of the association. The democratic procedure became an instrument of social spectacle (the proposal for a no-confidence motion to the board was rejected, and the Committee of Sixteen opponents passed a resolution condemning the act and the exclusion of rebellious and politicised members).[5] These political performances involved not only the actors themselves, but also the spectators and went far beyond the premises of the 'Szklane Domy' [*Glass Houses*] Association, thus covering the cooperative's whole area of activity. 'The Committee of Sixteen proposed that the members of the WHC management should only be composed of the Żoliborz estate residents. They wanted to remove the Cooperative founders from its management board: Tołwiński, Toeplitz, and Szwalbe (who lived outside the housing estate); it was, in essence, an attack on the myth of the founding fathers which was very important for the Association's active members. The residents also distributed leaflets 'directed against the payment of some part of the wages with the 'Cooperative Inn' retail vouchers', in order to undermine the principle of the social economy based on exchange, which is essential for cooperativism (Szymański 1989, pp. 100–101). This type of resistance exposed the network of social connections and ideological interdependencies. The Women's Club, involved in a socio-politico-ideological spectacle, acknowledged that 'excluding political issues from the cultural and educational actions run by the 'Szklane Domy' [*Glass Houses*] Association condemned it to 'sterility and atrophy' (Szymański 1989, p. 101) and that, in turn, entailed the decision to suspend the Club's activities, so women pledged not to raise political issues ever again. These formally made declarations at meetings and assemblies had a reverse side too. Women very often organised various types of manifestations to support the striking workers (in 1935—for the employees of the leather industry, in 1936—they organised an action for the amnestied political prisoners, and in March and April 1936—they collected money to help families of the workers killed in combat with the police in Krakow and Częstochowa).

Thus, it can be seen that the political situation in the housing estate clearly reflected the national mood. When at the beginning of 1935, the Polish Socialist Party (PPS) and the Communist Party of Poland (KPP) started negotiating the implementation of the slogans of the common ideological front, the petition for the readmission of the excluded four members (from the 'Committee of Sixteen') was approved within the estate. Stefan Purman (a KPP activist) was also elected to the Association's Management Board and he was entrusted with the function of the head of the artistic and literary section of the 'Szklane Domy' [*Glass Houses*] Association.

WHC was a neutral organisation in its assumptions, but it was difficult to expect a similar attitude from its members. Their political zeal manifested itself extremely clearly and in a frequently caricatural manner, despite WHC's statutory provisions and rules. And if today people talk about Żoliborz as about a socialist and secular intelligence, it is because ideological manifestos, calls and rituals were extremely

[5]This resolution was published in: Życie WSM [Life of the WHC] (1934a).

important there. The political disputes also meant that Żoliborz's democracy had a chance to become a better version of itself. It is possible only when it is constantly subjected to criticism, transformed in a dialogical manner, and the introduced order makes it more socially acceptable. Hence the extreme importance of the distance is towards democratic participation. If democracy is to consist in generating changes in social experience, self-management and in shaping the public space without excluding its participants, it needs distrustful and rebellious citizens.

The association worked in an extremely formalised but also versatile manner, creating and getting involved in the work of various institutions. Kazimierz Tołwiński, Stanisław Tołwiński's father, turned his private book collection into a library, and soon afterwards, settlement house activities began to be run. In 1928, at the initiative of several mothers, a preschool was set up in Janina Bierutowa's private flat, which was eventually handed over to the Żoliborz branch of the Workers' Society of Children's Friends (RTPD) in the 1928/1929 school year. The housing estate residents were actively involved in the work of these institutions, as there was no trained settlement house worker or kitchen help that could prepare breakfast for preschool children. The association members were on duty, and some of them (like 'Comrade Burkotowa' for example) were involved throughout the entire year. Thanks to the inspiration of Zofia Lubodziecka, a puppet theatre for children, later known throughout Warsaw as the 'Baj' theatre, was created (Fig. 7.2). A drama group was organised for young people and an orchestra (composed of brass musicians and mandolinists at the initiative of the residents who played these instruments). The 'Szklane Domy' [*Glass Houses*] Association delegated their committees to help and improve the activities of the workers' kitchen and attempted to establish cooperation with the cooperative stores kept at that time by the WSS.

The number of association members grew, though not as quickly as the WHC management would have imagined.[6] Not all the main tenants (i.e. those who were also cooperative members) joined the 'Szklane Domy' [*Glass Houses*] Association. Admittedly, in adopting the operational model of the French mutual assistance association, the principle of compulsory membership was sought, but the Commissariat of the Warsaw City Board did not express their consent. One could see in these attempts a violation of the main cooperativism principle which, after all, was based on the voluntary participation.

The association's management splits into two sections: an economic one (headed by Stanisław Szwalbe) with loan, work and care sections. The second section was headed by Adam Próchnik and included: music, lecture, spectacle and interclub committees.[7] The association's expanding administration required the hiring of a permanent employee who would be a member of the Bureau of the 'Szklane Domy' Association Board, acting at the same time as a secretary. Until 1933, Kazimierz Now-

[6]In 1934, 1001 people belonged to the association, including 730 main tenants, i.e. 60.7% of the total number. In January 1938, the percentage amounted to 73.9% and a year later it dropped to 70.2%. Cf. Szymański (1989, p. 95). The decline in the number of cooperative members was mainly due to the non-payment of contributions.

[7]The scope of activities of these committees was published in the 'Życie WSM' magazine (Życie WSM 1932).

Fig. 7.2 Scenes enacted by children from the Workers' Society of Children's Friends (RTPD) school, Report on the WHC activities for 1932. 3738, p. 154

icki fulfilled this function (succeeded by Emanuel Freyd, then Julian Hochfeld and eventually Henryk Jędrzejewski). The Board was composed of 11 people and consisted of 8 sections: work and financial support section, educational section, library section, youth section, artistic section, social life section, self-government section, and physical education section. As part of this activity, the hosts of the following 7 clubs and teams were selected: visual artists' club, radio club, women's club, 'Vitraj Domoj' Esperanto club, a mental games club, tourist club (with pedestrian, water and ski sections, and a glider club functioning briefly) and a brass orchestra club (Fig. 7.3). Among the numerous members of the association's management board, there were many representatives of the Warsaw Housing Cooperative management. Each of the sections had its delegates, as well as committees and subcommittees which appointed residents to specific positions and functions.[8] The Intercompany Consensus Commission (*Międzystowarzyszeniowa Komisja Porozumiewawcza*), made up of representatives of the cooperative management board, was established, along with the Tenant Council and delegations of individual social housing organisations. The Intercompany Consensus Commission coordinated individual events, projects and daily activities. These, in turn, were initially conducted in very modest housing conditions. The significantly enlarged preschool with numerous children was allotted premises in a liquidated non-alcoholic inn run by the Association of Working Women's Clubs. The reading room was used by children in the morning, and by the secretariat of the association in the afternoon, while the large meeting room managed by the cooperative management was used for gymnastics exercises and kids games in the morning, and for meetings and readings in the evenings. In this way, the association's daily work cycle was determined by the needs of all residents of the housing estate.

7.3 Transformation Performance and Self-study Work

It seems, however, that one of the most important educational institutions in Żoliborz was the library. It was in the library that open access to cooperative ideas was beginning to sprout, and young people were offered wide intellectual horizons. In December 1933, after Kazimierz Tołwiński died, it was decided during the Extraordinary General Assembly of the 'Szklane Domy' Association by the resolution of the Association's Board to name the Association after Kazimierz Tołwiński. The library worked very effectively and enjoyed a lot of interest. It is also worth noting that it was created earlier than the Public Library established by the municipal authorities in this district. With time, it turned into 'a kind of headquarters, serving other libraries' (Szymański 1989, p. 102): children's libraries of the Workers' Friends of Children Association (RTPD) in Żoliborz and Rakowiec, a reference library at the reading room of 'Szklane Domy' Association, a quiet room and a library of the Domestic Servants' Trade Union. 'On the premises of the 1st Colony, there was a library run by the *Grandfather*, i.e. Stanisław Tołwiński, one of the WHC founders and member of the Cooperative Management Board, and eventually, after the war, the mayor of Warsaw). His daughter, Hanna Tołwińska, helped him in the library for adults. There was also a library for children. It was run by Maria Arnoldowa, a warm-hearted person, full of interest for small booklovers, able to encourage them to read, advise on what to read, getting into the interests of specific children' (Nowicka 2009, p. 14).

[8]For instance, 19 people were appointed to the educational section, 22 to the artistic one, 4 people were appointed to the work and financial support sections, while the loan section was composed of 11 people, the library section consisted of 12 people, and the social life section was composed of 17 people, etc. Szymański wrote about it in detail (Szymański 1989, p. 98).

Fig. 7.3 Orchestra of the "Glass Houses" Association, Report on the WHC activities for 1931. The brass band composed of 26 members at the end of 1931, and the mandolinists' band – of 14 members. 3534, p. 155

The reading room subscribed to dailies and magazines of different political orientations and various themes: there were socio-literary, economic, cooperative, popular-scientific, women's and informative[9] magazines, selected by the reading commission appointed for this purpose. Readers regretted, however, a very modest foreign magazine section (merely two or three titles subscribed a year!). In the reading room, you could run a mind games club (chess, checkers, dominoes. Gambling was prohibited and, over time, card games as well) and listen to radio broadcasts (Fig. 7.4).

[9]Andrzej Szymański described it in detail, quoting the titles of subscribed magazines (Szymański 1989, pp. 103–104).

Fig. 7.4 Mind game club of the "Glass Houses" Association, Report on the WHC activities for 1932. 3740, p. 156

As befitted Żoliborz's atmosphere, the selection of magazines was also the subject of a feud at the general meeting of members of the 'Szklane Domy' Association ('H. Weldsztaub and B. Nowicka and M. Nowicki filed a demand for removing the right-wing tabloid press from the reading room. The demand was rejected by a large majority'[10]) as well as criticised in the 'Życie WSM' [*Life of the WHC*] magazine. In April 1937, Marian Nowicki demanded a reconsideration of the position of the General Assembly and asked rhetorically whether the WHC reading room was a scrap heap (Nowicki 1937), arguing that dailies other than the working-class press

[10] A recursively repeated dispute was described by Purman. In his work titled *Polemiki* [Polemics], he published an extensive article titled *Lektura do poduszki, czy oręż w walce?* [Bedtime Reading or a Weapon in a Fight?] (Purman 1937).

or information magazines with no clear political orientation were merely tabloids. In response to this complaint, Stefan Purman, an educational commission member (and an active KPP activist), formulated a demand for a comprehensive and diverse access to the press (Purman 1937). It must be admitted, however, that the left-wing magazines dominated: 'Myśl Socjalistyczna' [*Socialist Thought*], 'Tydzień Robotnika' [*A Worker's Week*], also a radical student magazine (e.g. 'Po prostu' [*Just Simply*]) had been subscribed. Readers could read magazines such as the 'Lewar' biweekly, or the 'Lewy Tor' [*Left Track*] monthly, in which the texts written by both residents and members of the cooperative's management, such as Edward Szymański, Wanda Wasilewska, Stanisław Tołwiński and Stanisław Ossowski, were published. The reading room together with the library was, therefore, a kind of self-study workshop for social activists, but also for participants of numerous courses organised by the Association. It gave the possibility of self-study and became a place of reflection and rest for the inhabitants of the overcrowded and cramped flats. And for booklovers, a real reading refuge (Fig. 7.5).

The Association took great care of the intellectual and educational atmosphere in the housing estate. In addition to the library, the Association conducted readings. Initially, Adam Próchnik chaired the reading committee (and then Emanuel Freyd, Julian Hochfeld and Henryk Dembiński). The readings were attended by many well-known social and political activists (including Adam Próchnik, Stefania Sempołowska, Wanda Wasilewska, Tomasz Nocznicki, Edward Osóbka, Stanisław Tołwiński, Marian Nowicki, Adam Zdanowski, and Stefan Zbrożyna) and scientists (Janusz Korczak, Stefan Baley, Zdzisław Żmigryder-Konopka, Władysław Gumplowicz, Stanisław Ossowski, Jan Nepomucen Miller, Antoni Bolesław Dobrowolski, and others). The readings concerned educational and upbringing issues, as well as new concepts in pedagogy. Current political events and social moods were discussed (e.g. Ossowski spoke about the war in Abyssinia, Halina Święcicka about 'The Role of Cooperativism in Socialist Movement', and Henryk Dembiński—about Italian Fascism. In 1933, Adam Próchnik gave a speech titled 'On the History of Dictatorship', and Antoni Zdanowski, having come back from war-torn Spain, gave a speech on 'The Fighting Spain'. There was also a speech given on the situation in France under the rules of Leon Blum and Chautemps—the governments of the victorious Popular Front. Maria Ossowska spoke about human characters. However, cooperativism was the most frequently discussed issue.[11] Residents were also interested in socialism and modernist avant-garde. When in 1934, Tołwiński returned from the Soviet Union, he gave a speech titled 'The Crimean Socialist Republic' (Tołwiński 1934) which enjoyed a lot of interest (it was attended by a record number of 317 people), similarly during the speech given by Stefania Sempołowska who spoke about 'Impressions from Moscow' (293 people attended the lecture). These data testify to the Żoliborz

[11] Ż. Kormanowa mentions a whole list of issues raised during the readings (cf. Kormanowa 1970, pp. 85–89).

Fig. 7.5 Reading room of the "Glass Houses" Association, Report on the WHC activities for 1931, p. 157

housing estate residents' significant interest in the topics and the overall intellectual climate. Szymański noted, however, that 'in 1939, the intensity of the reading campaign decreased. The reason for that was the listeners' waning interest caused by the excess and monotony of the lectures' (Szymański 1989, pp. 106–107).

7.4 To Participate or not to Participate: That is the Question

The association also organised meetings with proletarian literature authors and English avant-garde film screenings. At the 'Czapka Frygijska' [*The Phrygian Cap*] Visual Artists' Club, Stanisław Ossowski, among others, gave a speech titled 'Artistic creation and capitalism'. The Warsaw Group of Visual Artists, operating from 1934 to 1937 and founded by KPP supporters, also presented its exhibition there. Mieczysław Berman, one of the artists from this group reminisces about Żoliborz being the only place in Warsaw that took up this exhibition risk. 'The exhibition of left-wing visual artists could have been held at that time only after gaining the support of a legal and progressive cultural institution. 'Czapka Frygijska' [*The Phrygian Cap*] Visual Artists' Club was such an institution at the 'Szklane Domy' Tenants Association which shared their company's name and provided material help. Hence, the attractive name of 'The Phrygian Cap' was firmly attached to the humble colleagues from the Warsaw Group of Visual Artists, and although more than twenty years have passed since the creation of our group, no one has called us otherwise than by the name of 'The Phrygian Cap' ever since. The exhibition opened on 4 May 1936 and lasted until the first days of June. It was located in the premises of the WHC in Żoliborz, at Krasińskiego 10, in two rooms. (…) The exhibition attracted numerous visitors, probably around two thousand. On the inn's tables were catalogues and questionnaires for the audience to enter comments about the exhibited works. The audience filled out about 300 questionnaires' (Berman 1960, p. 72). In this way, 'Szklane Domy' as an association of left-wing and progressive radicals took part in the city's life, while educating at the same time its residents. However, the 'Phrygian Cap' Club primarily meant the activity of an estate club whose 'form of action was to initiate meetings with authors, publish works unpublished so far with a special emphasis on proletarian literature and poetry, organise artistic recitations, concerts, art exhibitions, stage and dance events, experimental film screenings, arrange visits to art galleries, and organise club evenings, etc. This club, like the Aesthetics and Beauty Propaganda Club (*Klub Propagandy Estetyki i Piękna*), was meant to cover all cultural and social activities within the estate, but failed to do so. However, it organised a number of successful events in the years 1936–1938' (Szymański 1989, p. 116). No wonder it failed! It is was a real participatory nightmare! One could get the impression that the estate itself is inhabited exclusively by artists, actors, and writers who organise these projects mainly for themselves. Are the inhabitants of the housing estate, who learn how to lead an urban life, spend their free time, rest and organise their space, able to bear this responsibility of 'cultural participation'? And what about the other forms of activity: social life within the housing estate, gardening activities, active participation in democratic management structures, social and political activities? After all, the Żoliborz residents had to work and fight for the material bases of their existence. In addition, one could get the impression of some neophytic attitude.

Although the activity and development of the estate was dependent on loans granted by the city's municipal authorities of the ruling 'Sanacja' Party (which obviously did not want to engage in political activities or be associated with them), the cooperative, along with the accompanying institutions, strongly emphasised its working-class, socialist and cooperative character. After all, the 'Szklane Domy' Association was established primarily to raise the 'awareness of the WHC residents in the field of science and ideology, common to the entire working-class liberation movement' (Życie WSM 1934b). Thanks to its institutions, the Association constantly organised 'actions' of a political nature, and was keenly interested in workers' matters. The inhabitants were often involved in various types of extraordinary grass-roots events. On the one hand, these events were political, and on the other hand, they always emphasised their solidarity dimension and educational character. This does not mean, however, that all residents of the housing estate were willing to participate in this type of actions. One of the readers of the 'Życie WSM' [*Life of the WHC*] magazine spoke up for a 'greater socialisation', claiming that when shoemakers went on strike and people decided to help them, some unpleasant incidents took place. 'It wasn't that much about the fact that the collectors themselves [the people appointed by the 'Szklane Domy' Association for collecting donations for the strikers—author's note] were treated in a highly rude and even offensive manner, or that some residents, for various reasons, refused to help, but rather about their attitude towards the manifestations of social struggle, struggle for existence, struggle for the conditions which can merely provide the most modest family existence. The Warsaw Housing Cooperative was made up of people who were convinced that only people, whether working physically or mentally, would live here. It seems to me that everyone who works should understand the struggle for existence, to meet the most vital family needs. The cases of total socialisation occurring from time to time in our area which should not happen at least here, within the area of social cooperative inhabited by the working class, seem at least understandable' (Wawrzyńska 1935, p. 17).

There were calls for not displaying state flags in cooperative houses as a sign of protest against the signing by President Mościcki of the constitutional law on 23 April 1935.

Emotions, views, separate opinions and collective actions were manifested in the housing estate in an extremely strong manner and were often performative in nature. In every issue of the 'Życie WSM' magazine, the so-called readers' voices appeared, and the 'Chronicle' section was replete with descriptions of various social actions and events. Apart from everyday ordinary behaviours in which the city dweller were already involved on an almost routine basis, one could see a lot of spontaneous, unusual, conscious and creative activities. A functioning entity dominates the action. It is feverish and anxious. Readers express their doubts, negative moods, dissatisfaction with the actions of the estate's authorities. The latter, in turn, constantly call for joint work, active involvement, voting, attending delegates' meetings, readings and discussions. 'Everyone to the meeting! Let there be no cooperative member who will not fulfil their electoral duty, who will not have their trusted representative at the General Assembly of the WHC Delegates'. In the same article, we can also read that 'WHC is not only a pioneer of mass housing construction for the most disadvantaged, but also a pioneer in the transformation of the conditions of neighbourly co-existence, so that this co-existence, drawing its strength from mutual help and a growing sense of social awareness, leads to full satisfaction of all economic and cultural needs of the inhabitants of the new housing estates, to educating new people who are struggling to transform today's social system. The gauge of our progress in this area is the increase in residents' social activity and the Tenant Council's development. The inhabitants' active involvement in the estates' self-governing life was presented as one of the duties of colony delegations in the Tenant Council's regulations. The newly elected delegations have nowadays a pretty rewarding task to fulfil: to move and engage all those members who have not yet taken part in the elections to fulfil their civic duty. The competition between the inhabitants of different colonies can be complemented by social competition between delegations—entities of new self-government' (Życie WSM 1935b, p. 1). The victorious colony would usually be given priority over equipping and decorating its internal court. The post-election celebrations were held regularly (usually in June, the day before the cooperative day), and people celebrated their first prize award when it came to tenants' participation in election meetings. The neighbourhood reality was associated with actions, events, pioneer activities and the transformation of living conditions. In the pioneer housing estate which did not have any specific tradition nor common past, much was done to make it emerge. The visions of the future, i.e. cooperative and socialist were equally important.

Is there a place in Żoliborz's democracy for 'counter-attacking participation', for a distance which allows you to think critically about it (Mendel and Szkudlarek 2012) or even distance itself from it? Could such reserved citizens be perceived as a threat to Żoliborz's democracy? Or on the contrary: distanced and counter-attacking activities were perceived as an expression of concern for its condition. The condition for the existence of democracy is, after all, to build places for aloofness, non-involvement and non-participation in collective life.

7.5 The Neighbourhood Rhythm of Celebration Practices—Preservation of Rituals

Annual celebration of the Labour Day and the Cooperative Day, laying of cornerstones for the subsequent new houses and putting new flats into service was often accompanied by a proper setting and ritual involvement, typical for the WHC (Figs. 7.6 and 7.7).

As I have mentioned before, the residents were involved and encouraged to take part in these celebrations. 'During the meeting of the colony guardians, it was unanimously adopted to call all the housing estate residents to manifest their participation in the Labour Day celebrations by decorating the windows and balconies with flowers, greenery, and the portraits of Socialist Heroes' (Życie WSM 1935a, p. 6). Thus, the housing estate residents were involved in the Labour Day celebrations in 1935, but the Labour Day celebrations had been held since the very beginning of the WHC's existence and lasted continuously until 1939. They took place in accordance with the already determined rhythm and specific ritual.

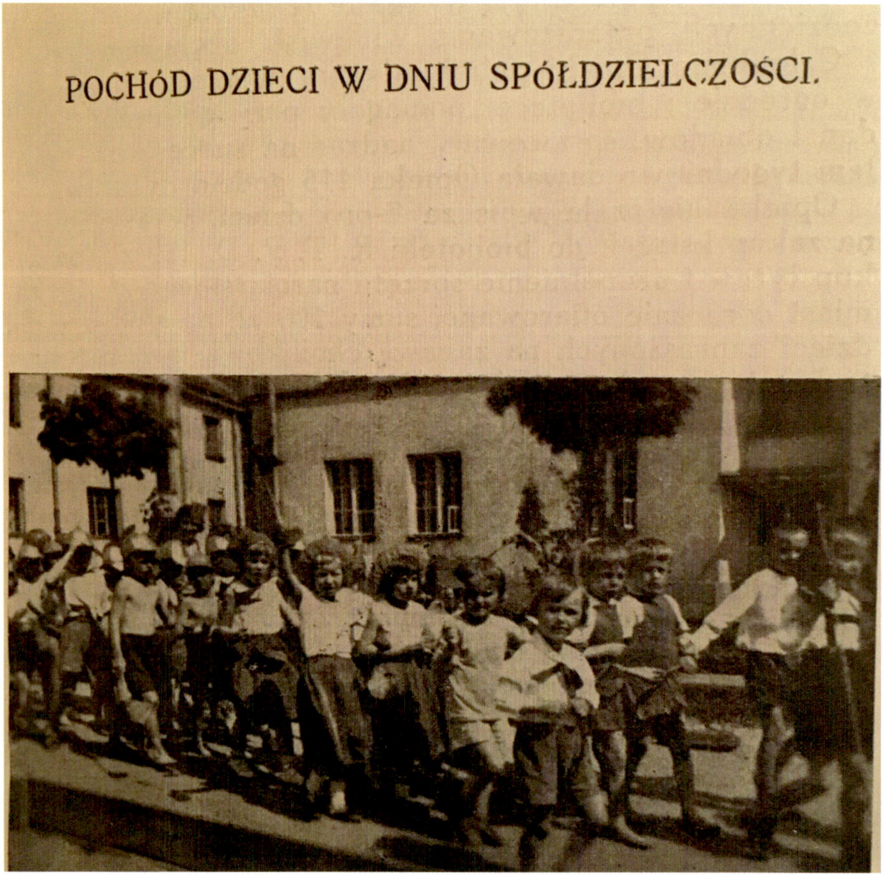

Fig. 7.6 Children's march on the Day of Cooperatives, Report on the WHC activities for 1933. 3771, p. 161

The scenario for such an event was described in the 'Życie WSM' magazine from May 1939:

> The evening bugle call on the eve of the Labour Day has become not only a tradition but also a citizens' right. It has also gained organisational practice which makes the course of the ceremony compact, deep and incremental. The orchestra sets off from the courtyard of Colony II. Trumpeters perform a bugle call on the stairs leading to the chimney. A. S.,[12] aligned in front of the courtyard, salutes the flag while it is being hoisted. Banners appear almost simultaneously in all colonies. The estate which—a minute ago—looked mundane takes on

[12] AS—Akcja Socjalistyczna, Socialist Action established in 1934 as a paramilitary formation modelled on the Austrian Republikanischer Schutzbund. Its task was to protect the structures and members of the PPS party from the radical right-wing militants, especially of the National Radical Camp, with whom violent clashes took place in Warsaw. Cf. Józef Żarnowski (1965, p. 371) and memories of Stanisław Sankowski (1981).

Fig. 7.7 Children's Academy, Report on the WHC activities for 1932. 3735, p. 161

a festive character in a blink of an eye. The orchestra marching in front, the A.S. marching lined up in rows, the colourful line of red scouts, and the Young People's Choir carrying their own banner are joined by the increasing number of people in each courtyard. The crowd is so large that it fills the courtyard of one colony and its end has not yet left the previous one. The last stage—the courtyard of Colony VIII is as crowded as a beehive. A concert is taking place in the courtyard. In the one-storey gallery, someone is accompanying a solo singer and the Young People's Choir on piano. The vigorous 'Szklane Domy' Orchestra is playing until late in the evening' (Życie WSM 1939, p. 129). During this ceremony, Stanisław Ossowski delivered a speech for young people, persuading them 'not to be obedient in their thinking'.[13] 'Do not let your youthfulness leave you when you grow up, this true juvenility which is sensitive to harm and oppression; do not lose the ability to rebel against what is evil

[13]Obviously, Ossowski did not use such a term. I am using this term in reference to the book by Jakub Karpinski bearing this very title, devoted to Ossowski and his political and social views (Karpiński 1989).

> and vile, a victorious glimpse into the future. Do not let petty personal matters obscure the brightness of your vision with dust; do not believe adults, if they tell you that an adult must come to terms with the world as it is, that universal brotherhood is a pipe dream; do not let the experience, which you shall acquire, become—as the poet says—the cotton which clogs your ears, so as not to hear human moaning (Ossowski 1939, p. 130).

Such a cyclical ritual constituted a kind of social framework in which the estate identity is enclosed, in which language is created, allowing participants to express themselves on what they usually tend to think and do. These are the liminal states in which the created community of experiences suspends everyday reality, occupying such space in which one is allowed to think and reflect on how one tends to think and act on a daily basis. This is the space and time in which the estate residents can feel what they experience in their daily lives. As if the rules became the voice of thought, and rituals—the model of experience.

The celebration liberates the desired state of reciprocity, aid and fraternity so often demanded insistently by the estate activists. One could say that such a celebration resembles ritual dramas that 'reach rather for spectacular than discursive means—so as to stimulate our senses and provide them with a whole mass of phenomenological evidence for the existence of the symbolic reality presented by the ritual. By imparting a dramatic structure to abstract, invisible concepts, the ritual makes ideas and dreams become vivid and tangible, and the order we experience blends with the order we dream of' (Myerhoff 1984).

Stanisław Ossowski, who saw in the development of mutual aid and the sense of brotherhood a chance to create an atmosphere of solidarity and kindness within the estate, was perfectly aware of the power of ritual. And 'the atmosphere of kindness facilitates the socialisation processes; socialisation develops mutual aid. The point is to break into this circle of causal interdependencies. (…) The methods that the Jesuits were able to exploit so skilfully in their socialist-theocratic Republic of Paraguay in the 17th century should not be underestimated in this respect, i.e. methods whose role in the socialisation of the community is completely underestimated, although primitive peoples had already known it perfectly well. I mean the collective singing' (Ossowski 1967c, p. 368)—a collective ritual engaging and teaching reciprocity, cooperation and common experience of social reality. These are important, performative, community experiences for Ossowski. That is why, when he mentions Stefan Czarnowski, he writes about an important event: 'During the great gala event of the 'Szklane Domy' [*Glass Houses*] Association taking place in Żoliborz and dedicated to the memory of Stefan Czarnowski, a month after his death, at the gala which gathered more than four hundred people in the packed meeting room of the Warsaw Housing Cooperative, the printers' choir sang both 'The Red Banner' song [*Czerwony Sztandar*] and 'The Internationale' [*Międzynarodówka*], i.e. the anthem of both workers' parties in pre-war Poland (Ossowski 1970a, p. 180). A song which reconciled the feuded parties, during the celebration, often carried their banners separately.

Fig. 7.8 Common room in Żoliborz—Workers' Society of Children's Friends (RTPD) eurhythmics classes, Report on the WHC activities for 1931. 3560, p. 163

When we start reading these estate celebration scenarios and public persuasions to manifest involvement in the ritual, they seem to us exaggerated, intrusive, overly socio-volunteering and 'terribly participatory'. But when we look at the memories of the estate residents, we can notice their performative power, tension and thoughtfulness which specifies each time the residents' identity. One cannot forget that the stylistics and poetics at that time were different (Fig. 7.8).

Maria Nowicka recalls: 'when Labour Day was approaching, we were taught commemorative songs, we made flags made of red tissue paper, we helped to sew costumes for the performance. (…) On Labour Day (1st of May), we marched with banners around the courtyards of our WHC colonies and sang; and later that day, there were great performances in 1st Colony or in the 'Tęcza' cinema which was packed to the last seat. It was a great day for all artistic groups from our settlement house: the rhythmic group, the choir and the percussive ensemble. There were also various role-plays, dialogues, short stage tricks (…) written especially for us by various people from among the WHC residents (…). The melodies were composed by Jan Wesołowski, Anna Osserówna and probably others, and the lyrics usually carried a powerful ideological charge. They talked about the fight for social justice, and about friendship between nations. I was really worried by the noble undertones of those performances. All of the above-mentioned activities were enthusiastically attended by many adult WHC residents and the RTPD school and preschool teachers. Probably there were many people willing to take part in numerous and various social initiatives. Some people devoted dozens of hours of their free time, finding enjoyment and satisfaction in such activities. And all of that was for free' (Nowicka 2009, p. 14). In these memories, the attention is focused on 'full socialisation', and devotion to the public cause. 'As if the reality reverted for a certain period of time to a 'conditional-complying' mode—just like during the *mardi gras* or Slavic carnival [*ostatki*]. The public liminality is governed by public power' (Turner 1984). The 'conditional-complying' mode connects 'what ifs', dreams, that which is desirable with a performative and materialising strength of the spectacle and preparations leading to it. It is all about connecting dreams with reality.

It seems that it is the lack of a common past, the unsatisfied desires of a life lived among others, of full socialisation, the swelling dreams of some socialist and social order that incite the cooperative members to the festiveness. It is completely different when 'socialised individualism' is a natural, obvious, 'gentle' attitude and not insistently desired. This is perfectly evident in Aleksander Ziemny's memoirs about Maria Ossowska, who was an extraordinary inhabitant of the Żoliborz estate. She was one of the authors of the 'programme of the people from Żoliborz' which 'had a dual dimension: a general social one and also a very personal one. They leaned toward the left, and organisationally were associated with the Polish Socialist Party (PPS) (its radical faction within the trade unions and cooperatives). (…) Workers' festivals were celebrated in a natural, unconstrained and—I'd say—homelike manner. (…) A Żoliborz resident is still an eager citizen; he/she does not shy away from projects considered by him/her as worthy and meaningful. (…) He/she wants to settle in, yes, but within the community, not at its expense. And without undue hassle' (Ziemny 1970, pp. 2–3). Simplicity, freedom from constraint and informalisation of everyday life, living in accordance with local rules in an unnoticeable, discreet and natural manner make some residents of this estate celebrate in a 'home-like' way.

I think that we can repeat after Sennett, that Maria Ossowska reached artistry in her everyday life on the Żoliborz estate in a natural manner; she was free to make various social gestures. It is hard to imagine, however, this 'home-like' nature of celebrations, when we look at these estate rituals. Perhaps it was all about a kind of familiarity which often accompanies community and its festive emotions. Such a 'family', 'homely' mood, however, was created by the Żoliborz residents on a daily basis. Despite the unusually formalised structure of the institutions belonging to the

'Szklane Domy' Association, their hierarchical and legal order, the extraordinary transparency of activities and the care for informing the public about their management methods, the characteristic features of structural *communitas* are really striking. Even the conspicuous ideological and political tensions do not disturb the atmosphere of homeliness, present even in certain language behaviours. For example, Kazimierz Tołwiński, the organiser of the estate library, was called 'Grandfather' (he was the father of Stanisław Tołwiński who could be considered 'the founding father'). Also, Zofia Lubodziecka, who offered the library many of its first books and was the inspiration for setting up a puppet theatre for the Żoliborz children, was called 'Auntie Zosia'. This extraordinary familiarity manifested in the ways of calling persons who performed essentially institutional functions testifies to genuine attempts to build a community in which 'institutional officials' were treated as important close ones, those who shaped their identity and were instrumental in the process of socialisation. A preschool operating initially in a private flat and organised from the bottom-up, and the theatre mentioned earlier testify to the spirit of commonality and devotion, and at the same time to the pioneering and spontaneous nature of the housing estate institutions. 'The 1st Colony was also inhabited by Szczepan Baczyński who, after the war, served as rector of the State Higher School of Theatre (Państwowa Wyższa Szkoła Teatralna). At the time, he was an official by profession, but an actor by predilection. He was also interested in educating children. One day he rallied a few people—including my mother who was very musical and sang beautifully—and organised a puppet theatre. (…) It was the beginning of the 'Baj' theatre. The adult members of the theatre team were back then amateurs—either neighbours or colleagues' (Nowicka 2010, p. 25). Therefore community-related and familiar, bottom-up activities aimed at organising everyday reality could be easily seen. In no way did they interfere with the demonstration of power which seemed to be visible everywhere, although it was eagerly delegated to the cooperative members. Periodic 'carnivals', on the other hand, revealed the one who dominated others. Exposing the paradoxes of the housing estate' institutions and the political enthusiasm of the important members of the cooperative was possible, for example, thanks to the 'WHC Satirical Performances'. The first satirical performance titled *W krzywym zwierciadle* [In The Distorting Mirror] was staged in April 1937, and the second two years later. The satirical criticism of the cooperative management members and of the 'Szklane Domy' Association and some of the estate activists and local government officials resonated with the exposing of the residents' passivity, and their apathy and aloofness towards social life.

Great Wanda	Chair
Our 'flame' is burning in the cloud, wanting to breathe miasmas, while I deprave children, as the IKC[a] claims…	I'm not old, although I'm already bald, many storms have passed over my head – my bald head is shining in sparkles, that is why my thoughts are enlightened
Fascism will be swept away like fluff proletariat etc. and 'Homeland' lacks the idea, as the KPP claims…	Chair of the Supervisory Board of the Warsaw Housing Cooperative – and construction director, who doesn't know his job
And now I know myself, whether I am or not the eastern miasma, or a bourgeois lady whether I'm going backwards or forwards	When I am at SPB (i.e. Social Construction Firm) then WHC is my client, and when I am at WHC, then it's SPB in turn
And I keep on writing and thinking, until the daybreak, how looks the close-up of my 'Face of the Day'	When SPB builds a colony and bungles something, I pity myself, for I am the president of both Cooperatives
[a]IKC, Ilustrowany Kurier Codzienny [*Illustrated Daily Courier*]	I am both a debtor and a creditor, a supervisor and a host, I am the president of all presidents and boss of all bosses!
	I run the cooperative movement with an iron fist – the project will pass through if I give my consent everyone is looking, everyone is talking The president, the president is walking the WHC's *beau*! – whispers and muffled buzz can be heard this word 'President' casts a spell on everyone
	They keep on whispering 'President' here, and they keep on whispering 'President' there, and the President is none other than me' (Goetling and Szymański 1937)

When reading the materials about the Warsaw Housing Cooperative one feels ambivalent all the time. Members of the Cooperative management are at the same time members of the authorities of almost all institutions and enterprises operating within the estate (mainly as part of the 'Szklane Domy' Association, but also as part of the Social Construction Firm, as the song from the satirical performance shows). One could say that power is scattered, capricious, but it does not mean, like in Foucault's works, that it is invisible or seductive. It manifests itself very clearly, we know exactly who initiates collective actions, improves and directs them. On the other hand, however, the management constantly demands from the estate residents their active participation, to take part in the elections and to stand as candidates, delegating various people to various positions.

At times, the power structure is strictly formalised, with a precise legal framework, official language, and structured and hierarchised estate reality. At other times, it is homely and familiar. At times, one has the impression that the estate residents fall into the 'participation hell', because they are constantly mobilised, encouraged to be involved in social actions, political events, into the socialised reality of everyday life.

On the other hand, however, Maria Nowicka's enthusiasm quoted in her memoirs was sincere and, after years of reflection, did not weaken and was not treated critically. Similarly, another new resident of the Żoliborz colony, after she moved in 1938, regretted the lack of interest in the residents' fate on the Cooperative's part. 'I take the liberty to suggest: Should not a permanent body be brought into existence, or should a continuously functioning institution be introduced to familiarise new members with the totality of WHC matters? Newly arrived people would thus undergo a kind of field training, thus having the opportunity to contact local activists and being immediately affiliated among 'active' tenants. If, after completing ordinary formalities, instead of stereotypically sticking a few pamphlets into people's hands, the colony guardian or a Tenant Council delegate devoted half an hour to 'socialise' each new cooperative member, I am sure that each of them would become an aware and active cooperativist' (Życie WSM 1939c, p. 122).

This ambivalence can be explained by an egalitarian and mixed (because of its working class and intelligentsia) character of the estate. The Cooperative and its accompanying institutions instil the educational and bell-ringing model in the workers who have just started learning how to lead an urban life, learn the principles of 'socialised individualism' and are incited to emancipation. The authoritative speech acts are directed to them, while the principles of 'cultural coexistence' are repeated and explained, likewise appeals, calls and encouragements formulated in the 'conditional-complying' mode'.[14] Festive spectacles are apparently organised for them; they are conducive to building an identity, not only the housing estate's identity, but also a civic one and one of the working class. For those who only search for affordable housing and decent living conditions, strict rules and carefully performed rituals are indispensable, because, according to Sennett, physical gestures invigorate social relations.

References

Berman M (1960) 'Czapka Frygijska' [The phrygian cap]. In: Barcikowski W et al, Księga wspomnień 1919–1939 [The book of recollections 1919–1939]. Czytelnik, Warszawa, pp 55–90

Czubaj M (2007) W stronę miejskiej utopii. Szkice o wyobraźni społecznej [Towards the Urban Utopia. Essays on social imagination]. Wydawnictwo SWPS, Academica, Warszawa

Filipczak S (1963) Moje wspomnienia pracy w ruchu spółdzielczym [My memories of working in the cooperativist movement]. In: Kossowski K (ed) Wspomnienia działaczy spółdzielczych [Accounts of cooperativist activists], vol 1. Zakład Wydawnictw CRS, Warszawa, pp 151–156

[14]Zofia Topińska wrote that the founders of the Żoliborz community wanted to serve the workers in accordance with the socialist ideals. 'Socialism was a dream come true for these people. According to its principles, they accepted or criticised the activities of the board, administration, formative institutions, laundries, boiler houses and everything that existed there. Socialism was a longed-for future in which everyone could find their field of action, an outlet for their energy, initiative and will to do something useful in life that would be useful for others (...)' (Topińska 1984, p. 10).

Goetling JR, Szymański E (1937) Szopka WSM. Fragmenty piosenek z Szopki WSM 'W krzywym zwierciadle' [WHC satirical performances. Fragments of songs from the satirical performance titled 'In the distorting mirror']. Życie WSM, May:98
Harvey D (2012) Rebel cities: from the right to the city to the urban revolution. Verso, London, New York. http://abahlali.org/files/Harvey_Rebel_cities.pdf. Accessed 9 Sept 2017
Karpiński J (1989) Nie być w myśleniu posłusznym: Ossowscy, socjologia, filozofia [Not being obedient in thinking: the Ossowskis, sociology, philosophy]. Polonia, London
Kormanowa Ż (1970) Postępowe tradycje inteligencji żoliborskiej 1918–1939 [Progressive traditions of Polish intelligentsia 1918–1939]. In: Anonymous (ed) Żoliborz wczoraj, dziś, jutro [The past, present and future of Żoliborz]. Książka i Wiedza, Warszawa, pp 67–101
Mendel M, Szkudlarek T (2012) Demokratyczna izolacja. Krytyczna partycypacja obywatelska [A democratic isolation. Critical citizen participation]. Zoon Politicon: Rocznik Naukowy 3:207–222
Myerhoff BG (1984) A death in due time: construction of self and culture in ritual drama. In: MacAloon JJ (ed) Rite, drama, festival and spectacle. ISHI, Philadelphia
Nawratek K (2012) Holes in the whole: introduction to the urban revolutions. Zero Books, Winchester, Washington
Nowicka M (2009) WSM – RTPD – Szklane Domy [WHC (Warsaw Housing Cooperative) – RTPD, Robotnicze Towarzystwo Przyjaciół Dzieci (Workers' Friends of Children Association) – 'Szklane Domy' (Glass Houses) Association]. Życie WSM, September:14–15
Nowicka M (2010) Wspomnienia Marii Nowickiej. In: Pałka J, Szamruchiewicz A, Szymańska-Szwąder M (eds) Warszawa międzywojenna w pamięci jej mieszkańców. Drugi zeszyt serii Z archiwum historii mówionej [Interwar Warsaw in memories of its residents. The Second notebook of the oral history archive], Dom Spotkań z Historią [History meeting house]: Ośrodek Karta, Warszawa, pp 22–31
Nowicki M (1937) Czytelnia czy śmietnisko? [A reading room or a scrap heap?]. Życie WSM, April:76
Orsetti M (1926) Robert Owen, wielki przyjaciel ludzkości (1771–1858) [Robert Owen, A great friend of humanity (1771–1858)]. Wydawnictwo Związku Spółdzielni Spożywców, Warszawa
Ossowski S (1939) Fragment z przemówienia ob. Prof. St. Ossowskiego [An excerpt from the speech given by professor Stanisław Ossowski]. Życie WSM, May:130
Ossowski S (1967) Organizacja przestrzeni i życie społeczne w przyszłych osiedlach [Spatial organisation and social life in future housing estates]. In: Ossowski S (ed), Dzieła [Collected works], vol 3 [Issues of social psychology]. Państwowe Wydawnictwo Naukowe, Warszawa, pp 351–370
Ossowski S (1970) Stefan Czarnowski. In: Ossowski S (ed), Dzieła [Collected works], vol 6. Państwowe Wydawnictwo Naukowe, Warszawa, pp 127–131
Próchnik A (1933) Współżycie ideowe czy walka polityczna [Ideological coexistence or political struggle]. Życie WSM [Life of the WHC], October:1–2
Purman S (1937) Lektura do poduszki, czy oręż w walce? [Bedtime reading or a weapon in a fight?]. Życie WSM, May:96
Rewers E (ed) (2014b) Kulturowe studia miejskie. Wprowadzenie [Cultural urban studies. An introduction]. Narodowe Centrum Kultury, Warszawa
Sankowski S (1981) Wspomnienie o Akcji Socjalistycznej [Reminiscences about the socialist action]. http://lewicowo.pl/wspomnienie-o-akcji-socjalistycznej. Accessed 21 June 2018
Szymański JA (1989) Warszawska Spółdzielnia Mieszkaniowa 1921–1970. Zarys dziejów [Warsaw Housing Cooperative 1921–1970. History outline]. Wydawnictwo Spółdzielcze, Warszawa
Tołwiński S (1934) Osiedla społeczne w ZSRR [Social housing estates in the USSR]. Życie WSM, December:1–3
Topińska Z (1984) Żoliborskie obietnice: o wychowawczym środowisku WSM [The promises of Żoliborz: the formative environment of the WHC]. Wydawnictwo Spółdzielcze, Warszawa
Turner V (1984) Liminality and the performative genres. In: MacAloon JJ (ed) Rite, Drama, Festival and Spectacle. ISHI, Philadelphia
Wawrzyńska W (1935) O większe uspołecznienie [For a greater socialisation]. Życie WSM, April:17

Żarnowski J (1965) Polska Partia Socjalistyczna 1935–1939 [Polish socialist party 1935–1939]. Książka i Wiedza, Kielce

Ziemny A (1970) Żyć aby świadczyć [To live in order to witness]. Ty i Ja [You and Me] 125 September:2–3

Życie WSM (1932) Życie WSM, June–July, Aug:1–8

Życie WSM (1934a) Uchwała Zarządu. Życie WSM Jan:8

Życie WSM (1934b) Walne Zebranie Stowarzyszenia 'Szklane Domy' [The general assembly of the 'Szklane Domy' association]. Życie WSM, July:3–4

Życie WSM (1935a) 1-szy maja w Osiedlu [The Labour Day in the housing estate]. Życie WSM, April:6

Życie WSM (1935b) Godnie zakończymy kampanię wyborczą [We will properly finish the election campaign]. Życie WSM, April:1

Życie WSM (1939c) Papierki nie wystarczą, ale nie wolno się konspirować [Tiny papers are not enough, but it is not allowed to conspire]. Życie WSM, April:122

Chapter 8
Models of Urban Cultures

Abstract The Żoliborz social experiment, however, took place with the active participation of 'organic intellectuals' (Antonio Gramsci). Hence, I describe the Żoliborz reformers with the help of modern categories, calling them activists or amateurs. It is important to show that urban lifestyles can only be created from the inside. This allowed me to exclude the category of external experts. In this chapter, I also present the knowledge transfer between society and academia on the example of sociologists such as Stanisław Ossowski and Maria Ossowska, cooperative activists who not only shared their knowledge on the estate, but also produced and tested it there as if in a social laboratory. It cannot be said, however, that they were external experts—they lived on the estate and were members of the cooperative, subordinate to its management. Knowledge was transferred in two directions. Researchers and social reformers inspired the estate's residents but also learned from them, verified the proposed solutions, constantly corrected the neighbourhood *praxis* and erred systematically just as they would in a real laboratory. The conditions, however, were not sterile: the point of the exercise was rather peripatetic experimentation and the sharing of knowledge.

Keywords Organic intellectuals · Experts · Transfer of academic knowledge · *praxis* · Laboratory · Stanisław ossowski

Let us return to the question posed at the beginning of the previous chapter. Which of Sennett's models, i.e. workshop or community centre, is more fitting for the Żoliborz estate? Is it possible to combine the two models? Or maybe an entirely different model could be applied? How should one go about analysing the estate's institutionalised, often oppressive order while not disregarding its care for fostering solidarity and reciprocity among all members of the cooperative?

Sennett goes back to the 1900 Paris Exhibition. Apart from presenting the newest industrial, technological and imperial developments, the exhibition showcased the outcomes of those achievements. All exhibitors prepared *musée social* (social museums), which they felt touched upon the 'social issue'. The question was: how can society be changed? Solidarity, understood as the interrelations between social ties and political organisation, was the answer. 'Cooperation made sense of this connection: the German's united labour union, the French Catholic voluntary organisation

M. Matysek-Imielińska, *Warsaw Housing Cooperative*, The Urban Book Series,
https://doi.org/10.1007/978-3-030-23077-7_8

and the American workshop exhibited three ways to practise face- to-face cooperation in order to bring about solidarity' (Sennett 2012, pp. 36–37). Leftist visitors of the exhibition split into two groups: those who wanted to impose solidarity from the top-down (as had been the case with centralised German labour unions) and those who saw it as a consequence of grass-root movements (American workshops). The division reached deeper and is still evident today. Supporters of the top-down approach see cooperation as a tool, as a means of achieving political goals. Therefore, interactions and exchanges have to be ordered, disciplined and formalised. On the other hand, supporters of the grass-root approach want 'as much free participation as possible within the parish hall or on the street, even if this meant sacrificing a certain amount of discipline' (Sennett 2012, p. 39).

Top-down organisations wish to project an image of a united ideological front, meaning that conflicts cannot be public and, instead, have to be resolved behind the scenes. Once agreement is reached through a dialectical process, it is presented to the public. Grass root strategies, on the other hand, involve a dialogic process in which conflicting ideological arguments are publicly confronted. Despite often coming off as incoherent quarrelling, dialogic exchange of opinion is considered to be creative and reminiscent of agonistic democracy, although it does not always result in concrete solutions.

Labour union representatives (e.g. Ferdinand Lassalle and Samuel Gompers), firmly grounded political realists that they were, did not believe that community centres could improve the life of the poor. As regards Owen's workshops, they considered them just another social and political dead end.

Sennett argues that social reformers and local activists of the early twentieth century were convinced that once social issues related to planned parenthood and the isolation that migrant workers experience in cities are resolved, grass-root initiatives will take off and take care of the rest. Providing access to education as well as affordable and comfortable housing will be enough to set the process in motion. The activists often drew ideas from cooperative movements, which did not view cooperation as a political tool but rather as a goal in itself (Sennett brings up Piotr Kropotkin's projects and Robert Owen's cooperatives to prove his point).

'The star exhibit in Paris for solidarity built from the ground up was the settlement house. In form the settlement house was a voluntary association, located in a poor urban community, where poorly skilled workers could receive education, get advice on everyday problems or simply find a warm, clean place to hang out' (Sennett 2012, p. 43). Although the settlement houses were run by middle-class women, who received no compensation for their work, and funded by private philanthropists, much of the work and simple repairs were carried out by the poor who frequented them. There is, however, a notable difference between such settlement houses and similar institutions in Żoliborz. Unlike the latter, the former were not charities meant to 'ease the bourgeoisie's conscience' which, in essence, further humiliated the poor. Żoliborz's institutions were in fact founded by the well-educated and often rich members of the elite (e.g. Teodor Toeplitz). Yet, they formed the backbone of the Polish cooperative movement and often made their living by the 'sweat of their necks and minds' (Adam Próchnik).

Both types of institutions, however, based their strategies on fostering cooperation among diverse individuals, thus trying to determine the best conditions for establishing cooperation and participation. According to Sennett, settlement houses did not struggle with this. The goal was to fully engage the poor in daily activities: raising children, education, shopping. Informal agreements, loose exchanges and free interactions are crucial for this to succeed. Solidarity may transform into sociability. 'For all this—Sennett concludes—informality always risks disorganisation. And even if it does rouse people inside its hallways and rooms, the settlement house risks becoming just a good experience they have occasionally, rather than a guide to life outside. That may be true more largely of communal cooperation: it offers good experiences but is not a way of life. (…) The results of bonding in the community have to lead somewhere; action needs a structure, it has to become sustainable' (Sennett 2012, p. 55).

Was Żoliborz's institutional life similar to such settlement houses?

There is no doubt that many social relationships were informal and many activities, even those that had been institutionalised, were based on cooperation, such as the estate preschool that initially was based in a private flat, or the library, which was a 'pleasant place' that brought people together and formed a hub of social life. As did the estate's clubs, hobby groups, countless public lectures and courses. Participants could practice cooperation and share even the most trivial of daily experiences. That process was made easier by the estate's layout, its 'life axis' and the boiler house which doubled as the 'heart of the estate'. The tenants' daily activities involved helping one another and taking part in preparing the skating rink in the winter and playgrounds in the summer, gardening and playing group mind games.

Workshops, on the other hand, were different. Their goal was to teach skills that could be used in the outside world. 'Each workshop was in part self-governing, involving special meetings where the student–labourers discussed their work without the presence of a teacher. The Rochdale Principles thus appeared in these ground rules: work open to all, active participation, the work in which people cooperated rethought But the Institutes were not free-form processes; each workshop had fixed productive targets, and the overall design of the Institutes was set by Booker T. Washington alone' (Sennett 2012, p. 56).

Complex social rituals had developed in the workshops and were performed publicly (codes of honour, pledges of allegiance) and not covertly. They were expressions of the mutual relationships and natural inequality between: masters, apprentices, students. The workshop institution has to combine long-term mutual benefits and loyalty with short-term flexibility and openness. Flexibility is essential for delegating various types of work and initiatives to labourers (Sennett calls it 'flexible networking').

We can be certain that Sennett's workshop involves collective work based on 'mobile solidarity' rather than a close-knit community. The skills learned may be used, tested, changed and modified outside of the workshop. The goal was to experiment with the newly gained abilities, be it physical labour skills, cultural or social competencies. Flexible cooperation is skill that is shaped in the workshop, but has to be practised and developed individually. It then serves the local community, the home and social groups, but most importantly the public sphere. Tenants are thus transformed into citizens who consciously shape their social relationships. Żoliborz's activists wanted to create a 'new man not only in the workshop, but in the flat, home and estate' (Szwalbe 1962, p. 17). Strict rules, formal relationships, hierarchical roles, norms and legal regulations—all of that teaches certain social life habits, which are then transformed, developed and used in diverse social contexts, leading to rituals being commonly and freely accepted.

Sennett provides us with several hints allowing us to answer the question posed earlier: workshop or settlement house? 'The settlement house took up the issue of sociality, (…) of living in a complex society full of difference; (…) [settlement houses] sought to convert inner and often passive awareness of others into active engagement. To make this happen, the strategy of the settlement house (…) emphasised informal contact, a principle organisers applied to themselves in 'Toll's Rule': advice rather than direct. But encounters on these terms could remain fleeting and shapeless long-term' (Sennett 2012, pp. 62–63). According to Sennett, workshops handled this issue by shaping cooperative activities. Washington's establishments, for example, focused on developing skills that would serve the community and could be applied in different contexts. They were based on Robert Owen's Rochdale Principles. 'But in practice these principles could produce a paradox: mutuality among members in a workshop, but still subservience to someone at the top about how they should live. Nonetheless, mutuality within the workshop was genuine in the Institutes: it made technical competence into sociable experience' (Sennett 2012, p. 63). How does that come to be? I will cover that later. Right now, it is worth adding that the Warsaw Housing Cooperative was a workshop insofar as it taught residents how to live, it instilled 'principles of cultured coexistence', leading to changing the identity of residents. It was a workshop that turned passive tenants into responsible citizens who were aware of their social and political rights, both inside and outside of the estate. The estate workshop was indeed overseen by formal and hierarchical authorities who imposed principles and rules of living (estate regulations, rules listed in the housing savings book, mandates and prohibitions), used authoritative speech, published articles in the 'Życie WSM' magazine encouraging active participation in the estate's cooperative life, including voting, acting as delegates, taking on the role of colony guardians. It was not a manifestation of a one-person entity, but rather the authority of the multiple-member Cooperative Board and the Tenant Council. Conflicts were resolved publicly, often using democratic rules. A similar institutional framework was adopted by the 'Szklane Domy' [*Glass Houses*] Tenants' Mutual Assistance Association. Its structure, consisting of offices, departments, committees and subcommittees, was stipulated by its constitution, but in practice it was a 'flexible network' allowing the delegation of various everyday tasks to tenant activists, regardless of their administrative title, without any issues. Relations with tenants were informal, free and familiar. 'The life of WHC—'Szklane Domy' [*Glass Houses*]—WFCA [Workers' Association of the Friends of Children] was rich and diverse. I cannot say with certainty which of the institutions were responsible for the various areas of estate activity (organisation, events, teams). What is more, many of the activists actually were « one in three people » , combining membership in and acting on behalf of various agendas. All of the institutions figured out ways to cooperate with each other, coordinate their activities and support each other, both in organisational and financial matters' (Nowicka 2009, p. 15).

The principles of cooperation and obeyance of the estate's formal rules and rituals of cooperation took on a form reminiscent of the workshop. The settlement house model, on the other hand, was applied to all activities associated with sociality and daily tasks. Yet, the matter was more complicated than that and we cannot rely solely on Sennett's models to do it justice. The two types of social behaviour and social ties were often interconnected. After a number of years had passed, Julian Hochfeld recalled: 'that self-governing network of institutions, which was able to function thanks to organisational efforts, the enthusiasm of its pioneering founders and activists and the direct material (and sometimes ideological) interests of its users-participants, was informally associated with the labour, political, educational, youth and cooperativist worker's movement; these informal ties were more important than some of the formal and organisational ones' (Hochfeld 1962, p. 9). And the Active Cooperatists' Club [*Koło Czynnych Kooperatystek*], despite being strongly formalised and institutionalised, often relied on sociality and proximity. Even in such trivial areas as grocery shopping, the Club's members, mostly housewives, established store commissions and committees, agitated and became involved in various aspects of public and economic life. It seems that the members of the Architecture and Urban Planning Studio were aware of the intertwining of the settlement house and workshop models. Also, it is very likely that they wanted to preserve the unequivocal nature of the two types of behaviour that those models shape. In January 1943, they wrote that 'the work of all clubs has to be based on voluntary participation. We must not forget that clubs are social gatherings, which strengthen neighbourly bonds, and that they should allow their members to relax, serving as an « extension » of their flats, allowing them to connect with others who share similar interests and, finally, serve their educational function. That is why club regulations should be flexible and not overly intrusive. (…) We cannot oppose clubs when they show initiative in organising trips for its members, provided that they are able to handle the organisation of such outings on their own and do not engage the trip section, tourist club, etc.' (Architecture and Urban Planning Studio 1976, p. 318).[1]

It was sometimes that case that certain political competencies and identities were not gained in the estate's public spaces, but in seclusion, for example in private flats (the 'Sanacja' regime did not take kindly to such activities). Yet, those were not intimate or personal encounters. Instead, we may refer to them as 'counter participation'. 'Somewhere around the early days of spring—Edward Kozikowski recollects—I got a call from Antonina Sokolicz to visit her for a social gathering on the next Sunday. (…) I went to Żoliborz at the appointed hour. Antonina lived in a flat in one of the buildings in the notorious Warsaw Housing Cooperative next to Krasińskiego Street [Antonia Sokolicz lived not too far from Ossowski—author's note]. (…) The small room that I was taken to was strikingly red. The walls and ceiling were covered with a crimson fabric and a memorial fire was on. Lenin's portrait hung on the wall

[1]The book covers the findings of the general studies team led by Stanisław Ossowski and the Rakowiec architecture and urban planning team led by Helena Syrkus.

across from the door, with two palm twigs adorned with black crepe beneath it. It was obvious that the room's solemn atmosphere was meant to honour the memory of the leader of the October Revolution whose life had drawn to a close a year earlier, in January of 1924. The attendees were numerous and quite diverse. (…) I was not familiar with anyone apart from the hostess and her husband, engineer Merkel. (…) I was not sure what impressed me the most: the courage of Antonina Sokolicz, who decided to hold Lenin's memorial meeting in her Warsaw flat, in the very heart of Poland, at a time when police agents were everywhere, or the perfect calm that permeated the room. The threat of being arrested was very real (…). I could feel that the participants of that secret gathering were aware just how serious the matter was; the act of paying respects to Lenin's memory, as well as the danger that everyone was in. The ceremony started shortly after my arrival. Each speech was carefully crafted. I could feel that the ceremony had been prepared by one person, that it had been thought through countless times and even the smallest details were taken care of. Each of the speakers introduced a new perspective on Lenin's life and achievements. Only later did I find out that it was Antonina Sokolicz who was behind it all and that she had planned almost the entire ceremony by herself' (Kozikowski 1960, pp. 185–186).[2]

For democracy to work, it is important for distanced attitudes, isolation or 'counterparticipation' to be more than just legal. Tomasz Szkudlarek claims that such forms of civic participation reflected by rituals and procedures, human behaviour and mentality, should be accepted (Mendel and Szkudlarek 2012, p. 211).

Antonia Sokolicz and other Żoliborz activists were perfectly aware of the fluid and permeable border separating private and public spaces. For private flats often were home to public events. And the social rituals that such events entailed allowed their participants to develop certain social patterns of secrecy, collaboration, trust and political emotions.

The workshop, therefore, teaches certain attitudes and imposes certain rituals. It is a public space in which etiquette, duty, trust and respect for 'distinguished leadership' are instilled, leading to a transformation of personalities. Yet, rituals cannot rely solely on restating rules if they are to remain lively and dynamic. Instead, they should be undergoing an incessant internal evolution. As should be all those who take part in the rituals, practising new social patterns and habits. To illustrate the above process, let me quote the account of female resident of Żoliborz who worked in a hosiery factory. 'In 1927, I had my first encounter with the collective. It took place in Żoliborz's WHC. I was 17 back then and looking for ways to improve my standing in life. Times were very hard. I desperately needed a place to live. My hopes rose when I heard that there was a collective that provides flats to its members. It really sparked my interest (…). We [together with a friend—author's note] decided to visit WHC's 1st Colony. We came into a big hall, which did not yet have a floor at the time (it later turned out that it was the settlement house), and saw two gentlemen inside. We introduced ourselves and explained our reasons for visiting the estate. They told me how to

[2]The information provided by Kozikowski is not precise. It could not have been 1925, because the first residents moved in only in 1926.

apply for membership and a flat at WHC. It later turned out that the person whom I had talked to was comrade Stanisław Tołwiński. He explained that only labour union members may join WHC, that members of cooperative should take part in the estate's social life, and he provided me with a brief description of the tasks and goals of cooperative members. Even today I can vividly recall how much of an impact that conversation has had on me and the new problems that it had introduced. Labour Unions? Social work? (…). At the time, my « social and cultural » life was limited to religious practice, and failing to attend service on Sunday or during holidays, even for the most serious of reasons, was considered a sin. Hence, I initially found all of the concepts that comrade Tołwiński talked about difficult to grasp. Yet, common sense prevailed. I was thrilled by the opportunity to improve my life (…). Initially, I was very reluctant to join a labour union and become involved in social work, which I had not really understood at the time, but after a period of internal struggle I let go of my misgivings. (…) I joined the Class Textile Workers Union. That is how the first major obstacle separating me from participating in political and social life had been torn down. (…) WHC's programme, requiring members to be actively engaged in the collective's social life, was what pushed me to do it. (…) In 1929, I finally moved into my long-awaited flat in Colony II of WHC in Żoliborz. (…) I was fired from work because I had been encouraging my colleagues to join the Class Textile Workers Union. That is how I lost my livelihood. (…) Having the freedom to use my time however I pleased, I joined the Żoliborz Cooperatists' Club in 1930, and had remained a member until 1939. Across the years, I held various positions in the Club: Audit Committee member, treasurer, vice-president. I also volunteered at store committees subordinate to the Cooperative Inn and at the Labour Office for Home Workers. (…) In 1932, I joined the Polish Communist Party and that really broadened my political horizons, giving them the right « hue » . From that point onwards, I was no longer clutching at straws, « using my intuition » as a guide, but instead had clear direction. (…) Between 1933 and 1936, I served as the colony guardian of WHC's Colony VIII. During that time, I took care of the various issues that WHC tenants were struggling with, relying on my position on the Board of WHC's « Glass Houses » Association' (Filipczak 1963, pp. 151–156).

Such engagement in the life of the estate and strengthening of social bonds was founded on, as we see above, political activity. Żoliborz was a very politically active district. Similar to Sennett's workshop, the WHC completely changed Stanisława Filipczak's life through practicing cooperation, engagement, trust and respect for 'distinguished leadership', which is evident in her memoirs, especially when she talks about Stanisław Tołwiński and Jan Hempel (the latter of whom explained the intricacies of religious and political matters to her). After a while, Stanisława Filipczak felt at home in the organisational structures of the Cooperative and in the political movement that she was part of, and the flat that had brought her to Żoliborz was no longer as important.

Maria Belsiger provides a different account stating that, despite serious efforts of numerous people, the social bonds between tenants were not growing. They had no motivation for common, personal or even economic engagement, usually because they associated their living standard with their work and not their place of residence.

Belsiger also makes clear that white-collar and blue-collar workers had no common interests. That prevented cooperation from developing and, thus, the estate's social and cultural life suffered. Therefore, claims that there was some kind of 'ideological common ground' among the residents may be considered unfounded. Most tenants wanted to make their daily life easier. According to Belsiger, any signs of closeness among the residents were only temporary and emerged during collective participation ceremonies, celebrations and meetings. Although the 'Glass Houses' clubs brought together people with fairly similar interests, membership was low, usually ranging from several to several dozen members. Their participation was often limited to passive usage of the library, reading room, lectures and courses. Despite the best efforts of estate activists, not all residents and members of the WHC took part in the election meetings organised by the delegates to the General Assembly. 'According to claims made by Belsiger, most residents were ideologically passive, and some were even openly opposed to the cooperative's ideological programme' (Szymański 1989, p. 89).[3] They most likely treated the WHC estate as a type of settlement house, a casual and informal sphere of social life with no major consequences. We may also assume that the passive attitudes and the refusal to commit to a non-individual, intersubjective sphere, which were mentioned by Belsiger, led to the estate's festivities becoming spectacles rather than rituals. A ritual morphs into a spectacle (of power, of the church, of a political institution) when people spectate rather than participate. And this gives rise to a hierarchy of passive spectators and active actors. The spectators do not participate, nor do they cooperate or create. They stand on the sidelines. They do not identify with the spectacle and do not cooperate with it.

Jan Hochfeld seems to have been aware of the various motives for joining organisations; therefore, he was able to provide an accurate description of the matter: '« Glass Houses » were joined by those who had money problems, those who needed 'tenant-insurance' which covered rent when tenants were unable to pay it, those who wanted to use the library and reading room, those who wanted to socialise with Warsaw's intellectual and working-class elite and those who were attracted by the activity of the Glass Houses clubs, especially « The Phrygian Cap » artistic club and « Vitraj Domoj » Esperanto club, both of which were immensely popular' (Hochfeld 1962, p. 9).

How did this relative mutuality take root in the estate when nearly all of its residents were subordinated to an authority that initiated and supervised their lives to a certain extent? Why did genuine cooperation, which morphed technical competencies into prosocial experience, come to be?

Rituals are external entities and their intensity is rooted in recurrence. As a result, certain patterns of experience are created. Although recurrence leads to stagnation, ingrained habits (or efficient methods of action) may be consciously analysed and transformed until they become effortless, natural and unintrusive. 'Rituals go stale if they remain stuck in the first stage of learning, that of a habit; if they go through the full rhythm of practice, they self- renew' (Sennett 2012, p. 91): the habit is formed, ques-

[3] Szymański does not provide a source for the quote. We only know that it was included in a report prepared for the Cooperative management.

tioned, analysed, transformed and ingrained. 'Directions are first given us, which we ingrain as habit; these directions dissolve into evocations we try to pursue more consciously; the pursuit is not endless; we recover our sense of direction in an enriched habit, re- ingrained as tacit behaviour' (Sennett 2012, p. 91) which then becomes dense in meanings. 'Życie WSM' was the actor that gave directions, instructions on how to behave, reminders and appeals through regulations and instructions as well as written and oral obligations. In September 1939, 'Życie WSM' published a set of instructions for new tenants: 'Those who have become members of WHC and those who, more importantly, have become residents of the Estate will not only be expected to pay all of the required fees and perform social work, but also take upon them the moral obligation to cooperate with the Cooperative's social and formative initiatives' (Życie WSM 1931a, p. 1). All of the above may be considered 'dramatic expression' meant to evoke emotions. Such expressive performances free individuals from the obligation of representing and speaking for themselves. They enter a multiple sphere of shared (intersubjectively) reality and play roles, embody norms and cross boundaries of that which is public and private. They enter a sphere of interaction. And that sphere was very broad; built by the authorities and maintained by the residents. The city interacted with its users, tested them (as the users did the city) and allowed them to experiment with their habits in an environment that allowed for natural performances. That, in turn, led to the emergence of 'Żoliborz customs', the Żoliborz attitude, and even 'Żoliborz style'. Kazimierz Brandys included an expressive, slightly academic, but very sentimental description of the typical 'Żoliborz' native in his novel entitled *Listy do Pani Z* [Letters to Mrs. Z]:

> I suggest Żoliborz not as a district, but as a worldview and a set of customs. This city of Warsaw has long been home to working intelligentsia; it has secular, cooperativist and democratic roots. (…) Being aware of one's origins and place in the city is the primary trait of a Żoliborz native. He does not live in Warsaw—he lives in Żoliborz (…). A prosocial attitude is the second trait that Żoliborz natives have. Żoliborz has always had its own approach to running the country, it had a district-wide social programme, and it had its own ideological campaigns. It created an authentic, very European lifestyle and did it without resorting to imitating: the customs of unionised intelligentsia members of modest means and considerable cultural needs, readers-debaters, critical and informed viewers. (…) Żoliborz schools have their own social-secular character. (…) Finally, the individualism of the district is evident in the fact that its residents are christened with its name. Try saying « ochociarz » or « czerniakowianin »—it just does not sound right. (…) So, now you have the general idea. « Żoliborz » is a model of culture that Poles may adopt. Let me repeat: possible, meaning that it may be wilfully adopted and become a mass phenomenon (Brandys 1999, pp. 26–27).[4]

Brandys goes on to describe Żoliborz not as a reality, but as a certain cultural construct:

> Take a look at theatre halls, for example: « Żoliborz » makes up a part of the audience. Because « Żoliborz » did not give up. They want to plug social holes and diluted living spaces with culture. (…) Compared to the West, their civilisational amenities are very modest. They have no television sets, refrigerators, scooters or *air conditioning*… They have nothing but good

[4]Brandys wrote the novel between 1958 and 1961. "Ochociarz" is a resident of the "Ochota" district and "Czerniakowianin" is a resident of the "Czerniakow" district.

> will. They lack the modern instrumentation that injects culture with rhythm; they lack the backdrop, the decorations of everyday life, the efficient system of services and attractions. They are left with work, books, bridge, some art, some sports and travel, as long as they do not leave the country. They are also free to discuss how things are and how they should be. They have that much less, but they are that much more. (…) You may bet that they will make a stand every time civic disobedience and social justice is infringed upon. They will support you, show tolerance and stand with you against fanaticism; they will not, however, let go off the principles that a certain part of humankind has set for itself in the fight for freedom. They stand guard over national culture and keep abreast of contemporary art. (…) « Żoliborz » is home to the socialised civic class which may end up building the Polish socialist average (Brandys 1999, pp. 30–31).[5]

8.1 The Art of Cooperation, or Who Shapes the Local Culture: Organic Intellectuals, Amateurs or External Experts

Almost all of WHC's founders, including Stanisław Tołwiński, Stanisław Szwalbe, Teodor Toeplitz, Adam Próchnik, Maria Orsetti, Antoni Zdanowski, Jan Hempel, Bolesław Bierut among others, had their roots in left-wing institutions such as Uniwersytet Ludowy [People's University], Wszechnica Polska [Polish Free Educational Centre] and Związek Robotniczych Stowarzyszeń Miejskich [Union of Urban Worker's Associations]. Their agenda was based on Western European projects (Vienna, Paris, Frankfurt am Main) (Syrkus 1976, pp. 97–104, 131–161). Their 'leftism' was very broadly defined. They all had one trait in common; however, they wanted to improve the living conditions of the impoverished, foster civic attitudes in society and promote the idea of cooperativism, which was meant to support civic attitudes or, in the words of Julian Hochfeld, 'everyday democracy'. Today, we would rather use the term 'direct democracy'. It should be noted that the cooperative's founders were not its sole creators. The estate's residents also made a big contribution, among them sociologists Maria Ossowska and Stanisław Ossowski, Julian Hochfeld and Nina Assorodobraj (Nina Kula), whose involvement in 'cooperative culture', in the culture of a new form of society, they referred to as 'sociology in action'[6] and

---

[5]That was not the only literary depiction of Żoliborz. In 1935, Edward Szymański wrote a poem titled *Trzy Miasta* [Three Cities] (Szymański 1935) about the estate. Jarosław Abramow-Newerly's interesting memoirs covering the times of German occupation (Abramow-Newerly 2000) and the post-war period (Abramow-Newerly 2003) are two other important titles. As an aside, I should mention that Jarosław's father, Igor Newerly, was a Glass Houses canoeing coach and worked closely with Janusz Korczak.

[6]Bauman's *Visions of a Human World. Studies on the Social Origins and Purpose of Sociology* were published in 1964 (Bauman 1964). One of the book's chapters was titled *Antonio Gramsci's Sociology in Action*. Bauman's analysis of Gramsci's thought uncovered aspects that I found the most striking about the WHC educational project, that is the strong ties between the intellectuals and the masses. Gramsci saw sociology as a means of reconciling the emotions of the masses with the knowledge of the intellectuals which 'aimed to explain emotions using the language of knowledge and, likewise, translate knowledge into the language of emotions, thus becoming an important

what today we would call 'performative sociology'. They treated Żoliborz not only as their place of residence, but were also deeply aware of its pioneering nature. Thus, they could approach it as a social laboratory in which the participants, i.e. residents, learn how to use equipment, internalise social habits, experience boundaries and transformations.

Marek Rapacki, who was born in 1938 and could only remember Żoliborz in its post-war form (it seems that he tended to mythologise it too), recalls that the founders of WHC, 'regardless of their views, have retained a certain social sensibility, an inability to treat others with ruthlessness' (Rapacki 2009, p. 60). Rapacki's legend of Żoliborz is built around 'Żoliborz intelligentsia'. 'Legends are mostly about people, so let us recall a few of the most notable figures. They were visionaries, social activists, social activists, educators, physicians. Others—those residents who were active in the cultural, artistic and scientific spheres—also contributed to that tradition. Their work and heritage exemplifies the values that WHC was built on. (…) They dreamed of creating a community modelled after the phalanstère, a concept created by the French socialist utopian Charles Fourier, and they almost managed to achieve it for a short time during Poland's interwar independence. Among those responsible were Joanna Brzezińska-Landy, Janina Cygańska, Jan Libkind, Aldona Lipszyc and Feliks Zelcer. The Żoliborz estate and two smaller WHC estates built in the late 1930s were designed by the avant-garde architects of the day who were enamoured with the idea of social housing: Barbara and Stanisław Brukalski, Helena Syrkus and Szymon Syrkus, Bruno Zborowski. Maria Ossowska and Stanisław Ossowski deserve special recognition for their efforts on behalf of the WHC. It was them that laid the foundations for the Żoliborz intelligentsia ethics: selfless social activism, collective forms of neighbourly coexistence, and respect for the fellow human being regardless of social status. Among the distinguished academics with a similar intellectual perspective were professors Witold Kula and Nina Kula—him a world-class historian, her a sociologist who cooperated closely with professor Stefan Czarnowski. (…) The Żoliborz estate was also home to some of the best literature historians: Juliusz Wiktor Gomulicki and Zdzisław Libera. Writers also lived there: Maria Kownacka, who had written children's books that remained popular for generations, Irena Jurgielewiczowa, Igor Newerly and his son Jarosław Abramow-Newerly, who chronicled the history of Żoliborz's community, and a representative of the younger generation Antoni Libera, a noted writer, critic, theatre director, son of Zdzisław Libera' (Rapacki 2009, p. 58). The author goes on with that sentimental list, naming directors, theatrology experts, sculptors and politicians. The many famous and cherished figures served as the protagonists of the founding myth that no community can do without. Żoliborz's urban legend, presenting it as an enclave of the intelligentsia, cannot be maintained without ritualistic rejuvenation.

---

identity component of the « historical block » , an element of creative historical activity. It was to combine unconscious suffering with a conscious historical process, transforming history into a conscious process that would bring on the end of suffering. It was not simply engaged sociology, as defined by Mills, but rather sociology in **action**, a sociology which, as Marx put it, becomes a material force once it gains control over the masses' (Bauman 1964, p. 336). Bauman compares sociology in action with C. Wright Mills' engaged sociology and his notion of 'sociological imagination'.

In the previous chapters, I referred to those social reformers, creators and architects as 'critical spatial practitioners' (Marcus Miessen). They may also be thought of as 'organic intellectuals' who live close to the estate's residents and together they shape their social reality and create conditions for the emergence of democratic experiences. One cannot practise politics without that organic connection—'history without this passion, without this sentimental connection between intellectuals and people-nation' (Gramsci 1971, p. 767).

Yet, they were not law-making intellectuals, experts or professionals. The act of creating the Żoliborz estate was a novel experience for all of them. What they did have, however, was enthusiasm. They were visionaries or, as Merrifield would put it, 'amateurs'. Their pursuit of new social forms was motivated by curiosity, innovation, imagination, dissidence against the ruling order; being institutionally independent, they took full responsibility for their creation. Amateurs create an elaborate network of relationships with municipal entities (residents, grass roots movements, and self-governing organisations) as they are forced to rely on other amateurs. And that requires taking a stand against urban experts and 'real professionals'. Merrifield's amateurs may topple the brutal and socially regressive tyranny of 'professionals' (Merrifield 2017). Amateurs prove that the city may be governed by its residents, thus promoting a different a view on how we think and act.

The 'amateur', visionary and passionate process of building the Żoliborz estate resembled another phenomenon which Colin McFarlane, a contemporary urban academic, refers to as translocally 'learning the city' (McFarlane 2011). Similar to Merrifield's concept, it was based on a network- and solidarity-based sharing of knowledge between residents and the creators of urban policies. McFarlane is convinced that such atypical, firmly rooted and autonomous 'expertise' is 'amateurish' in nature for it relies on knowledge produced by the activist community and urban movements which create the city and its policies in cooperation with the citizens. There are no ready-made recipes and professionals who are in possession of be-all and end-all solutions. Amateurs learn from residents, and residents benefit from the vision and enthusiasm of urban reformers.

8.2 Stanisław Ossowski—A Tenant, Researcher and Social Reformer

Naming Ossowski as one of the chief performers of the Żoliborz estate would not be an exaggeration. In his afterword to Ossowski's *Works*, Stefan Nowak outlines three areas of Ossowski's activity: an in-depth and authentic interest in the life and work of Professor Stefan Czarnowski, an uncompromising leftist and diligent cultural scientist (Ossowski continued in Czarnowski's footsteps during his time at the University of Warsaw Cultural History Department). Nowak writes that Ossowski hung Czarnowski's picture above his desk alongside a reproduction of Zygmunt Waliszewski's *Don Quixote*). Ossowski, recalling the memories of Czarnowski, called his master 'a pioneer'. Ossowski claimed that Czarnowski had been guided by a clear ideology, but was far from a propagandist (Ossowski 1970a). The second aspect mentioned by Nowak is Ossowski's activity in the Warsaw Housing Cooperative which 'attempted to bring to life Żeromski's dreams of building glass houses for the working class. Numerous speeches at the WHC Cultural Centre, a strong focus on the issue of integrating people from various social classes using common ideological and cultural values' (Nowak 1970, p. 444). Finally, Nowak emphasises Ossowski's lively and committed involvement in the social life of Warsaw's Left and his attempts to integrate it. For it was in those circles that discussions concerning the possibilities and forms of socialist reform took place.

When describing the WHC, Stanisław Ossowski often used the term experiment, as he was aware of being immersed in it himself and, to a certain extent, of initiating it. He described it as a conscious attempt at creating a social environment: 'the WHC Żoliborz estate not only created a territorial community, but also shaped its own collective lifestyle. New cultural values were born in the estate. The community was influenced by those residents who were bound by a common ideology. The estate ceased to be only a local matter' (Ossowski 1947, p. 186). What is curious, Żoliborz residents formed their identity by reading pieces similar to the one quoted above which were published by the estate's newspaper. One example of the estate residents' identity and reflection upon their place in the world was Wanda Jakubowska's and Józef Cękalski's documentary about the WHC titled *Budujemy* [We are Building].[7] The residents were informed that they would be the 'subject' of a film. 'Życie WSM' published a lengthy article titled *Film ilustrujący działalność WSM* [A Movie Illustrating the Activity of the WHC]: 'The ceremony of laying the cornerstone for WHC's Colony XII, which took place on 21 May of this year, gave the filmmakers an opportunity to capture a lot of film footage in our estate which will be used to create an original movie documenting the housing, construction and cultural activity of the WHC. The present film, directed by comrade E. Cękalski and comrade Wanda Jakubowska, members of the Start group, ties into another documentary that was released in 1928 and aims to present the progress that has been achieved since then. After progressing through a synthetic overview of WHC's various buildings, the cameras focus on the cultural work of our institution. The film as a whole is truly interesting. When it was first aired at the Cooperativist Academy on June 5, the audience received it with loud applause. The film was « screened » for quite some time as an « extra feature » at the « Żoliborz » cinema at Wilson Square, and during the summer it will be screened at the construction exhibition on Bielany Fields' (Życie WSM 1932, p. 6).[8]

The WHC authorities also ensured that the participants of the aforementioned experiment were well aware of the novel and pioneering character of their everyday lives in the estate. The goal was to confront the Żoliborz experience with the rest of the world, contrast it with other urban experiments. The city of the future was not invented in Żoliborz, nor was Żoliborz a perfect incarnation of the concept. That is why it was decided to also screen the avant-garde movie entitled *Miasto Jutra*

[7]For more information, see Chap. 4.

[8]See also Chap. 4.

[The City of Tomorrow]. As an aside, we should mention that cinema itself was an avant-garde means of expression at the time. Thus, modern urban solutions were presented using equally modern methods. 'Modern residential housing does not stop at catering to the individual needs of tenants, or even the needs of entire estates, but rather attempts to explore the principles on which cities and even entire regions are built. The city should not grow chaotically, as is the case now, but rather follow a certain plan based on the universal needs of the present and future generations of city dwellers. This theme is explored by « The City of Tomorrow » , an important film created by German's Housing Reform Society. The film was loaned by WHC and screened in the estate cinema on October 23 of this year. One hundred people interested in all matters urban turned out to see it' (Życie WSM 1931b, p. 11).

Due to its novel and avant-garde character, the estate was also a long-lasting exhibition piece.[9] Even then, the estate's founders and residents were well aware of the experimental, laboratory-like character of their work. Such initiatives prove that even though Żoliborz was an 'enclave of collectivist life', an island within a city, or as Stanisław Szwalbe put it, an 'oasis isolated from the Sanacja-run reality', it ceased to be a local matter. Its pioneering, novel nature resonated across the city, making it the topic of many analyses as well as public and academic debates.

8.3 Methodical Wandering Under Sterile Social Conditions

Contrary to what Rapacki has stated in his memoirs, WHC was not planned to become a closed community, commune, or something resembling Fourier's phalanx. WHC members were to become fully fledged citizens (not only workers, as was the case with patronage estates, not only socialists, as stipulated by the Party, not only community members, which would give them a feeling of solidarity and connection, but deprive them of individuality and demand their unwavering loyalty). Żoliborz is not a 'genetically pre-urban oppression, sectarian oppression', which Nawratek fears may emerge in urban enclaves. Thus, we may say that pre-war Żoliborz was a multitude of singular entities building that which is common (*multitude*). It was not an anarcho-syndicalist social utopia, nor did it resemble Le Corbusier's vision of perfect society. Therefore, I do not find the utopia metaphor fitting in this case. Utopian thinking relies on the lack of appreciation for the historical context of social change. The 'total innovation' myth, based on the notion of starting social life anew, or the revolutionary myth, that is 'completely abandoning historical continuity as the single step towards excellence, as an instant creation of ideal society' (Szacki 1968, pp. 150–168, cf. 1971, pp. 25–31), were not among the plans and intentions of the estate's founders. The Żoliborz community was not founded on the periphery of the society that rejected it. In fact, the opposite is true: it was designed to form a symbiotic relationship with the city, which was especially evident in the concepts of 'centre of attractions' and connection with the outside environment, in avoiding backwardness,

[9]For more information, see Chap. 4.

all of which can be attributed to Brukalska. Ossowski emphasised opening the estate up to the world, creating a connection between the estate's institutions and the city. 'Let us feel the pulse of the world in the estate just as the pulse of the heart may be felt in each local artery' (Ossowski 1967, p. 369). For the personality profile of a social activist and the individual elements of a Żoliborz native's cultural ethos were not planned and set in advance, nor was the estate's urban space. They are not based on a coherent and integrated pattern. The typical Żoliborz resident was not a 'product' of a top-down, planned socialising undertaking of ideologues, but rather the consequence of incessant and ongoing creation of everyday estate reality, which often proved as far from ideal as the reality that surrounded it.

Mariusz Czubaj attempts to find some utopian elements in the WHC project. He notes that the Żoliborz colonies were built on the periphery, far from the urban life of the city. Furthermore, he claims that Żoliborz replaced the anonymity of the metropolis with a sense of community and cooperation. The estate-wide gambling ban, motivated by the fact that gambling may be associated with capitalist competition, brings to mind More's utopian ideals. Although Czubaj draws some parallels between the WHC and utopian thinking, he does not go so far as to consider Żoliborz a utopia. 'The social estate project is an expression of conscious utopian critique. I say so because the WHC writers focused on the interrelations between the concept of utopia and the plan of the « cooperativist township » . The question they ask, however, is not what makes their estate similar to a utopia, but rather in what ways does the estate differ from a utopia. At the design level, the critical utopian awareness becomes apparent in the cultural pattern built around social values' (Czubaj 2007, p. 91). The will to oppose bourgeois reality, an egalitarian approach to sharing common urban goods and the conviction that it is the residents who create the city are all reflections of pragmatic and reformist thinking. It was by no means revolutionary or escapistic, which is typical for utopias. That does not mean, however, that we should disregard the aforementioned methodological suggestions of Ruth Levitas. She proposes inducting normative thinking into the sociological discourse (or, more accurately, cultural discourse) as to 'not let go off that which most of us can see as realistic, doable or achievable' (Levitas 2013, p. 201). Let me reiterate that the author of *Utopia as Method* considers descriptions of worlds that are plausible even for a short while and instate their own democratic institutions, thereby fostering a lively involvement in public matters and creating alternative scenarios for the future, as fundamental for engaged humanities practised as a critical analysis going against the anti-utopian discourse to unleash the transformational potential inherent in the discussed reality.

A sense of exceptionality, difference and innovativeness can be felt in the accounts of Żoliborz's creators and residents. The ideological designers of Żoliborz often use the experiment and laboratory terms in their writings. Even if they did treat their undertaking as an educational project, or an urban experiment in civics, they were aware that in the end it was interrupted by state institutions. Stanisław Ossowski knew that the collective movement had died in post-war, communist Poland, as it was taken over by the authorities who centralised it and made socialist attitudes obligatory. No longer were socialist views a reflection of one's empathy. In 1947,

Ossowski wrote that the 'WHC's role is less apparent today and that has led to the collective losing its lustre and with it its status as an ideological island: it no longer opposes reality as it did before the war. (…) [At the time—author's note] it was an important experiment, an attempt to build a social environment on a temporary spatial foundation. An experiment whose results could be of particular value at a time when cities and estates are being designed' (Ossowski 1947, p. 187).[10] Julian Hochfeld saw the WHC estate as a special kind of sociological laboratory of cooperativism and self-governance. He knew that state centralisation was coming and decided to stand guard over the notion of cooperativism during the inaugural proceedings of the District National Council of Northern Warsaw in January of 1946: 'It was here that the efforts of PPS members [Polish Socialist Party—author's note] led to the creation of an exemplary self-governing workers' institution founded on the right ideals, and it was their active involvement that allowed it to function and make an impact. I am talking about the Warsaw Housing Cooperative along with its vibrant accompanying institutions, including « Glass Houses » , RTPD [Robotnicze Towarzystwo Przyjaciół Dzieci—Workers' Friends of Children Association], « Cooperative Inn » , SPB—Social Construction Firm [*Społeczne Przedsiębiorstwo Budowlane*], Warsaw Bookstore Cooperative [*Warszawska Spółdzielnia Księgarska*], First Cooperative Laundry, RKS « Marymont » [Robotniczy Klub Sportowy—Workers' Sports Club]. Everyday work on concrete matters taught socialists and communists how to reach a compromise. It was here, in the cooperativist community, that « Żoliborz socialism » , both united and PPS-leaning, was born. We were also able to join forces with representatives of the peasant movement and the democrats. This was a laboratory of PPS-led self-governance' (Hochfeld 1946, p. 262).

Stanisław Szwalbe expressed a similar opinion on the matter in 1970. 'It was easier to organically manage WHC when it had 2000 or 3000 members, to shape the ideology, foster and organise social life. That has now changed, as the WHC currently houses more than ten thousand residents. Also, WHC used to be an enclave isolated from the reality created by the 'Sanacja' regime. That had brought the collective's members together, pushed them to act, as they were forced to rely on each other' (Szwalbe 1970, p. 213).

If we were to browse the writings of WHC's founders in search of laboratory metaphors, we would find them used to describe the pioneering nature of the estate, its novel architectural solutions and interior design meant to teach tenants how to live instead of allowing them to passively mimic bourgeois tastes.

Such metaphors were also used to describe the shaping of a new/novel urban lifestyle, or even a Żoliborz native model (as described by Brandys). Unlike the bourgeois lifestyle, it was founded on solidarity, cooperation and socialised individualism. Those were the traits that modern citizens, who are making their life 'by the sweat of their necks and minds' (Próchnik), should have had. The estate transformed

[10] Julian Hochfeld wrote in 1945: 'Cooperativism is not an organ of the democratic state but rather an essential supplement greatly contributing to its democratic contents—so long as the state is not replaced by a self-governing, all-encompassing organisation of farmers and manufacturers. (…) Such cooperativism cannot be bureaucratic. (…) We will definitely not abandon the idea of « Glass Houses »' (Hochfeld 1945, p. 260).

personalities which is evident in the aforementioned recollections of the hosiery factory worker. The transformation consisted in abandoning the role of an excluded worker whose association with traditional institutions only reinforces their passive behaviour. It shaped engaged attitudes among educated citizens who were ready to fight for their right to the city, of receiving their equal share of the city's goods. Żoliborz's practician-researchers used dialogue as a tool of shaping personalities, as life's multifacetedness is formed by discussions on values, ideas and dreams. Identity is not only shaped individually but also through participating in community life. Yet, self-education, built on engagement, perseverance, devotion and submission to leaders, is the primary means of achieving the abovementioned transformation.

The third context in which laboratory metaphors were used concerned the estate's island-within-a-city nature. Żoliborz was considered to be an isolated zone in pre-war Warsaw ruled by the 'Sanacja' regime: including its autonomous economic status, ideological coherence and totalist nature. I use the word totalist to emphasise the estate's holistic approach to controlling all aspects of life: hygiene, local horticulture, 'cultural coexistence', living and shopping habits and education. One may say that the founders of the Żoliborz estate had the means to control the conditions inside it, much the same as scientists control their experiments in a laboratory. The Żoliborz experiment was planned and allowed the researchers to influence those areas of social life that were of interest to them (Sułek 1979, p. 16).

8.4 Social Laboratory and Alchemical Knowledge

If we are to discuss the laboratory, we should take some time to outline the source, nature and possible 'transfers' of the knowledge 'produced' therein.

In December 1932, the 'Glass Houses' Educational Committee organised the Free Workers' Educational Centre which offered systematic paid courses (including the humanities and social sciences, economics and natural science, foreign language courses and, for those lacking elementary education, an elementary course). Although the courses were led by professionals living in the estate, they generated considerable losses and had to be discontinued. Nevertheless, the Active Cooperatists' Club organised paid reading and writing courses, two-tiered math courses and lectures in economics between 1934 and 1935. The courses were free of charge for the unemployed. Both of the above initiatives convinced the Municipal Education and Culture Department to launch an adult elementary education project in Żoliborz, which reduced the significant illiteracy rate in the estate.

Very popular Social Worker Courses were held in 1934 and 1935, and in 1936 they included a General Socialisation Course, which I would like to discuss in more depth. The Social Worker Courses were a special institution reflecting the self-awareness of the estate's founders and residents who were well aware of the pioneering and experimental nature of the estate and their activity within it. The courses included lectures and workshops on the issues encountered by the WHC's various institutions. Thus, it provided a space for broader reflection, often academic in nature, attempting to create a theoretical framework for the estate that would reflect the fact that it was a phenomenon that not only 'happens', but also should be constantly improved upon and consciously planned (lecturers included S. Ossowski, M. Nowicki, A. Próchnik, S. Tołwiński, S. Czarnowski, S. Rudniański, Z. Szymanowski).

During the course, 'the impact that these institutions had on the nationwide workers' movements was discussed, shortcomings were outlined and plans for the future were presented. Numerous lectures provided information on recent developments in sociology and socialist philosophy, participants were educated on working class movements' (Szymański 1989, p. 107).

The General Socialisation Course consisted of three subsequent cycles. The first cycle was a series of lectures. Stefan Czarnowski lectured on the sociological implications of socialism, whereas Kazimierz Czaplicki and Stanisław Rudniański covered the sociology and philosophy of socialism. The second cycle concerned working-class social movements (and their political, professional, cooperativist and cultural engagement), with lectures given by Próchnik, Szczucki and Tołwiński. The third cycle consisted of lectures on the Warsaw Housing Cooperative.

Among the course's participants were employees of the various WHC institutions and members of the 'Glass Houses' Association, who not only were involved in organising the estate's life, but also were able to reflect upon it using their broad humanistic knowledge. The course was very modern for the time: the lectures were transcribed and each listener received their own copy of the transcription. Upon the conclusion of each cycle, participants were given problem sets to solve and then took part in three tests. On average, 80 students out of a total of the 120 that enrolled regularly attended classes.

The Social Worker Courses can be approached as a certain form of social knowledge laboratory which brought down the boundary separating action and research, theory and practice, in which social activists carry out research on their own initiatives. For social reformers and experimenters have a lot in common. Also, it seems that the courses inspired Stanisław Ossowski's idea of founding the Stefan Czarnowski Research Institute, which he planned during the German occupation of Poland. We may assume that the courses were useful and organisationally efficient given that in January of 1943, during a working session of the Architecture and Urban Planning Studio, a group of 'urban activists' led by Ossowski put forward the following suggestion: 'A pioneering estate could build a pioneer school—an educational institution for the estate's social workers. Given its role, the culture centre should be able to accommodate all manners of training courses' (Architecture and Urban Planning Studio 1976, p. 313).

The organisational and educational duties were delegated to other institutions, including ones unaffiliated with the WHC. In 1939, 'Życie WSM' provided an overview of several courses. 'Courses organised by the Workers' Organisations Association have been held in our estate for the last few years. Last year, we had 5 Courses. Four of them drew in people from all over Poland, whereas the fifth one, 'On Collectivist Work', gathered crowds from Warsaw and the neighbouring towns. The latter Course, which was organised by the Central Cooperativist Section of the Workers' University Society in cooperation with the Society for the Support of

Cooperative Work, had a record-breaking attendance of 80 participants and spanned 4 days. Participants of the other courses: the Collectivist Course (Central Collectivist Section of the Workers' University Society), PPS Youth (PPS Central Executive Committee), Red Scouts (Main Council of the Central Executive Committee Red Scouts)—were accommodated in a number of flats in the Żoliborz estate, which were prepared specially for the occasion, and dined at the 'Cooperative Inn'. The lectures were held in the 'Glass Houses' Association office. Course participants were also free to use the 'Glass Houses' library and reading room. Among the lecturers were the following members of our Cooperative: A. Próchnik, A. Zdanowski, M. Nowicki, R. Froehlich, H. Dembiński, S. Tołwiński, S. Szwalbe, W. Schayer, H. Jędrzejewski, M. Zdanowska, H. Purman, J. Hochfeld, O. Haubold, K. Haubold, L. Lenk, J. Święcicka, E. Osóbka. In all, 174 people participated in the Courses. The lectures spanned 46 days. Each course gave participants an overview of the WHC's activities and a tour of the Estate's facilities' (Życie WSM 1939, p. 113).

The Polish humanists listed in the article (historians, sociologists, pedagogists) were not only living in the estate, but also took active part in public life by initiating numerous social events, disseminating academic knowledge in Żoliborz and leading General Socialisation courses. For example, Maria Orsetti practically abandoned her academic career and devoted herself fully to the women's cooperativist movement by founding the model Active Cooperatists' Club in Żoliborz. Julian Hochfeld,[11] on the other hand, started his academic career in Warsaw's higher education institutions only in 1948. Whereas Adam Próchnik (who wrote a doctoral thesis on history entitled *Kościuszko Democracy*) did write academic papers on occasion, but most of his efforts were devoted to social, political and cooperativist initiatives. In addition, he taught history at the Żoliborz school. I could go on and on about the academic backgrounds and theoretical interests of the WHC's social life designers. My goal, however, is to demonstrate that the Żoliborz estate was not an entirely spontaneous and grass-root initiative, the idea for it did not naturally develop in the minds of workers. Instead, the estate was designed using certain theoretical frameworks as an attempt to implement academic knowledge in political, cooperativist and tenant initiatives. To use the terms devised by the estate's founders, the Żoliborz laboratory may be described as a space of methodical meanderings in which appropriate conditions are created (prepared) and potential results are influenced. And knowledge is transferred in both directions. As I have mentioned earlier, the WHC was regularly visited by University of Warsaw students (especially members of the Sociology Club). The *Third International Conference on Social Work* was held in London

[11] He studied Law and Administration at Jagiellonian University. Later, he continued his studies at École des Sciences Politiques in Paris. In 1937, he defended his doctoral thesis on social insurance issues at Jagiellonian University. In 1948, Hochfeld decided to pursue an academic career as a professor at Main School of Planning and Statistics and Main School of Foreign Service. He lectured on social growth theories as well as dialectical and historical materialism. In 1951, he became head of University of Warsaw Historical Materialism Department. One of his notable achievements in that role involved initiating research on the living conditions and identity of the Polish working class, which ran from 1952 to 1956. At the time, it was one of the relatively few empirical studies in Polish sociology.

between 13 and 17 July 1936, where Maria Belsiger and A. Minkowska gave a lecture on the WHC. Their research was based on surveys, descriptions of social initiatives and their impact on shaping various forms of collective life.

Ossowski often used his academic expertise to analyse the activity of Żoliborz's institutions. *Urbanistyka i socjologia* [Urban Planning and Sociology], in which he focuses on the issue of neighbourly ties and social relationships in the new, post-war conditions by analysing Żoliborz's institutions, is one example of such works. When Ossowski declares that he 'is interested in institutions operating on an estate-, district- or region-wide scale', he uses examples of successful solutions implemented by the WHC. 'We may distinguish—writes Ossowski—institutions meant to create social ties, e.g. clubs, discussion halls, and other institutions, such as laundries and the Cooperative Inn, which are meant to address practical personal matters unrelated to neighbourly relations but which nevertheless contribute to the growth of social cohesion. The difference is that, in the latter case, the growth of social cohesion is a side effect' (Ossowski 1970b, p. 347). Ossowski goes on to compare various institutions encountered in London and Żoliborz. The latter, he writes, serve very important social functions because they are not anonymous.

As a researcher and social reformer, Ossowski saw great importance in all kinds of research institutes and free educational centres, both of which could lead to breaking down class divisions in post-war Poland and allow individuals to pursue 'self-learning with one's intellectual development in mind'. The aim was to 'bring down class monopoly on academic knowledge' (Ossowski 1970b, p. 347), which we could thusly rephrase in modern terms: to make cultural common goods available to all citizens, and not only the educated elites. Cultural common goods should be evenly distributed. Therefore, space needs to be assigned for social institutions when designing a new socio-scientific life of society. Likewise, designs of new estates and reconstruction of old ones should include research institutes. Ossowski's perception of such field academic units is very curious. He does not, however, ignore the possible issues that they might face. 'When organising such units, precautions need to be taken to prevent the precious time of scientists being wasted on conversations with individuals with insufficient intellectual capabilities, especially ones with psychotic tendencies' (Ossowski 1970c, pp. 148–149). Ossowski wanted special schools and universities to operate alongside the aforementioned academic units, which would offer courses tailored to the needs of the local population and conduct research on local issues to discover possible solutions. The ultimate goal would obviously be long-term. 'Only a collaboration between academics, pedagogists and the people representing the aspirations of the masses, for whom the institutions of higher education are being established for, will lead to closing the cultural divide between the social classes. The collaboration would be beneficial for both parties, as academics often lose sight of what is important when they have no direct contact with the lives of the masses, and being isolated in the circles of the cultural elite hinders their perception and understanding of certain issues that could inject excitement into their endeavours' (Ossowski 1970c, pp. 156).

It is evident that Ossowski the academic saw the importance of a two-way transfer of knowledge. He was, it should once again be emphasised, a social reformer as well. When designing new estates and new urban culture, which could be introduced during post-war rebuilding of the country, Ossowski remembered that estates should not be schematic, that institutions should not be replicated in each one of them. In addition to district and estate institutions, which by definition have to be similar across the city, each estate should have its own institutions that would give it a unique character. They would be the 'attraction centres'. Ossowski assumes that 'they would be founded based on the needs, resident breakdown, imitative of cooperative members and random events. Thus, one estate could have a workers' educational centre, or something resembling Ruskin's Oxford workers' college, another could have a research institute, which could resemble the unit that was to be built in pre-war Żoliborz in honour of Stefan Czarnowski, and another yet could have sports facilities, an aquarium or a botanical garden' (Ossowski 1967, p. 365).

On the one hand, the goal was to democratise the 'higher forms of intellectual culture' to create a desire for knowledge and understanding of the world, to create the aware citizen. On the other hand, a two-way transfer of knowledge would refresh academic thought and provide empirical data for research on pressing social issues. When considering, perhaps prematurely and loftily, how the collaboration between academics and social reformers should be organised, Ossowski once again brings up specific solutions that were tried and tested in Żoliborz. 'We have little experience in Poland. One example would be the Warsaw Housing Cooperative estate with a population of four thousand tenants and the Glass Houses association that is based there', and as a footnote he adds: 'This experience, involving the kindergarten, schools and adult courses, libraries and reading rooms, discussion, art and sports clubs, academic lectures and literary/artistic events, is covered by several volumes of yearly reports and several years' of issues of 'Życie WSM' monthly' (Ossowski 1970c, p. 155) (Figs. 8.1, 8.2, and 8.3). Thus, they were adopted as empirical data for further research.

The mutual benefits brought on by a critical, scientific reflection on Żoliborz's educational/emancipatory project, the collaboration between residents and academics as well as the theoretical reflections upon the WHC's reality cannot obscure the fact that Ossowski himself, who was a resident of the estate and not an external observer, was perfectly aware of the importance of everyday artefacts, notes, memos and the estate chronicle published by the local newspaper. He knew that the cooperative reports were not only empirical data describing the organisation of the estate's social life, but also served as a discursively argued narration in favour of a certain spatial, ideological and educational condition. They serve as a meta-commentary for the planned and implemented actions. They are, finally, an expression of the documentary and practical research on one's own practice of daily life. The articles published by 'Życie WSM', covering rules of coexistence, collective work and living principles, often bore resemblance to political programmes, manifestos or appeals. They are accompanied, however, by a common identity expressed by the 'we' pronoun (although that may be in part due to the pre-war writing style).

Fig. 8.1 Cover of the report on the WHC activities for 1934, issued in 1935. 3773, p. 195

8.5 Can We Live in a Laboratory?

The laboratory is not a new construct augmenting attempts at describing the city. Such a perspective was adopted by Robert E. Park who, at the start of the twentieth century, observed social behaviours in the urbanised and relatively small space that cocreates modern society. As a representative of the Chicago school, his treatment of the city as a laboratory relied mostly on examining how the organisation of cities leads to changing not only the environment, but also the people who inhabit it; partly because individuals experiment with adapting to new environments. 'The city—writes Ewa Rewers on Park's findings—was used as a laboratory in which the individual's ability to take part in supra-local cultural strategies was tested.

Fig. 8.2 Cover of the report on the WHC activities for 1935, issued in 1936. 3803, p. 195

Furthermore, those changes were combined with the social pressure of experimenting, consisting in combining different cultures within the city: local, metropolitan, rural and urban' (Rewers 2014, p. 25).

When the city is thought of as a laboratory, it becomes a space in which individuals experiment by learning how to be citizens (traditional sociology or anthropology dictionaries would define this process as socialisation or enculturation), but the researcher treats the city as a specific kind of observatory where he may conduct research, find patterns and build theories. In 1941, Stanisław Ossowski also analysed Chicago's urban studies and found the researchers' scientific approach particularly interesting. The laboratory is a space of controlled experiments, testing theories, searching for interrelations between urban nature and culture. Ossowski thought that this approach loses sight of the dynamics of urban culture, and that elaborate

Fig. 8.3 Cover of the report on the WHC activities for 1936, issued in 1937. 3812, p. 195

theories may only apply when the city enjoys stable social and political conditions. He was perfectly aware that the laws of human ecology were nothing like the laws of natural science. The laws devised by American sociologists obscure the dynamics of urban culture and its often spontaneous and unpredictable nature. Ossowski also was hard pressed to find the issue that fascinated him the most about Żoliborz in Park's works, i.e. 'the participation of the conscious will in building society'. For he was searching for evidence proving that 'conscious anticipation is the essence of urban planning' (Ossowski 1970b, p. 344). That would explain why he was so driven to design estate scientific centres and had such an affinity for educational practices aimed at already engaged 'organisers of social life' and for conducting conscious field work aiming to turn residents into the reformers of their own urban lifestyles. That is how Ossowski defined the role of the social sciences.

In his approach to Żoliborz as a laboratory, Ossowski seems to have been more forward-thinking than Park, mostly by showing scepticism towards his lofty scientific ambitions. Ossowski saw the estate as a laboratory because it experienced ongoing construction, testing and transformation of new urban planning, architectural, social and educational solutions. We may say that it resembled present-day city laboratories which emerged at the crossroads of academic thought, the activities of various types of organisations (Workers' Friends of Children Association), cultural institutions (the estate's 'social facilities') and activist initiatives. Residents participate consciously and actively, not by submitting to the various innovative undertakings but by cocreating them. Similar to an ideal laboratory, all parties participating in the experiment may discuss its results and methodology. The estate's discussion clubs and chronicle published by 'Życie WSM' were the means for doing that. In that sense, we may say that dialogical, rather than dialectical, solutions were used.

It is here that a multi-dimensional, diverse transformation of lifestyles, manners of dwelling, raising children and developing the public sphere takes place. 'This approach [i.e. the modern approach created at the turn of the 20th and 21st centuries—author's note] towards experimenting forms the basis for the existence and activity of emancipated urban entities. Their experiments involve not only the division and use (inhabiting) of urban space, but most importantly assign meanings to it' (Rewers 2014, p. 57). Ossowski deemed it necessary to adopt a critical perception of this process, a theoretical encircling of the issue, research in action. That is why he became engaged in the activities of the underground Architecture and Urban Planning Studio, which was an interdisciplinary group of 'independent practicians', including architects, urban planners, sociologists and pedagogists. Those 'critical spatial practitioners' posed questions regarding their role in shaping neighbourly ties and the possibilities of invoking bottom-up cooperation. Not only that—they did not stop at designing social ties within the estate, but went on to draw up the entirety of post-war urban culture.

Being convinced that experimenting is among the most common activities of city dwellers and a part of their daily practice, Ewa Rewers wondered whether that metaphor could prove useful for urban studies. She divides the city-laboratory concept into two categories: interventionist concepts and founding concepts.

Interventionist approaches to the city-laboratory concept view it as a space in which new ideas are implemented, a place consisting of spaces shaped in the past that are presently undergoing a crisis. 'They are forced to serve the new goals of cities, new visions are imposed on real spaces, therefore experiments are conducted on the old flesh of the city. The subjects of such « surgical » interventions or, as others would have it, urban acupuncture are professionals, activists and local communities. Despite being widespread, such activities do not always converge with the opinions and interests of citizens' (Rewers 2014, p. 23). The Haussmanisation of Paris and the founding of artists' districts at the turn of the twentieth and twenty-first centuries would be examples of that.

Founding concepts, on the other hand, find their expression in designing new cities in line with sustainable development principles (e.g. the Masdar city in Abu Dhabi), allowing residents to shape new lifestyles. 'The push towards sustainability encompasses all aspects of urban life. It is pretty obvious, however, that technological, urban planning and social experiments are at the forefront of such initiatives' (Rewers 2014, p. 23). Founding concepts are thus based on strategies of subordinating, repairing and healing. And external experts (or perhaps technocrats) are the driving force behind them. Therefore, innovative changes resulting from bottom-up, everyday practices of residents become less noticeable.

Ewa Rewers' typology of modern-day city-laboratory implementations is inspiring because it draws attention to the involvement of external experts. If we go back to the Żoliborz estate, we may see that even if external experts had an impact, their ideas had to be negotiated with the residents, leading to conflicts and social friction. Although the experts often made use of their academic experience, their actions were similar to those of 'amateurs' (Merrifield). For introducing innovations and principles of cooperation requires participation and not a passive acceptance of ready-made solutions. Such participation relies on the never-ending setting and ordering of rules. Yet, as is true for all laboratories, that procedure has to have an open-ended nature leading to concrete outcomes. The aim is to participate in the development of certain cooperation rituals, 'avoiding the fantasy of 'settling matters' once and for all' (Sennett 2012, p. 234).

Stanisław Ossowski followed a similar path when he wrote about the reorganisation of social life in both new and existing estates. Founding new estates resembles the logic of experiments. Creators and designers have a say over the tenant selection process and thus contribute to the shaping of neighbourly ties. Because all tenants move in at the same time, facing new social and urban conditions, chances of changing social attitudes, shaping new habits or perhaps new modes of cooperation are higher. Thirdly, organising new estates facilitates coordinating urban planning with the organisation of social life. Also, it allows adapting new forms of coexistence to established spatial forms.

The biggest difficulties encountered when designing new estates, however, are a result of the lack of tradition, for there is not one set of ready-made forms of cooperation and living. 'There is no questioning—writes Ossowski—the role that common history, or a common capital of memories, plays in the life of a community. New estates will be inhabited by individuals coming from various places, having various habits, differing expectations; individuals who will have social ties with various groups outside of the estate. Abramowski's friendships were doomed to fail because they assumed that strangers would be able to form bonds of true friendship. That only happens, and seldom, in times of group enthusiasm. In normal conditions, however, no one can force people to make friends. (…) Thus, factors allowing to mitigate the lack of common tradition, by either uniting for a common future or building a new estate tradition, have to be taken into consideration when planning the social life of new estates' (Ossowski 1967, p. 361).

If we were to regard the WHC as an experiment consisting in creating new living conditions by external experts, then filling those forms with the contents of social life, building habits and rituals of estate living would only take on the form of a settlement house if it involved 'spreading common interests', or the form of a workshop if it involved 'hastening the creation of new traditions'. That is why rituals of cooperation and celebration are so important.

If we were to adopt Ewa Rewers' typology of founding and interventionist concepts, we would find that the Żoliborz project would not fall into either of the two categories, proving such typologies useless. That is mainly because the Żoliborz estate, even though it seems to have been 'awfully participatory' and overly oppressive, was based on mutual cooperation, trust, 'respect for authority' consisting of the radical intelligentsia which had the residents on its side. The radical reformers were not 'external experts'. I would rather call them 'organic intellectuals' or 'amateurs'.

Using the laboratory metaphor in urban studies, and especially when researching the WHC, is tempting because the most serious of social problems arise in urban environments. Not because the residents that were taking part in this educational-civic project were only learning urban living, thus undergoing a transformation of identity. In my opinion, Sennett's definition of the workshop better fits the nature of learning the patterns of urban living. The only reason why we could find the laboratory metaphor useful is because it makes apparent the transfer of academic knowledge into the estate and makes possible the analysis of the practices of its residents in academic discourse (in the areas of urban planning, architecture, social sciences, pedagogy). The metaphor adds a practical dimension to the vast theoretical framework of critical urban studies. Furthermore, all participating entities may engage in dialogue to exchange opinions, views, ambitions and dreams. The organic estate solutions proposed by the abovementioned intellectuals did not always work in a natural social environment.

8.6 The Engineering Purpose of Social Knowledge

Both the laboratory and the experiment bring to mind undefined effects, manipulating phenomena and conditions, observing the outcomes of those manipulations and formulating new interdependencies. The workshop, on the other hand, has set rules and the participants know what their cooperation will produce. Sennett brings the workshop metaphor up to date, mainly because of its increasingly heterogenic social and cultural nature and possibilities of transferring knowledge. He is aware that, similar to the collaborative work of the creators and users of the Linux operating system, the outcomes are not entirely predictable. He is also aware that the modern laboratory does not rely on the principles of truth and falsehood which have been replaced by efficiency and organisational effectiveness. Therefore, he boldly combines the two metaphors. 'Another modern variant of the workshop is the scientific laboratory, which Owen explicitly foresaw. 'Factory-style science' appeared to him as the mechanical testing of hypotheses; a more innovative laboratory engages in true experiment, open to surprise—which is to say, discovery. Good laboratory work should run like an experimental workshop' (Sennett 2012, p. 58).

Experiments in social sciences often evoke questions regarding ethics, the limits of manipulation, the goals of science and the scientists themselves, the practical and hidden motives of sociologists. And when sociologists want to save the world, they turn the laboratory into an alchemist's chamber. The above issues are not the main topic of this work, which aims to explore the role that Żoliborz strategies had on shaping the city. Yet, it is worthwhile to consider the engineering purpose of sociology as a side note to the main topic. For sociology has shown its performative face in the modernist WHC project.

We can also see just how original Julian Hochfeld's and Stanisław Ossowski's ideas were for the time and why they had a lasting impact on post-war debates on the role of social sciences. The two Polish sociologists saw them as engaged sciences in which the choice of issues that undergo analysis is impacted by the researchers' world view, as are the ways in which those issues are presented. Social sciences are based on the intent and belief that academic activity may change social reality. Such an approach was not limited to academics under the influence of Marxist ideas (Julian Hochfeld, e.g., was both an academic and a politician), but also those who were critical of Marxism but adopted an active social reformer attitude (Stanisław Ossowski was not ashamed to identify as a socialist).[12] The belief that humanistic thought may be used to shape the world was evident in the academics' engagement in reconstructing social institutions, creating new cultural patterns and modifying social relationships to make them more egalitarian and democratic.

In 1964, Zygmunt Bauman's *Visions of a Human World* was published, a book which made an enormous impact at the time as the first published work presenting an interpretation of Antonio Gramsci's thought to Polish readers. In his book, Bauman supports Gramsci's view of rational knowledge forming the foundation for action. The purpose and rationality of human action (including cognitive activities, philosophy and science) becomes apparent only through relating it to historical processes and the social reality in which such action takes place. The conclusion being that both social and scientific knowledge as well as philosophy have pragmatic meaning: they order human experience allowing individuals to find their place in the world and act accordingly. Bauman concluded that knowledge has an engineering purpose. The term engineering was understood very broadly as 'exploiting the knowledge of the interrelationships between phenomena to achieve a desired sequence of phenomena' (Bauman 1964, p. 36). Thus, engineering would be understood as a practical activity making use of theoretical knowledge but not entirely identical to it. The 'engineering' term (both outdated and suspect) contains the notion of combined theoretical and practical activity, a combination of thought and action, knowledge and practice. In *Visions of a Human World*, Bauman contrasts engineering through rationalisation with engineering through manipulation. Antonio Gramsci and Charles W. Mills represent the former approach, whereas Talcott Parsons and George Lundberg took the latter approach. I will not go into detail on the Bauman's interpretation of engineering through manipulation. Let me just say that Bauman claimed that it took on the form of 'behavioural science' and encompassed nearly the entire scientific and empirical social sciences paradigm. To put it simply, Bauman defines engineering

[12] I should emphasise that I do not consider all social reformers to be socialists. For example, I see no reason why Christian conservatives could not be social reformers.

through manipulation as influencing people to behave in a way that is desired by the manipulators. Behavioural science is not a type of engineering that would attempt 'to rationalise mass tendencies, but instead aims to elaborate a set of efficient techniques allowing individuals with sufficient resources to make their subordinates behave in a way that they please' (Bauman 1964, p. 61). A socially engaged sociologist would regard science as emancipatory and critical, a form of social consciousness that frees people from the oppression of the ruling class thereby becoming a means of resistance. For rulers/ruled, those who have power/those who do not have it, is the main axis of class divisions. If we were to apply engineering knowledge in that area, we would find that sociology (understood as engineering through rationalisation) gives power to those who have not had it before. It makes possible formulating social definitions of situations by setting goals and providing direction, giving purpose to human activity. Sociology as engineering through manipulation has no practical use for people who wield no power (but are subject to power). Influential people use it to gain information allowing them to adopt more effective means of exerting their influence. Therefore, knowledge benefits the rulers as it enables them to manipulate the behaviour of the ruled and thus strengthens their power. In that sense, it is socially engaged but stands in stark contrast to the social engagement of engineering through rationalisation. Its engagement relies on practical, utilitarian function benefiting only one side of the social relationship (the effectiveness of Mackenzie's performance also relies on that principle). And because it does not set any values, goals, orientations in addition to the means of achieving arbitrarily set goals, it cannot be considered an ideology. In that sense, it seems to have a neutral approach to ideological systems. Leaving all moral aspects aside, Bauman's critique of engineering through manipulation was rooted in his strong conviction that thought is inextricably linked with acting, a notion which he had drawn from Gramsci. Theory is not a servant to practise and important social processes are successful thanks to the emergence of social consciousness, thanks to knowledge and ideas. Social activity and the consciousness that accompanies it are the human *praxis*, the philosophy of practice. There is, however, a different way that we could look at the social engagement of social sciences.

Such an approach would not involve social knowledge functions that are autonomous from the researcher, but would instead focus on cognition as a process in which the researcher plays an active role using his or her extensive historical, social and cultural experience.

Although Ossowski and Hochfeld shared Mills' and Gramsci's enthusiastic view of the important role that humanities should play in the process, they adopted different ideological approaches to social sciences (the main difference being Marxist views). We can trace the parting of ways of those two 'organic intellectuals' back to the years 1947–1948 when both of them wrote important ideological pieces. The first of the two articles, considered today to have laid the foundation for such polemical forms of expression, is Stanisław Ossowski's *Doktryna marksistowska na tle dzisiejszej epoki* [Marxist Doctrine and the Modern Era], whereas the second is Julian Hochfeld's response *O znaczeniu marksizmu* [On the Meaning of Marxism]. The discussion between those two researchers and social reformers made apparent the subtle albeit crucial differences in their opinions regarding the claims and expectations towards the engineering role of social sciences, and especially towards experimenting on social relationships. Ossowski saw Hochfeld's joint political and academic activity as a foray into the precarious domain of alchemy. Hochfeld, on the other hand, claimed that the 'golden age' of Marxist theory was a 'time in which the worker's movement had not yet developed practical and theoretical-academic specialisations'.[13] Thus, Hochfeld's vision is characterised by his conviction that sociology should be creative and performative. It should not be limited to devising laws and making predictions, but rather changing the world in accordance with the sociologist's vision. That is why Hochfeld needed Marxism.

At the same time, the Ossowskis were organising weekly meetings with Tadeusz Kotarbiński during which Jan Strzelecki's concept of humanistic socialism was being fleshed out.[14] During those seminars, works of Abramowski, Georges Sorel, Marx, Emmanuel Mounier and Jacques Maritain were discussed. Axiology, a critical analysis of socialism's past, was the main subject of discussions which, according to Strzelecki, 'was a result of the universal distrust towards obscure and optimistic theories, a category under which all kinds of Marxism undoubtedly fall'.[15]

References

Abramow-Newerly J (2000) Lwy mojego podwórka [The lions of my courtyard]. Wydawnictwo Twój Styl, Warszawa

Abramow-Newerly J (2003) Lwy wyzwolone [Free lions]. Rosner i Wspólnicy, Warszawa

Architecture and Urban Planning Studio (1976) Organizacja życia zbiorowego w osiedlu spółdzielczym [The organisation of collective life in a cooperative housing estate]. In: Syrkus H, Ku idei osiedla społecznego 1925–1975 [Towards the idea of the social housing estate 1925–1975]. Państwowe Wydawnictwo Naukowe, Warszawa, pp 313–319

Bauman Z (1964) Wizje ludzkiego świata. Studia nad społeczną genezą i funkcją socjologii [Visions of a human world. Studies on the social origins and purpose of sociology]. Książka i Wiedza, Warszawa

Brandys M (1999) Listy do Pani Z. [Letters to Mrs. Z]. Prószyński i S-ka, Warszawa

Czubaj M (2007) W stronę miejskiej utopii. Szkice o wyobraźni społecznej [Towards the Urban Utopia. Essays on social imagination]. Wydawnictwo SWPS, Academica, Warszawa

Filipczak S (1963) Moje wspomnienia pracy w ruchu spółdzielczym [My memories of working in the cooperativist movement]. In: Kossowski K (ed) Wspomnienia działaczy spółdzielczych [Accounts of cooperativist activists], vol 1. Zakład Wydawnictw CRS, Warszawa, pp 151–156

---

[13]He provided an interesting perspective on the subject in his study of Oskar Lange: *Oskar Lange as a Theoretician of Historical Materialism* (Hochfeld 1964, p. 210).

[14]Jan Strzelecki has a special place in Polish humanities. He was the founder of 'Płomienie' [Flames], a magazine which published the writings of almost all representatives of the young socialists. His concept of humanistic socialism laid the ground for many polemics and important academic discussions concerning the shape of post-war sociology in Poland (including the Ossowski-Hochfeld debate of 1947–1948). It should be noted that Strzelecki considered both socialism and Christian thought as possible saviours of humanistic values.

[15]Strzelecki's private notes for the son, cited in Neyman (2002, p. 44).

Gramsci A (1971) Selections from the prison notebooks of Antonio Gramsci (eds, trans: Hoare Q, Nowell-Smith G). Lawrence & Wishart, London. http://abahlali.org/files/gramsci.pdf. Accessed 7 Sept 2018

Hochfeld J (1945) O istotną treść spółdzielczości [On the fundamental contents of the cooperative movement]. Przegląd Socjalistyczny [Socialist review] 1(XI); reprint in: Hochfeld J (1946) My socjaliści [We, socialists]. Wydawnictwo Spółdzielnia Wydawnicza 'Wiedza', Warszawa, pp 259–260

Hochfeld J (1946) P.P.S. wobec samorządu dzielnicowego [PPS and the district government]. In: Hochfeld J (1946) My socjaliści [We, socialists]. Wydawnictwo Spółdzielnia Wydawnicza 'Wiedza', Warszawa, pp 261–266

Hochfeld J (1962) Więź społeczna wuesemowców [The social ties of WHC activists]. Życie Osiedli WSM, January, p 9

Hochfeld J (1964) Oskar Lange as a theoretical of historical materialism. In: On political economy and econometrics: essays in honour of Oskar Lange. Polish Scientific Publishers, Warszawa, pp 207–232. https://www.surplusvalue.org.au/Marxism/hist%20mat%20lange.pdf. Accessed 7 Sept 2018

Kozikowski E (1960) O Antoninie Sokolicz [On Antonina Sokolicz]. In: Barcikowski W et al, Księga wspomnień 1919–1939 [The book of recollections 1919–1939]. Czytelnik, Warszawa, pp 181–189

Levitas R (2013) Utopia as method. The imaginary reconstitution of society. Palgrave–MacMillan, New York

McFarlane C (2011) Learning the city. Knowledge and translocal assamblage. Wiley–Blackwell, Malden, MA and Oxford

Mendel M, Szkudlarek T (2012) Demokratyczna izolacja. Krytyczna partycypacja obywatelska [A democratic isolation. Critical citizen participation]. Zoon Politicon: Rocznik Naukowy 3:207–222

Merrifield A (2017) The amateur. The pleasures of doing what you love. Verso, London and New York

Neyman E (2002) Słowo wstępne [Foreword]. In: Neyman E (ed) Intymny portret uczonych: korespondencja Marii i Stanisława Ossowskich [An intimate portrait of scholars: the letters of Maria and Stanisław Ossowski]. Sic!, Warszawa, pp 5–82

Nowak S (1970) Posłowie [Afterword]. Ossowski S, Dzieła [Collected works], vol 6. Państwowe Wydawnictwo Naukowe, Warszawa, pp 432–465

Nowicka M (2009) WSM—RTPD—Szklane Domy [WHC (Warsaw Housing Cooperative)—RTPD, Robotnicze Towarzystwo Przyjaciół Dzieci (Workers' Friends of Children Association)—'Szklane Domy' (Glass Houses) Association]. Życie WSM, September, pp 14–15

Ossowski S (1947) Zagadnienie domu społecznego w osiedlu [The issue of the communal house in the estate]. Życie WSM, July–August; reprint In: Ossowski S (1970) Dzieła [Collected works], vol 6. Państwowe Wydawnictwo Naukowe, Warszawa, pp 184–188

Ossowski S (1967) Organizacja przestrzeni i życie społeczne w przyszłych osiedlach [Spatial organisation and social life in future housing estates]. In: Ossowski S, Dzieła [Collected works], vol 3 [Issues of social psychology]. Państwowe Wydawnictwo Naukowe, Warszawa, pp 351–370

Ossowski S (1970a) Stefan Czarnowski. In: Ossowski S, Dzieła [Collected works], vol 6. Państwowe Wydawnictwo Naukowe, Warszawa, pp 127–131

Ossowski S (1970b) Urbanistyka i socjologia [Urban planning and sociology]. In: Ossowski S, Dzieła [Collected works], vol 6. Państwowe Wydawnictwo Naukowe, Warszawa, pp 337–349

Ossowski S (1970c) O drogach upowszechniania kultury umysłowej na wyższym poziomie [On the means of popularising high culture]. In: Ossowski S, Dzieła [Collected works], vol 6. Państwowe Wydawnictwo Naukowe, Warszawa, pp 144–157

Rapacki M (2009) Czwarta ćwiartka [The fourth quarter]. In: Chałasińska E, Gawecka A (eds) O pięknym Brzegu. Żoliborz, ludzie, wydarzenia [About the Beautiful River Bank. Żoliborz, People and Events]. Towarzystwo Przyjaźni Polsko-Francuskiej [French Polish Friendship Society], Fundacja 'Żoliborski Fundusz Lokalny', Warszawa, pp 56–62

Rewers E (2014) Miejska przestrzeń kulturowa: od laboratorium do warsztatu [Urban cultural space: from the laboratory to the workshop]. In: Rewers E (ed) Kulturowe studia miejskie: wprowadzenie [Cultural urban studies. An introduction]. Narodowe Centrum Kultury, Warszawa
Sennett R (2012) Together. The rituals, pleasures and politics of cooperation. Yale University Press, New Haven
Sułek A (1979) Eksperyment w badaniach społecznych [Experiments in social sciences]. Państwowe Wydawnictwo Naukowe, Warszawa
Syrkus H (1976) Ku idei osiedla społecznego 1925–1975 [Towards the idea of the social housing estate 1925–1975]. Państwowe Wydawnictwo Naukowe, Warszawa
Szacki J (1968) Utopie [Utopias]. Iskry, Warszawa
Szacki J (1971) Tradycja [Tradition]. Państwowe Wydawnictwo Naukowe, Warszawa
Szwalbe S (1962) Sylwetki [Silhouettes]. Życie osiedli WSM [The Life of WHC Estates] 1:17
Szwalbe S (1970) Jeszcze o WSM [More on the topic of the Warsaw housing cooperative]. In: Anonymous (ed) Żoliborz wczoraj, dziś, jutro [The past, present and future of Żoliborz]. Książka i Wiedza, Warszawa, pp 211–214
Szymański E (1935) Trzy Miasta [Three cities]. Życie WSM, November, p 9
Szymański JA (1989) Warszawska Spółdzielnia Mieszkaniowa 1921–1970. Zarys dziejów [Warsaw housing cooperative 1921–1970. History outline]. Wydawnictwo Spółdzielcze, Warszawa
Życie WSM [Life of the WHC] (1931a) Do naszych nowych mieszkańców [To our new tenants]. Życie WSM, September, p 1
Życie WSM (1931b) Miasto Jutra [The City of Tomorrow]. Życie WSM, December, p 11
Życie WSM (1932) Życie WSM, June–July, p 6
Życie WSM (1939) Kursy Towarzystwa Uniwersytetu Robotniczego [Courses by the Workers' University Society]. Życie WSM, April, pp 113–114

Chapter 9
'Total Pedagogisation'?

Abstract The analysis of life on the housing estate and its educational and emancipatory character can lead one to suspect some form of total pedagogisation. One cannot ignore the fact that these modern social reformers and founders of the cooperative used the neighbourhood newsletter to diffuse the model of 'a new man in the new housing estate'. In the context of the criticism of modernity, I wonder whether education really had an emancipatory character. I use Jacques Rancière's concept of education and stultification. This is why the categories proposed by Gert Biesta, who studied the culture of learning and education, turned out to be vital for my analysis of the Żoliborz estate. The founders and teachers of the WHC school (Wacław Schayer and Stanisław Żemis) also pointed out to the difference between learning and education. The Żoliborz model of education, based on John Dewey's concepts and the assumptions that social practices are experienced by the body, confirmed Gert Biesta's thesis that civic education is the matter of organising democratic development conditions and a democratic living environment, and not the matter of implementing models of a perfect citizen, specific skills or competencies. I ask, however, the question whether the rather specific, experimental and hermetic educational model was beneficial for children and youth? Wasn't this model too detached from the Polish social and cultural context?

Keywords Model of a new resident · Experimental school · Culture of learning · Culture of education · Emancipation

One can have the impression that the 'teaching how to live' (teaching urban living) phrase implies that the estate pushed for an unmitigated 'pedagogisation' of the community. Yet, I decided to choose the workshop instead of the school out of the various metaphors that could be used to describe the estate. The settlement house metaphor fits some aspects of the estate's social life, whereas the laboratory reflected the mutual transfer of knowledge between the 'organic intellectuals' and the residents.

A 'pedagogised society' would resemble an enormous school whose goal would be to civilise the working class by working on the assumption that there is a relationship of inequality and distance between those who require teaching and those who know better. This inequality can be seen in the attempts to describe the perfect society, the concrete teaching model (Jacques Rancière's stultification) (Rancière 2009,

M. Matysek-Imielińska, *Warsaw Housing Cooperative*, The Urban Book Series,
https://doi.org/10.1007/978-3-030-23077-7_9

pp. 1–23), and viewing emancipation as waking up from a state of social lethargy and passiveness. The inequality gains legitimacy through the 'good citizen'—'bad citizen' opposition which lays the groundwork for lawfully excluding from the public sphere all those who do not fit the desired citizen profile. Therefore, inequality results from the notion that an individual has to possess certain prerequisite skills and characteristics to be allowed to participate in democracy and the public sphere and be treated as a respected member of the community. Appropriate, personal competencies required to become a citizen, which are acquired through education, are the key. Many academics, including Jacques Rancière and Gert Biesta, treat all relationships between the citizen and the political body (society) as forever tainted by domination and see them as an entrenchment of social hierarchy. The legality of domination uncovers an evident 'sensible distribution' of two kinds of people (those with refined or plain taste, those who are active or passive, those who discover and those who spectate) (Rancière 2004, p. 53). Rancière called this the police model, based on 'partition of the perceptible' (Rancière 1999, pp. 21–42). The model results in political decisions being made over which activities should be visible and which should not, which voices should be considered as part of the discourse and which are merely noise (Rancière 1999, p. 29). Such an approach leads directly to the emergence of governance forms that are referred to as post-democracy or a consensus based on the ability of the ruling class (including intellectuals) to reduce all conflicts (mainly class-based) to irrational and emotional demands of the masses (i.e. noise).

The issues with 'citizen education' may be interpreted in other ways, however. A democracy based on discord brings on a clash of statements coming from various individuals, statements which are not treated as noise but rather as a manifestation and creation of subjectivity. Democracy thus defined leads to the ongoing creation of new subjects which give meaning to the world that they live in and make collective statements, making them heard, seen, corporeal, present and included. That approach makes each part of life, no matter how mundane, a political matter and leads society from stultification to emancipation based on the assumption that all subjects are equal in a world of differences and struggles. Emancipation and equality are prerequisite conditions for changing reality; they are an assumption, a point of entry, and not the ultimate goal of political activity (Biesta 2010, p. 51; Rancière 2010, pp. 9–15). All of the above means that the essence of politics lies in creating structural conditions that make possible social change and shaping a world of common experience. As a result, a disjointed society of people who work towards establishing equality is formed. Such a social, collective definition of emancipation relies on cooperation and collective habit forming, involving a collectivisation of abilities and taking advantage of the skills of each subject involved in the process.

Undertaking educational and emancipatory activities while disregarding the wider social, political and economic environment inevitably leads to unpolitical activity, a tyranny of intimacy, which means that individuals will start blaming themselves for the exclusion and inequality that they suffer. And that only serves to strengthen and reproduce the existing domination structure. If an activity aims to make one's voice public and question the existing cultural, social, economic, etc., order, then such an activity should be considered political.

Emancipation therefore does not consist in teaching individuals how to live in the established democratic system. If such an approach is adopted, then emancipation would be reduced to a simple cultural reproduction of people who are required to acquire certain competencies allowing them to take part in the public sphere. Gert Biesta proposes replacing teaching democracy, which relies on implementing the 'good citizen' model, with learning through doing and creating (Biesta 2011a, p. 6, 2011b, p. 83). The lack of democracy is not a result of a lack of competencies, but rather a lack of structural conditions. The crisis of citizenship stems from the unavailability of democratic experience and the limited opportunities to take part in democratic practice. The only way to counteract the lack of democracy is with more democracy which can be achieved through making attempts to create it, organising self-governing institutions that would allow making statements in the public sphere. Stanisława Filipczak's (the hosiery factory worker) account and the activities of the Active Cooperatists' Club are a perfect example of this approach in action.

Although the Żoliborz estate was by no means a perfect environment in which the above emancipation-based education practices could be introduced, and despite the fact that certain areas of the estate's life were governed using the 'law-making model', its creators genuinely intended to overcome the structural conditions that undermined equality. That would also explain the estate's total, holistic nature. For Żoliborz residents collectively created citizenship wherever there was none. The imperfections and incoherence of the project, its ongoing revisions and repairs prove that Żoliborz's creators and activists had no initial vision of the 'correct citizen'. They were, however, thoroughly familiar with the economic, political, social and cultural factors that led to oppression, inequalities and lack of democracy. And that required imagination. We can therefore say that the cultural context in which they acted was not static. Rather, it evolved with time; it was a process capable of setting off events, ensuring existence but also change.

Janusz Ziółkowski made note of the importance of 'sociological imagination' when describing the architectural and urban planning achievements of Żoliborz's founders, who developed the social estate concept (Szymon and Helena Syrkus, Barbara Brukalska) (Ziółkowski 1965, pp. 227–248) and Teodor Toeplitz's 'social urbanism'. Ziółkowski saw the entire Warsaw Housing Cooperative project as an attempt at building common material and social conditions in which democracy could develop. He was obviously referencing Mills (1959)[1] who, in the late 1950s, created the term 'sociological imagination' which allows analysing how human lives are connected to the broader social, political and cultural contexts and vice versa.

---

[1] Bauman was also hugely inspired by Mills' sociology. He wrote: 'Readers are fascinated by two facets of Mills' sociology (...) the enlightening humanism of Mills' sociology, a ruthless war waged against the notion of transforming humans into servomechanisms, (...) pure hatred towards « eager robots » and their managers, as well as warm, humane and, frankly, « unsociological » emotions permeating his writings on human abilities, which are human freedom, and on freedom itself, which is the independence of intellect and human action (Bauman 1964, pp. 290–291). In Bauman's view, Mills' approach was based on the conviction that all people are irrevocably free and equal. The contemporary concepts of educating through practicing democracy are also based on that conviction.

'Sociological imagination' is not solely in the hands of academics. Each member of a structured social environment has it. Similar to common knowledge, it grows spontaneously and allows leading a purposeful life and experiencing intersubjective reality. In that sense, we may say that it does not belong to those who know better. Instead, it is egalitarian. Therefore, it reinforces the assumption that all subjects are irrevocably equal.

Mills' methodology raises some suspicions about those discourses which blame individuals for their misfortune and internal fears, stripping them of their historic and social context, drawing their attention away from the public scene by focusing on mentality and conscience, which breeds political apathy and unwillingness to participate in public life. In order to preserve one's mental health, feeling of self-worth and internal peace, people are ready to accept all of the requirements that are pushed upon them from the outside with increasing force. People do not see the links between their own anxiety and the social world. That is why Mills called for placing the life stories of specific individuals into a broader social context. In doing so, their concerns, fears and worries would be elevated to the status of public issues and would be recognised in social, structural and political categories. Transforming personal fears into engagement in common causes is the 'promise' made by Mills' vision of sociology.

Maybe that is why Zofia Topińska decided to title her book on the educational model of the Warsaw Housing Cooperative *The Promises of Żoliborz*, which is a perfect reflection of Mills' concept of imagination (be it sociological, anthropological, pedagogical) making a promise.

Although the project had educational and emancipatory plans for its participants, mainly adult residents of the estate, from the very beginning, it is worth looking into the vision of education that Żoliborz residents had for their children. After all, they were to become the future residents of the city within the city which aimed to build democratic living conditions. Thus, the ultimate goal was to educate 'a new human in a new housing estate'.[2]

[2]The educational approaches introduced in the WHC have been widely covered in Polish literature. They were analysed on an ongoing basis and many publications were released between the 1960s and late 1980s. Therefore, there is no need for me to go into excessive detail on the philosophical and pedagogical roots of those practices ('The RTPD was undoubtedly influenced by « new education » trends (Montessori, Decroly, Freire), the diverse « labour schools » theories (Dewey, Kerschensteiner and Błoński), Adler's individual psychology, Rowid's « creativity school » , Radlińska's social pedagogy, (…), Spaskowski's « school of the future »') (cf. Kuzańska-Obrączkowa 1966, p. 18). Also, I will not provide an in-depth description of the functioning of educational institutions, although it is a very interesting topic due to the novel, experimental methods of organising children's lives that were used in the estate. The topic deserves its own paper which would present the WHC's educational achievements from today's perspective and analyse it using both late-modern educational discourse as well as contemporary performative concepts. Academics will discover that numerous 'events' took place there, all of them built in accordance with John Dewey's pragmatic philosophy and the extremely innovative approach of Aleksander Landy (cf. Demel 1982). Maria Kuzańska-Obrączkowa provides an extensive list of titles on the subject.

9.1 A New Human in a New Housing Estate: From the Culture of Learning to the Culture of Education

The estate's Workers' Friends of Children Association (RTPD) established the first Polish secular coeducational school with a distinct ideological, moral, social and, I dare say, political curriculum. Here is what Aleksander Landy, a paediatrician, had to say about the school's experimental character: 'our school does what it can to bring up the children in accordance with the ideals of the working class, to prepare them in both morals and physical abilities for reconstructing the political system in line with socialist principles'.[3] The WHC preschool (Figs. 9.1 and 9.2), school and gymnasium all were strictly secular (for that reason the authorities initially refused to allow running a coeducational gymnasium in the estate). Religion was replaced by lessons in morality, active engagement on behalf of others, instilling principles of pacifism and a creative attitude. Just as with any other issue that the estate faced, Stanisław Ossowski spoke out on secular education. In 1936 'Lewy Tor' (Left Track) published an article in which he wrote: 'We want a secular school, (…) which will not turn out anxious intellectuals who are too afraid to thoroughly analyse reality, but rather individuals with a brave, honest attitude towards life and social issues (…). We want a school that will not serve the egoistic interests of the bourgeoisie (…)' (Ossowski 1936, p. 114).

Żoliborz's founders adopted a strict anti-church stance. Ossowski and others, including Stanisława Filipczak, saw the Catholic Church as a source of inequality and social passiveness, the latter of which reinforced and reproduced the social and political status of workers and women. On the other hand, their approach exhibited the broad social and political objectives of education. Those were also evident in the statements of the estate's residents published by 'Życie WSM'. After the closing of the school year in July of 1932, RTPD held the annual meeting of the association. Despite its vigorous activity and overwhelming support from the community,[4] only 27 people turned out. An 'observer' of that event wrote a short article titled *A Sad Case* [W przykrej sprawie] which was published by 'Życie WSM'. In the article, the observer outlined the causes for the abysmal attendance and lack of engagement in RTPD's activities. He remarked that both are caused by 'the close proximity of the presbytery and church, and the stately house of the Sisters of the Resurrection. Those institutions are one of the sources of the mistrust that women citizens–residents have in the RTPD school and, in consequence, in the Association as well. As we all know, the RTPD school does not pollute the minds of children with unscientific fairy tales, nor does it warp their spirit with hypocrisy and prudishness (false shame). That is why the clergy claims that the school is immoral and spreads bolshevist

---

[3]An interview with Aleksander Landy] titled *Czy szkoła nasza jest eksperymentalna?* [Is Our School Experimental?] (Landy 1934). Aleksander Landy, a paediatrician, social activist and pedagogue, is the epitome of a pre-war Polish intellectual.

[4]In 1931, RTPD ran a preschool—for 40 children, school with 5 departments—for approximately 100 children, doctor's clinic, summer vacation programme, after-school club, music school, puppet theatre and nutrition programmes for children (Warszawska Spółdzielnia Mieszkaniowa 1931).

Fig. 9.1 Kindergarten of the Workers' Society of Children's Friends (RTPD), Elective classes, Report on the WHC activities for 1931. 3546, p. 209

propaganda. Alas, this competitive envy is not the only source of mistrust. A long-standing misconception about the methods of teaching children has also made its mark. There are those people who remember how they were taught at school and wish that their children be taught using the same methods. They know perfectly well that there still are plenty of schools where children are punished by being told to stand in the corner, hold out their hands to be hit with a ruler, are assigned homework and asked to memorise long passages from books. Children often do not see such institutions as schools but as prisons, and they treat learning as hard labour. They are forced to behave and submit, nothing more. The RTPD school, on the other hand, has made it its mission to raise children in a cheerful, morally and physically healthy atmosphere of group work. Such an approach requires using novel methods, and that proves too hard to comprehend for some. (...) The lack of interest has to be overcome.

Fig. 9.2 Kindergarten of the Workers' Society of Children's Friends (RTPD), Clay modeling classes for kids, Report on the WHC activities for 1931. 3497, p. 209

The estate's residents, working-class people, should be informed in detail that: (1) the RTPD school and preschool belong to them and are meant to raise their children not as dimwits with warped minds who feel ashamed of their roots, but as healthy youngsters who will have a purpose in life and will be well prepared to pursue it, (2) that in times of overwhelming crisis, which hit the working people the hardest, RTPD helps workers' children, and given that the number of children in need may grow, working-class people should support the RTPD if they know what is good for them, (…), (3) that RTPD is not a charity whose sole task is to make wealthy ladies feel good about themselves by feeding the poor. The Association is a workers' self-help institution' (Życie WSM 1932).

The school, which was designed by Warsaw's radical intellectuals, was truly special at the time. Jan Szymański claimed that student–teacher relationships were based on trust and directness, and students addressed teachers by their first names.

Novel teaching methods were used (Montessori, Dalton, and Dewey). Primary school students had only 2–3 h of lessons daily, and during the rest of the school day, they took walks, played, cared for the school's animals, worked in small workshops and in the school garden (children often helped cultivate the estate garden as well) (Mazur 1993, p. 147).

When writing about the educational policy of the WHC in 1982, Maciej Demel emphasised the social sensitivity that had been instilled in students. The school's 'secularity, coeducation and high teaching standards drew in children from outside of the WHC and from outside of Żoliborz as well. It truly had to have a magnetic reputation given that the gymnasium was not recognised by the state authorities, meaning that graduating students had to take an additional external examination before they received an official diploma' (Demel 1982, p. 84). Ossowski wanted the Żoliborz estate and the youth educated within it to create the culture of a future society 'in which not only the economically privileged elites could participate'. It would not be a proletarian culture, however, as, according to Ossowski, in the future 'there will no longer be a proletariat as we know it today' (Ossowski 1970, p. 143). He assumed that it would be possible to create structural conditions that would counteract the exploitation of workers, allowing them to manifest their subjectivity, and make all people equal.

And Żoliborz self-help institutions did consider education an important factor, and the vision of a new human was made possible by building a new estate, i.e. through building democratic institutions. By setting ambitious political and social goals for the educational system, Żoliborz's founders and activists wanted the 'new human' to possess more than natural knowledge and consider the wide spectrum of 'social problems' to be a part of his or her personal concerns.[5] It was due to the scholarship and material student aid aimed at working-class children that the proletariat was disappearing. By not charging fees, the Workers' Friends of Children Association (RTPD) was making an effort to convince workers to the idea of equal access to gymnasiums. The Association wanted to break the pattern of workers almost automatically and somewhat naturally sending their children to vocational schools, thus effectively preventing them from receiving further education. Paradoxically, only one in three children was paying tuition in 1937. Forming an altruistic attitude started at the very beginning, encouraging youths to take active part in the estate's life, cooperate with the local government and Cooperative Inn, write for the 'Życie Młodych' ('Youth Life') magazine published by 'Życie WSM' and be active in many other institutional forms supplementing the activity of estate's residents.

There was little difference between the manner in which adults and children were educated. The youngest residents were actively engaged in the functioning of estate institutions; they were constantly 'introduced to politics' and were perfectly aware of the social and economic environment in which they learned. They witnessed collective performances and were involved in the practice of norms. That is why

[5] I consciously referenced Alfred Schütz's 'enlightened citizen' and Mills' sociological imagination because both have much in common with some aspects of humanistic knowledge which has the ability to transform human practices.

children strongly identified with the activity of 'Glass Houses' and even went as far as to create an ideologically heavy anthem. If we examine the lyrics, we will come to realise that children knew what the cooperative's role was and were not ready to accept the political order that they had been born into. The lyrics also prove that WHC children understood the purpose of social housing.

The Glass Anthem
Damp basements
smoky, dark, stuffy attics
gloomy backhouses
the bane of years gone, years terrible.
Rising up to the sky
are the Glass Houses of our dreams!
Warsaw's people join together
fighting the despair of homelessness
flying the flag of the WHC... (Życie WSM 1939).

In September 1938, 'Życie WSM' published a new issue of *Nowy człowiek w nowym osiedlu* [A New Human in a New Housing Estate] which was devoted solely to matters of education. The issue included statements from Żoliborz's foremost education experts, and each of them could be treated as a political manifesto or programme. They tied into the pedagogical and urban-hygienic discourse that was going on in Europe at the time.

In the foundational article, Henryk Dembiński, who at the time was a journalist for 'Życie WSM' magazine under the pen name Henryk Kora,[6] demanded a *Socialised Manufacturing School* [*O szkołę uspołecznioną i produkcyjną*]. In the article, he identified the most important areas for shaping a new civic culture and new educational challenges. His views, typical for many activists involved in WHC education, can be summarised in several points (the author himself divided them into subsections). The chapter titled *Prawo dżungli a wychowanie* [The Law of the Jungle and the Upbringing] was founded on an outright objection to the capitalist reality perpetuating social inequalities. According to Dembiński, the idea of socialised school and capitalist society is simply irreconcilable. Both of the above-mentioned concepts, considered individualistic and selfish, create conditions for competition, 'desocialising all areas of life and isolating the school from the surrounding reality. Without being immersed in the social environment, the school cannot educate at all, but when it opens up to the influence of a socialised environment, then the school will educate egoists, but will not be able to give them a social upbringing' (Dembiński 1938, p. 170). For Dembiński, the systemic context and structural conditions were therefore a fundamental educational issue. He does not force the reader to choose between capitalism and socialism. The socialised school is also a solution *for those who are not socialists*. Here, Dembiński refers to Bogdan Suchodolski, for whom (in Dembiński's interpretation) love for human beings can be instilled in people '« only when intransigent demands are made in order to rebuild the social and economic

---

[6]Henryk Dembiński's biographical note was provided by Szymański (1989, pp. 98–99).

system »' (Dembiński 1938, p. 170).[7] The views of Stanisław Witkiewicz for whom, in turn, social justice was an incarnation of 'love connecting humanity' also support this argument. Thus, using the Poland's intellectual leaders, who were far removed from socialism, Dembiński determined tasks pertaining not only to the domain of politics but also pedagogy. But can we be certain that his ambitions were not political? He wanted 'young people to be able to experience together with the socially conscious working class its everyday burning problems, its aspirations towards establishing a new social order, its assessment and understanding of the current political system, its heroic historical traditions and its methods of collective work and democratic discipline'. For Dembiński, it was a form of socially engaged pedagogy, based on disagreement with 'oppression of the nation, social class or an individual and the harm inflicted upon the nation, social class or individual'. Here starts Gert Biesta's politicalness. The focus is on the fundamental principle of equality, a sense of solidarity and responsibility, an important moment for practicing democracy.

Although it can be assumed that the programme is quite general, it also provides methods for achieving these social goals. *Do not advocate, but rather lead by example*—this is yet another subsection of Żoliborz's pedagogical programme. 'Socialising our youth depends on the socialisation of our housing estates and ourselves'—writes the author. The aim was to foster authentic, vivid, everyday practice of cooperation, establish 'collective culture', concern for the public good and build a democratic experience. The aim was to foster collective rituals and common celebration, social gestures and methods of upbringing 'which would give vent to creative instincts that would strengthen fraternity ties between humans through everyday situations, in the course of socially organised work and efforts' (Dembiński 1938, p. 171). Children in the housing estate could learn all that only by implementing the principles of coexistence and getting involved in the organisation of housing, self-government and cooperative institutions. As we know, there was no shortage of such institutions within the Żoliborz housing estate, and the pedagogues actively sought to arrange space for the public activities of the WHC youth. These historical experiences prove that Gert Biesta was right when he wrote that 'the cultural theory of learning is a theory that sees learning as practical, embodied and social. Seeing learning as practical means that it is understood as something that is done—that is, intrinsically connected to our actions and activities—and thus not simply as something that happens in the human mind, but as something that is thoroughly embodied' (Biesta 2011c, p. 203).

Although I am aware of the dangers arising from interpreting old, interwar visions of a creative school using contemporary pedagogical reflections, such an approach nevertheless allows us to better understand these experiences. It also highlights the point that practicing democracy from an early age and building civic structures is an indispensable condition for modernity.

Participation in productive work—this is the third factor of social education and at the same time the last point of Żoliborz's educational programme. Dembiński

[7]Dembiński probably refers here to Suchodolski's idea expressed in *Uspołecznienie kultury* [The Socialisation of Culture], which was published in 1938.

formulates in it the general principles that Stanisław Żemis had successfully used for many years, first as a teacher and then as the chairman of RTPD [Workers' Friends of Children Association]. School children actively participated in the life of the housing estate, often putting in work at the Warsaw Housing Cooperative, Social Construction Firm, Cooperative Inn, Warsaw Bookstore Cooperative, First Cooperative Laundry, and were interested in the principles of organising social life, observing the work of the 'Glass Houses' Association and the Tenant Council. They participated in practical classes, helping the kitchen staff in preparing breakfast and cooking dinner, and helped out at the preschool, cooperative store and school office. In the school workshops, they repaired school equipment and teaching aids and, during spring, helped cultivate the school garden. Stanisław Żemis' ambition was to send higher grade students to industrial plant internships. 'Social problems, problems with the organisation of work and the entire production process were more clearly expressed there. Unfortunately, under those conditions, we did not have access to factories'—recalled Żemis (1959). In Żoliborz school, as Dembiński described it, the necessity of political changes was discussed, and teachers were trying to get students emotionally engaged in the workers' struggle for a decent life. 'In our school, everyone discussed life issues and the workers' struggle with capitalism. I remember—Żemis recalled—when the issue of flooding two unprofitable coal mines in Silesia was raised and thus the threat of losing jobs by several hundred mining families became imminent, and when the miners staged a sit-in protest in order to defend these mines, our students, having heard about the protest, started spontaneously collecting donations for the strikers and managed to raise about 130 zlotys' (Żemis 1959).

The programme-related issues concerning productive work carried out according to the 'workshop' methods of operation, which were broadly outlined by Dembiński, are not surprising. They are particularly interesting for me because of the concept of creative work and its performative and corporeal character, the combination of body and mind, hand–head coordination. Dembiński emphasised that the WHC school had abolished the opposition between intellectual and production work. 'Whoever performs an intellectual work or studies at a school does not perform manual work and does not participate in processes which create a material basis for the existence of society' (Dembiński 1938, p. 172)—these popular beliefs stand in contrast with the view that 'collective production work nurtures creative instincts, quenches the instincts of greed, desire of acquiring and possessing, teaches a man to master his inner chaos, gain perseverance and social discipline, teaches people to interact with each other, mutual interdependence and mutual respect. All states of consciousness, all processes of understanding and thinking are the result of action and strive to reveal themselves in action' (Dembiński 1938, p. 172).

In these reflections, very expressively articulated by Dembiński, one can see the performative nature of the entire Żoliborz project. Here, Sennett's analysis of craftsmanship and workshop is manifested, which, unlike most contemporary proposals for the conceptualisation of immaterial labour, does not favour in any way the field of symbolic or communication activities.[8] It can even be said that it is a nostalgic incentive to renew artisanship, which is simple, efficient and reliable, carried out in a workshop, based on the economic use of resources and care for the material. While modern researchers of immaterial labour approach this topic from the point of view of the result (effect) of work, criticising commoditisation and capitalisation, Sennett focuses on the work process, on the artful work which is the only chance to deal with the alienating mechanisms of every economic system (not only capitalism). Sennett's proposal, however, is not an affirmation of the individual, lonely work of an expert, but rather a positive perspective in which he formulates two main principles: cooperation (practicing cooperation understood as a goal and not as a tool—social capital taken over by market logic), and coordination of conceptual and physical work with efficient coordination of strong and weak elements (e.g. weak hand–strong foot), but also matching various social temperaments and skills. That forms the basis for building an intersubjective world in which everyone is a causative agent equally participating in resisting the proliferation of alienating mechanisms.

If one analyses the practice of democracy in Żoliborz by means of categories proposed by Sennett, then its extremely important potential becomes apparent, which was underestimated in contemporary debates about the common good and metropolis. It is also worth reminding that both contemporary sociologists and interwar social activists (including pedagogues) were inspired by John Dewey and the pragmatic conviction that practicing democracy and cooperation makes human work an ever better experience.[9] Dewey proposes that education be understood as a process that involves the individual as a whole (including the body and mind, habits

[8]Contemporary humanities which conceptualise the category of work usually relate it to three issues: work as employment, work as a basic human activity (the anthropological approach), and finally as a political subject. As part of the critique of cognitive capitalism, as well as within the interesting arrangements of representatives of the Italian school of autonomous Marxism, and especially in the theoretical ideas of Antonio Negri and Michael Hardt regarding biopolitics, work is understood both as the object of a new form of exploitation and the subject of new political and social movements. Initially, their concept of immaterial labour basically did not take into account physical work and its material effects. It was only after a wave of criticism that it was extended to biopolitical labour as well. The same is true of the concept of Maurizio Lazzarato's immaterial labour; cf. Lazzarato (2010), *Immaterial Labour–Robotnicy opuszczają miejsca pracy* [Workers Leaving the Workplace]; Catalogue of the exhibition titled 'Workers Leaving the Workplace' held at Muzeum Sztuki in Łódź in 2010. Although it is an interesting theoretical-political-emancipatory project, it does not refer to the entire humankind, as its author would wish, but only to a specific group of people involved in the so-called creative industry or cognitive capitalism. It would therefore refer to the most-privileged employees.

[9]In Poland, John Dewey's philosophy became known very shortly after being published, thanks to the numerous translations at that time: *Moral Principles in Education* were published in 1921. A year later, *My Pedagogic Creed* translated by Józef Pieter and *The School and Society* translated by Róża Czaplińska-Mutermilchowa (published by Książnica ATLAS) were released, and in 1922, *The Child and the Curriculum* was published, translated by Helena Błeszyńska, with an introduction (written in 1913) by Édouard Claparède. In 1934, *How We Think* was published, with a short introduction by Z. Mysłakowski. Before the war, *Philosophy and Civilisation* was published, translated by Stefan Furman. It is also worth noting that the City School of Work, also known as the Experimental School or the Empirical School, was founded in Łódź in 1923. Romuald Petrykowski (one of the school's headmasters) described his experience in organising this school in the years 1923–1929 (see

and reflections on them) in interacting with the community. Dewey perceives the mind not as something separate, but as a function of human action. As a result, habits and social ways of being are shaped by the body. Learning takes place on a practical and physical level. In addition, thanks to carrying out constant 'transactions' with society, it is also a social phenomenon (Dewey 2008). Therefore, developing cooperation mechanisms and reaching consensus always require establishing coordination with other people. That is necessary because interactions with society create conflicts and various tensions when we encounter something that resists us.

Sennett can therefore be part of the contemporary debate within the framework of the so-called anthropo-sociology.[10] Like others, he is concerned about the 'culture of the new capitalism' (Sennett 2006) and what it has done to human work. It shows 'the personal consequences of working in the new capitalism', leading to 'the corrosion of character' (Sennett 1998). Testing Sennett's concept using Żoliborz's historical experiences turns out to be fruitful and proves that the analysis of the Warsaw Housing Cooperative still provides many important instructions regarding both past and ongoing struggles for the recovery of urban space through civic education. The Żoliborz story thus interpreted helps one notice a 'methodology' that allows to rethink the political and formative potential of organising city dwellers in both creative and democratic terms.

Many interwar pedagogues emphasised the difference between learning and education. Wacław Schayer, the director of the RTPD [Workers' Friends of Children Association] school, proposed a concept of a school that must defend itself against the 'invasion of bourgeois and petty bourgeois culture'. This dominance is opposed not so much by the set of principles that a human being adheres to, but by their behaviour and methods of operation. The ideological and political role of the school and great attachment to the operational, efficient and cooperative character are conspicuous and cannot be underestimated. The shaping of a moral stance takes place not in the way of inculcating certain rules, but also through rejecting traditionally maintained principles by the force of habit and unconscious routine. That is why the proposed rules of behaviour should 'not overtake the real needs and possibilities of the milieu' and 'absolutely avoid verbalism, that is not to introduce into our plans any such rules and demands for which we lack practical executive recommendations' (Schayer 1938, p. 173). The school concept proposed by Schayer justifies the demand for abandoning cultures of learning and replacing them with educational cultures 'Unlike 'learning'—which is in a sense a rather empty or neutral term—education always entails content, purpose and relationships. This implies that it is important to make a distinction between learning cultures and educational cultures. The latter are learning cultures framed by purposes. While in many cases it can be said that it is

---

Petrykowski 1963). About Dewey's reception in Polish pedagogy, cf. Sobczak (1979), Radziewicz (1989).

[10] 'Anthropo-sociology', a concept proposed by Lazzarato, is a field of reflection that analyses non-material work outside of business theories and examines it as 'a radical synergy of the energy generation' (Lazzarato 2010, p. 90).

the responsibility of educators to articulate and justify such purposes, this does not mean that students are necessarily excluded from doing so' (Biesta 2011c, p. 207).

Schayer covers *The Cooperative Educational Plan of 'Bolesław Limanowski schools'* in great depth. The plan itself may be considered as an attempt to formulate goals. It is an expression of both the needs and possibilities of the WHC housing estate which adopted shaping the forms of collective life 'by eliminating individual and collective conflicts of interests' as its main goal. Therefore, the school is subordinated to the idea of a social housing estate and is supposed to implement not only some pedagogical ideal, but rather to raise the future inhabitant of the housing estate (let me once again mention that Ossowska set a similar goal when she wrote *Wzór demokraty* [The Ideal Democrat]). Specific conditions, a cooperative environment in which the rules of coexistence, enacted and respected, and not just declared, become a living experience for students and are subsequently reproduced and implemented in practice. In this sense, we can quote Erica Fischer-Lichte, who stated that Żoliborz educational performances are unreferential, do not refer to something that existed earlier, to some substance or entity that should manifest itself in these acts (e.g. the model of a 'good citizen') (Fischer-Lichte 2008). The meaning of performative acts is generated as they happen. Schayer knows that 'the shaping of a moral stance takes place not in the way of inculcating certain rules, but through the participation in specific life situations'. Hence, the role of the cooperative itself and the life of the housing estate are inseparable from the pedagogical undertaking. 'This dynamic character of the cooperative housing estate gives it particularly prized educational values, thanks to which *cooperativism in present conditions is probably the only basic value of educational work* [highlighted in the original text]. The process of shaping personality, forming the new human, is conditioned by the simultaneous transformation of the individual and the environment, and thus the development of personality in harmony with the social environment cannot take place in an ossified, static environment. But it finds extremely favourable conditions in a cooperative environment. (…) If the Warsaw Housing Cooperative principles had no equivalents in the individual lives of the people constituting our community, if these people were not able to rise to their level in their private life, then these principles should be considered as a doctrine detached from reality and the whole experiment should be considered as a failure' (Schayer 1938, p. 173).

When summarising the achievements of the WHC, Zanna Kormanowa emphasises the educational nature of the project: educational for both adults and children. The secularisation of education, elimination of the influence of the clergy and 'fideism' and the introduction of school laboratories and workshops, combining intellectual and cognitive labour with physical work in numerous workshops, introducing didactic methods based on debates, focusing on physical exercise, sport and tourism, instating principles of self-management of students and democratic rules in school, while at the same time strengthening self-discipline and internal balance—all this proves the experimental and pioneering character of education. The WHC schools proved to be 'an institution operating in a capitalist environment, charged with nationalism and institutional religion, an institution developed in isolation and nearly laboratory like conditions' (Kormanowa 1970, p. 81). It is worth emphasising and reflecting upon this insularity and social isolation. One can have the impression that children educated in a WHC school should be open, confrontational pupils who would not be afraid to express his/her views, who are cheerful, ready for various cognitive works and participation in 'collective life'.

Yet, the analysis of the available materials and school memories reveals that the WHC's 'new human in a new housing estate' functions perfectly well within the cultural and, above all axiotic, housing system, benefiting from the institutions offered to him, making the best use of the WHC city (is a *plug-in* citizen, as Krzysztof Nawratek would say). The conviction that all of this provides the best possible education, however, makes the observer of Żoliborz education suspect that the school brings up self-satisfied, self-centred and uncompromising citizens, focused on their own happiness, who—looking down on people with a sense of superiority and paternalistic indulgence—might supply new members to the individualist and capitalist society in the future. This suspicion was raised by one of the tenants who published her fear-charged memoirs in the May 1939 issue of the 'Życie WSM' magazine. 'The people who praise the great social and political awareness of our school's children were amazed at their lack of socialisation in everyday life. Children do not have a deeper sense of obligation towards mutual help, and they are not adequately integrated and trained to perform work involving joint effort. I was told that on collective trips, for example, everyone resisted carrying backpacks with common food, and nobody offered to help their weaker colleagues. They accuse our children of having an undemocratic attitude towards people who work physically—especially to those who are involved in so-called personal services (kitchen, cleaning). Children like to be served and show no respect for physical labour, and tend to offer no help in it. Children's attitude towards the school itself, its institutions and people devoting their time and strengths to building it, can often be summarised by the sentence 'we deserve it.' Their demands are disproportionate to their duties. To sum up—the general attitude of children from our school is the attitude of spoiled children from upstart families, children who disregard the tender care they are surrounded with, children who despise poverty and its effects—cheap clothes, uneducated behaviour, etc. I do not mean to say that the above-mentioned attitudes result from our school's educational model. They may result from the children being isolated in an environment secluded from the outside world, without being sufficiently aware of the environment's distinctness and the efforts that were required to create such an environment. Should everything I wrote be correct, the school's task for the future should be to devote more attention to solving this problem. If these opinions are the result of isolated incidents among children—they should be submitted for the judgment of the children themselves. (…) Also, it is not my task to suggest what methods the school should use to fight the above-mentioned symptoms. I just think that if there is at least some reason behind my argument, the WHC management should pay close attention to it' ('Lokatorka WSM' 1939, pp. 121–122).

Besides, there were many critical remarks and parental concerns expressed about the school model and its educational methods. They were expressed mainly by the parents. The 'Życie WSM' magazine published many of these distrustful opinions, but they were less common than optimistic letters on the matter. 'Żoliborz pedagogues' believed that avoiding compulsion, not threatening pupils with giving them bad grades, notes or punishment was aimed at developing their initiative, independence and internal discipline. Perseverance, order and efficient organisation in action resulted simply from the needs of life within the community, 'collective culture', 'socialised individualism'. The memoirs of former students that are being published currently indicate that the school prepared them for practicing democracy, and many of them were active opponents during the 'Solidarity carnival' of the 1980s.[11]

References

Bauman Z (1964) Wizje ludzkiego świata. Studia nad społeczną genezą i funkcją socjologii [Visions of a human world. Studies on the social origins and purpose of sociology. Książka i Wiedza, Warszawa

Bełkowska K (ed) (2007) Na pięknym brzegu. Żoliborz, ludzie, książki. Towarzystwo Przyjaźni Polsko-Francuskiej, Warszawa

Biesta G (2010) A new logic of emancipation: the methodology of Jacques Rancière. Educ Theory 60(1):39–59

Biesta G (2011a) From teaching citizenship to learning democracy. In: Biesta G (ed) Learning democracy in school and society. Education, lifelong learning, and the politics of citizenship. Sense Publishers, Rotterdam, Boston, and Taipei, pp 5–16

Biesta G (2011b) Towards the learning democracy. In: Biesta G (ed) Learning democracy in school and society: education, lifelong learning, and the politics of citizenship. Sense Publishers, Rotterdam, Boston, and Taipei, pp 71–84

Biesta G (2011c) From learning cultures to educational cultures: values and judgements in educational research and educational improvement. Int J Early Child 43(3), 203–207

Chałasińska E, Gawecka A (eds) (2009) O pięknym Brzegu. Żoliborz, ludzie, wydarzenia [About the Beautiful River Bank. Żoliborz, people and events]. Towarzystwo Przyjaźni Polsko-Francuskiej, Fundacja 'Żoliborski Fundusz Lokalny', Warszawa

Dembiński H [alias 'Kora'] (1938) O szkołę uspołeczniona i produkcyjną [The socialised and manufacturing school]. Życie WSM, September, pp 170–173

Demel M (1982) Aleksander Landy: Życie i dzieło [Aleksander Landy. His life and works]. Państwowe Wydawnictwo Naukowe, Warszawa

Dewey J (2008) Democracy and education: an introduction to the philosophy of education. Project Gutenberg. https://www.gutenberg.org/files/852/852-h/852-h.htm. Accessed 7 Sept 2018

Fischer-Lichte E (2008) The transformative power of performance: a new aesthetics (trans: Jain SI). Routledge, London, and New York. https://dr2014theatretheory.files.wordpress.com/2015/09/fischer-lichte-c1-c2.pdf. Accessed 7 Sept 2018

[11]This view is based on the memories of Żoliborz residents which, of course, would require a broader analysis that would take into account the possible mythologisation and sentimentalisation of reality (cf. Chałasińska and Gawecka 2009; Bełkowska 2007; Marczykowa 2011). The last book [Workers' Friends of Children Association School no 1 Named After Bolesław Limanowski in Żoliborz. 75 Years of Its Existence] provides a different perspective, stating the need to build a pluralistic society. One of the authors, Wiktoria Załęska-Śliwerska, writes that 'under the Third Polish Republic—unlike under the Second Polish Republic—there has not been a single secular school: there is simply no one to stand up for it, as the party of true social activists has never been reactivated, true social activists that would follow in the footsteps of PPS members about whom I found out before the war, during the occupation and immediately after the end of war' (Załęska-Śliwerska 2011).

Kormanowa Ż (1970) Postępowe tradycje inteligencji żoliborskiej 1918–1939 [Progressive traditions of Polish intelligentsia 1918–1939]. In: Anonymous (ed) Żoliborz wczoraj, dziś, jutro [The past, present and future of Żoliborz]. Książka i Wiedza, Warszawa, pp 67–101

Kuzańska-Obrączkowa M (1966) Koncepcje wychowawcze Robotniczego Towarzystwa Przyjaciół Dzieci 1919–1939. Analiza założeń i funkcjonowania placówek [Educational concepts of the Workers' Friends of Children Association 1919–1939. An analysis of school principles and functioning]. Zakład Narodowy im. Ossolińskich, Wrocław, Warszawa, and Kraków

Landy A (1934) Czy szkoła nasza jest eksperymentalna? [Is our school experimental?]. Życie WSM, November, pp 5–6

Lazzarato M (2010) Praca niematerialna [Immaterial labour] (trans: Biskupski Ł). In: Sokołowska J (ed) Robotnicy opuszczają miejsca pracy = Workers leaving the workplace [catalogue of the exhibition]. Muzeum Sztuki w Łodzi, Łódź. http://www.generation-online.org/c/fcimmateriallabour3.htm. Accessed 7 Sept 2017

'Lokatorka WSM' (1939) Dzieciom i nauczycielom do rozwagi [For children's and teachers' consideration]. Życie WSM, May, pp 121–122

Marczykowa I (ed) (2011) Żoliborska 1: Szkoła Robotniczego Towarzystwa Przyjaciół dzieci im. Bolesława Limanowskiego. 75 lat istnienia [Workers' Friends of Children Association School no 1 Named After Bolesław Limanowski in Żoliborz. 75 years of its existence]. Wydawnictwo Hogben, Szczecin

Mazur E (1993) Warszawska Spółdzielnia Mieszkaniowa 1921–1939. Materialne warunki bytu robotników i inteligencji [Warsaw Housing Cooperative 1921–1939. The living conditions of workers and intelligentsia]. IAiE PAN, Warszawa

Mills C (1959) The sociological imagination. Oxford University Press, New York

Ossowski S (1936) O świecką szkołę [The fight for secular schools]. Lewy Tor [Left track] 9–10; reprint In: Ossowski S (1970) Dzieła, vol 6. Państwowe Wydawnictwo Naukowe, Warszawa, pp 110–114

Ossowski S (1970) Kultura robotnicza [Workers' culture]. In: Ossowski S (ed) Dzieła [Collected works], vol 6. Państwowe Wydawnictwo Naukowe, Warszawa, pp 139–143

Petrykowski R (1963) Szkoła eksperymentalna w Łodzi 1923–1929 [The experimental school in Łódź in the years 1923–1929]. Państwowe Zakłady Wydawnictw Szkolnych, Warszawa

Radziewicz J (1989) Szkoła uspołeczniona i uspołeczniająca [The socialised and socialising school]. Zespół Oświaty Niezależnej, Warszawa, Wydawnictwo Społeczne KOS

Rancière J (1999) Disagreement: politics and philosophy (trans: Rose J). University of Minnesota Press, Minneapolis and London. http://abahlali.org/files/Disagreement%20Politics%20and%20Philosophy.pdf. Accessed 7 Sept 2018

Rancière J (2004) The politics of aesthetics: the distribution of the sensible (trans: Rockhill G). Bloomsbury Academic, London and New York. https://pl.scribd.com/document/360491804/RANCIERE-Jacques-The-Politics-of-Aesthetics-The-Distribution-of-the-Sensible-pdf. Accessed 7 Sept 2018

Rancière J (2009) The emancipated spectator (trans: Elliott G). Verso, London and New York. https://imagemdissenso.files.wordpress.com/2010/07/the-emancipated-spectator-2009.pdf. Accessed 7 Sept 2018

Rancière J (2010) On ignorant schoolmasters. In: Bingham C, Biesta G, Rancière J (eds) Education, truth, emancipation. Continuum, London and New York, pp 1–24

Schayer W (1938) Spółdzielczy plan wychowawczy [The cooperative educational plan]. Życie WSM, September, pp 173–175

Sennett R (1998) The corrosion of character: the personal consequences of work in the new capitalism. W. W. Norton & Company, New York

Sennett R (2006) The culture of the new capitalism. Yale University Press, New Haven and London. http://asounder.org/resources/sennett_culture.pdf. Accessed 7 Sept. 2018

Sobczak J (1979) Recepcja idei 'Nowego Wychowania' w polskiej pedagogice okresu miedzy wojnami [The reception of the idea of 'New Education' in Polish pedagogy in the interwar period], vol 2. Wydawnictwo WSP, Bydgoszcz

Szymański JA (1989) Warszawska Spółdzielnia Mieszkaniowa 1921–1970. Zarys dziejów [Warsaw Housing Cooperative 1921–1970. History outline]. Wydawnictwo Spółdzielcze, Warszawa
Warszawska Spółdzielnia Mieszkaniowa [Warsaw Housing Cooperative] (1931) Sprawozdanie [1931 WHC report], Warszawa
Załęska-Śliwerska W (2011) Szkoła imienia Bolesława Limanowskiego. In: Marczykowa I (ed) Żoliborska 1: Szkoła Robotniczego Towarzystwa Przyjaciół dzieci im. Bolesława Limanowskiego: 75 lat istnienia [Workers' Friends of Children Association School no 1 Named After Bolesław Limanowski in Żoliborz. 75 years of its existence]. Wydawnictwo Hogben, Szczecin, pp 63–72
Żemis S (1959) Praktyki rzemieślnicze w szkole RTPD na Żoliborzu (wspomnienie) [Craftsman internship at RTPD school in Żoliborz (Memoirs)]. Przyjaciel Dziecka 6–7:23
Ziółkowski J (1965) O socjologicznej wyobraźni urbanisty [On the sociological imagination of the urban planner]. In: Ziółkowski J, Urbanizacja, miasto, osiedle. Studia socjologiczne [Urbanisation, city, estate]. Państwowe Wydawnictwo Naukowe, Warszawa, pp 227–251
Życie WSM (1932) W przykrej sprawie [A sad case]. Życie WSM, August, pp 3–4
Życie WSM (1939) Hymn 'Szklaniaków' [The Glass Anthem]. Życie WSM, May, p138

Chapter 10
State of Emergency and Everyday Life in Żoliborz

Abstract World War II, the Germans invasion of Poland and the occupation of Warsaw was a time labelled by anthropologists as a liminal situation, and philosophers (e.g. Giorgio Agamben) as a state of exception/emergency. The existing institutions and standards collapse. Social life takes on new forms. So I ask the following question: was the Warsaw Housing Cooperative, which developed its own social life institutions (legal, cultural, and economic)—alternative to the state ones, in such a liminal state, in the state of crisis? The cooperative organisational structures tested before the war proved to be efficient and effective under occupation, and the residents proved that 'Żoliborz socialism' was not only an idea, but a social practice of siding with those who suffer. The briefly described post-war period ends this narrative, because the WHC was taken over by the state and centralised power. From a residents' bottom-up, self-governing initiative, it evolved into a state-controlled housing enterprise. There was nothing left from the idea of cooperativism.

Keywords World War II · Institutional order · Saving Jews · Secret learning · Underground resistance · Centralisation

10.1 First Days of War and Everyday Life in the Żoliborz District

In Warsaw, the last days of August 1939 had been marked by the feeling of impending war. Although there was certain nervousness, people generally believed in the strength of the Polish army and the Polish state, as well as in the power of international alliances—signed between Poland and the UK, and Poland and France. Undoubtedly, however, the Polish authorities underestimated both the power of Nazism and Germany's overwhelming military superiority over Poland.

People were hurriedly returning back home from their summer holidays. Holiday leaves were officially cancelled, men were drafted for military service, and those who were not drafted waited for general mobilisation. In March 1939, an air defence system was put in place in Warsaw, and in June the City Provisioning Departments were set up to establish public food reserves.

M. Matysek-Imielińska, *Warsaw Housing Cooperative*, The Urban Book Series,
https://doi.org/10.1007/978-3-030-23077-7_10

In squares and public places, people were digging ditches and trenches as part of the anti-aircraft defence system. Windows were covered with paper in order to strengthen the glass panes from the effects of the exploding aerial bombs. There were long queues everywhere—in banks, as people rushed to withdraw their deposits, and in shops, as people were gathering food supplies in case of war. Numerous social and political organisations joined the preparatory action.

The experience of September 1939 in the Żoliborz district was similar to the rest of Warsaw. The activity of the civilian population increased: people were undertaking various self-help and self-defence activities. The Association of Żoliborz Residents [*Stowarzyszenie Żoliborzan*] established the Warsaw Social Self-Help Committee [*Stołeczny Komitet Samopomocy Społecznej (SKSS)*], and on the initiative of the Polish Socialist Party (PPS), the District Workers' Social Assistance Committee [*Okręgowy Robotniczy Komitet Pomocy Społecznej (ORKP)*] was established, with its seat at 10, Krasińskiego Street. A vigilante group [*Straż Obywatelska*] was also established, having its seat in the premises of the Warsaw Housing Cooperative at 10, Krasińskiego Street and 8, Niegolewskiego Street. Before the German aggression, the WHC established the Emergency Defense Commission [*Komisja Pogotowia Obronnego*]. In order to coordinate the provisions, the SKSS was entrusted with running the eatery and took over the Cooperative Inn canteen where, as of 5 December 1939, hot soup was served to Żoliborz residents.

During the siege of Warsaw in 1939, the mayor of Warsaw, Stefan Starzyński, appointed Stefan Zbrożyna as a district delegate for the Northern Warsaw District (which included Żoliborz). Zbrożyna was well known in the Żoliborz community as a socialist activist and committed philanthropist, and he enjoyed great authority and trust among Żoliborz residents. Żoliborz youngsters eagerly started working for self-help organisations. They were mobilised to the military training battalion. In addition to civil activities, the Polish Socialist Party (PPS) created the Workers' Battalions for the Defence of Warsaw [*Bataliony Robotniczej Obrony Warszawy*].

Numerous evacuations of civilians, air raids and electricity blackouts caused disorganisation of the city, exacerbated by the hysterical appeal of Colonel Roman Umiastowski calling on all men capable of bearing arms to leave the capital in order to create a new line of defence east of the Vistula. Those who responded to the appeal and had to leave Warsaw were the cooperative management members (Tołwiński, Szwalbe, Nowicki) and newly appointed members of the Emergency Defense Commission. Therefore, women briefly took control of the WHC estates during the first days of the war (including Ewa Wudzka—head of the Horticultural Centre, Janina Święcicka—activist of the Active Cooperatists' Club, Jadwiga Okorska—chief accountant).

On 28 September 1939, the Command of the Defence of Warsaw signed a capitulation treaty with Germany. Some of Warsaw's inhabitants were relieved to learn the capitulation, exhausted by continuous air raids and artillery fire. The Volunteer Workers' Brigade for the Defence of Warsaw, unwilling to accept Warsaw's capitulation and surrender, threatened with rebellion. While laying down their arms at Wilson Square, the soldiers of the Volunteer Workers' Brigade for the Defence of Warsaw organised a rally during which they openly manifested their reluctance to surrender their weapons and uniforms. 'They felt painfully disappointed and even betrayed when they finally could hold, if only for two days, the adequate weapons in their hands' (Poterański 1970, p. 106).

Neither the war nor the urban chaos managed to destabilise the estate structures and the well-established and solidified rituals of social cooperation in Żoliborz. Already in the first days of September, the meeting of the colony guardians was arranged in the WHC premises, who agreed to take care of children, reservists, the sick and infirm, if necessary. They also had to ensure that the residents themselves would keep order and cleanliness in the colonies, courtyards and streets on their own. In this state of exception/emergency, the Żoliborz multiplicity of institutions, which was obviously an advantage, had to be reorganised and coordinated. A Coordination Commission was established, and its task was, among others, transporting wounded residents of the housing estate to hospitals, taking care of children without parental care, providing residents with food, fuel, as well as controlling sanitary conditions and renovation works. This Commission also decided to replace the 'Życie WSM' magazine with a noticeboard which was also issued after the Siege of Warsaw and throughout the occupation of Poland. As of November 1939, the WHC management posted 'communiqués' on green paper in each staircase and at the entrance to each colony. Regardless of whether these were notices and orders coming from civil, military, police authorities or cooperative announcements, they would bear the WHC logo. It was forbidden at that time to place private announcements, mainly in order to make the official communiqués sufficiently legible. Thus, for example, in December 1939, a notice appeared about the possibility of subletting flats to tenants (this procedure was forbidden before the war for fear of speculative nature of sublease) but only after obtaining the consent of the cooperative management board.

The Commission's task included the coordinated protection of abandoned flats or those left open after fires to prevent looting and plunder, as well as the organisation of the WHC Court. The colony guardians, on the recommendation of the Commission and the outpatient clinic of the Workers' Friends of Children Association (RTPD), kept a register of children under the age of 14. The staff of the outpatient clinic prepared a linen tag for each registered child, with the name and address on it. The dairy kitchen was still working, although its manager, Janina Bierutowa, had to go outside of Warsaw to fetch milk from the farm, and when the occupation policy hardened, she prepared it from dried milk. The already prepared milk was distributed to individual houses by the Żoliborz youth.

Although since the capitulation of Warsaw the convening of district meetings was forbidden, in April 1940, the WHC Supervisory Board authorised the Bureau of the Main Ad Hoc Financial Assistance Committee for the Housing Estate Tenants to appoint and dismiss the main colony guardians in the WHC housing estates. In each colony, there were two guardians and one delegate responsible for food provisions. They were at the same time social (and tenement house) workers of the Warsaw Social Self-Help Committee (SKSS).

Due to the efforts of civil and military authorities, and with the huge support of social organisations throughout Warsaw, the order and life in the city's individual districts were somehow restored. In Żoliborz, there was also a spirit of readiness to restore order, clear away the debris and return to the activities so far pursued by the existing institutions. However, in the summer of 1940, the occupation authorities restricted the operations of the 'Glass Houses' Association and the Workers' Friends of Children Association (RTPD). In their place, the Social Welfare Department of the WHC was established. Its actual activity under the German occupation consisted in efficient crisis management within the housing estate. People gathered information about who was arrested by the Gestapo and when, and who must go into hiding. They probed what were the possibilities of releasing someone (buying out) from jail in individual situations, i.e. what kind of help the family of the arrested needed. The family was instructed how to prepare parcels to jail. It was not easy, because the trust towards new residents was limited in occupied Warsaw. People who did not know personally the employees of the WHC Social Welfare Department avoided them and had to search for information themselves. And here, the organised network of colony delegates and guardians, whom the WHC residents trusted, proved extremely helpful. The colony guardians always knew what had happened last night in the housing estate, who the Gestapo was coming for or whose flat was searched. In the morning, during daily briefings, they reported everything to the WHC Management Board. At that time, the collection of money and food was organised, and other help for the family of the arrested was provided.

During the state of emergency, the basic functioning rules were adopted by the Cooperative: 1. Institutions and tenants should provide help and social support as much as they can. 2. No tenant may be deprived of a flat during the war due to their poor material conditions. 3. Nobody within the WHC estate can suffer from hunger. 4. The cooperative and tenants help each other find a job or obtain social benefits within or outside the cooperative (e.g. the Warsaw Social Welfare Committee, Municipal Cooperative Care, Insurance Fund). 5. Children and adolescents should be given special care (the most important point).

In January 1941, the draft of a new substitute regulation of the Tenant Council of the residents of the WHC estates was discussed and subsequently voted. It was also planned to organise the WHC youth council. As the state and municipal institutions and power structures did not function, these housing estate institutions and structures proved to be efficient, superbly organised, and allowed for coordinated assistance activities to be undertaken as far as possible. These, in turn, concerned both important, broader defence and conspiracy activities, and those seemingly trivial, which nevertheless allowed to maintain a sense of normality in the state of exception/emergency. The cooperative helped WHC residents, anticipating acts of war and providing them with food and essentials.

In cooperation with the Warsaw Bookstore Cooperative, the WHC organised the collection of various goods desperately needed at that time. For example, the WHC provided its residents with first-aid kits. In total, 198 orders were accepted, of which 133 were completed. The rest was not completed, because the owner of a dressing plant, J. Chodkiewicz, took money and disappeared. During the war, the publishing houses usually did not operate, so the Warsaw Bookstore Cooperative (which had, before the war, two paid workers and under German occupation employed a dozen people) created an antiquarian bookshop, and Stanisław Kosiński (real name

Warhaftman, a Jew who had to go into hiding in German-occupied Poland) was one of its suppliers (and also advisors and informal 'employees'). Kosiński was an extraordinary erudite, editor of the Popular Science magazine titled 'Mathesis Polska', a mathematician by education, with an extensive knowledge of physics, astronomy, music and humanities. The Warsaw Bookstore Cooperative gave him shelter in a room above the store, where he lived through the war in hiding until the Warsaw Uprising. Unfortunately, I did not find any information about his fate after the Warsaw Uprising. Throughout his stay, he served both the Warsaw Housing Cooperative and the Warsaw Bookstore Cooperative by giving advice, offering help and taking action. He also became a vital link for initiating the social life of cooperative employees. He had a rich collection of records, and he played them, giving unforgettable concerts (Topińska, 1984, p. 116).

The Warsaw Bookstore Cooperative tried hard, during the pre-Christmas period of 1939, to make the first Christmas at war as normal as possible, so that children could celebrate festive rituals, such as making Christmas decorations or gifts together. Therefore, in the notice of 6 December 1939, WHC residents could read: 'We kindly ask all WHC residents who are involved in knitting gloves or intend to knit gloves, sweaters, or crochet napkins, etc., as well as toys and Christmas decorations, to come to the Warsaw Bookstore Cooperative store in the 1st Colony. The Warsaw Bookstore Cooperative is ready to take these products on consignment and possibly to initiate, in the future, the creation of proper worker cooperatives and expand customer market' ('Communiqué of 6 December 1939', archive material of the WHC History Chamber).

Although cooperatives did not come into being, the 'toy department' developed to considerable proportions, going far beyond the boundaries of the estate.

Władysław Baranowski, a teacher of drawing and handicrafts, managed to save the complete carpentry workshop by carrying it out from the 'Poniatowski' Middle School located in Żoliborz (neither a WHC nor RTPD institution) before the Germans occupied it. Together with the writer Igor Newerly, they started to produce wooden toys, delivering them to the store of the Warsaw Bookstore Cooperative. The shortage of toys was also felt by the pupils of the RTPD kindergarten. Therefore, Hanna Rembowska started painting herself board games and lottery boards for her pupils. Incited by her friends and the Warsaw Bookstore Cooperative Board, she handed her works to the production plant and thus she and the Warsaw Bookstore Cooperative started selling board games, thus earning money. Witold Miller who, after the war, worked as a professor at the Academy of Fine Arts in Warsaw, who had so far passionately designed puppets for the local 'Baj' theatre, and was a devoted social worker and enthusiast of Żoliborz youth, also contributed to the extension of the toy production at the Warsaw Bookstore Cooperative. He fabricated dwarfs (the size of a matchbox) from papier-mâché and fitted them in the tiny buildings made of toy blocks or toy cars. 'The production needed a lot of workforce, but was highly profitable. Newspapers for papier-mâché had to be passed through huge meat grinders, placed in different moulds, and painted according to the designer's pattern, etc. So at the end of the first settlement period, it was clear to everyone that income must be varied and that Witold Miller must earn more than the others. But Miller was irritated: 'How come? As if it wasn't enough that a human being has a talent, should he also get paid for it?" (Topińska, 1984, pp. 116–117). The production of toys continued to grow. One part of the WBC store was allocated solely for the toy production. Toys were also given away to stores located in Warsaw's downtown thanks to a cooperative trading agent.

On December 7, a notice was posted informing the residents about the possibility of working off rent arrears. Every tenant who was in arrears on rent due to job loss could work 2/3 of the overdue rent by working for the benefit of the cooperative, and the remaining part (1/3) had to be paid in cash. There were also calls on young people aged 14–18 to participate in the collection of everyday objects for the Rakowiec residents who suffered as a result of the fire which consumed their building.

It seems, however, that under wartime conditions, everyone preferred to work off their rent, rather than paying it in cash, which was at that time extremely necessary. A week after the publication of the notice, the WHC management had to correct and explain that 'in the interest of the cooperative members, the WHC management had to limit the possibility of working off the overdue rent only to the unemployed and those who lost jobs as a result of the war, and their financial situation significantly deteriorated. The WHC will organise this form of employment wherever possible and necessary, but the WHC is not able to guarantee permanent work to everyone. Thus the WHC counts on cash payments that will allow to buy fuel for the boiler house and pay for the WHC functioning' (Communiqué of 14 December 1939). Heating during this time was the most difficult. The fuel supplies prepared for the winter of 1939/1940 began to run out in January 1940, and at the end of the month, the heating was gradually switched off. Stanisław Szwalbe travelled to the coal mining region of Sosnowiec, which was already incorporated into the Third Reich at that time, in order to organise coal supplies for the WHC. They managed to survive through the winter. But because of the lack of fuel in June 1941, the laundry room and the bathhouse had to be closed.

During the following years of the war, winters became increasingly difficult. The WHC notices (e.g. from March 1942) informed that heating costs in the 1942/43 season could be covered only if the residents of the Żoliborz housing estate paid for this purpose PLN 1 per month per square metre of their flat. At the beginning of December 1942, the Tenant Council carried out temperature measurements in flats—it ranged from 0 to 8 degrees Celsius (Communiqué of 12 December 1942).

As a result of frequent bombings, residents had to constantly reglaze broken windows in their flats. The cooperative helped residents in collecting advances for window glass panes and organised their collective purchase and transport to the estate. The WHC management also encouraged residents to keep receipts, believing that after the war, the WHC would be able to claim war reparations. A glass replacement workshop was organised for over 40 residents of the housing estate, which lasted from 31 October to 9 November 1939. It was attended by attorneys, judges, teachers and workers, writers and the WHC youth. After completing the workshop, it was easier to get a job and thus avoid being deported to work in Germany. Jarosław Newerly recalls that his father—a writer—boasted about having replaced window panes in almost the entire WHC estate in Żoliborz.

The Cooperative Inn organised for its registered members food rations (potatoes, flour, barley groats.) The Cooperative Inn repeatedly reminded us about the need to maintain order and observe the WHC rules and regulations.

The WHC residents and the cooperative members, acting out of panic and uncertainty caused by the state of emergency, requested the WHC Board to withdraw housing contribution and monetise their membership share. The state of war, uncertainty as to what will happen tomorrow, and difficult financial situation forced them to look for any way to earn money.

The Cooperative Board, however, guarding its constitution and guaranteeing at the same time the transparency of property management had to explain in these short notices addressed to the estate residents the basic principles on which the Cooperative based its activities. The quotation shows, on the one hand, that the communiqués played the role of cooperative reports, and on the other hand, they illustrate the right to property which concerns the cooperative as whole and not individual members.

The representatives of the WHC management called upon the cooperative members to 'understand each other well and establish a common position. The members of the Warsaw Housing Cooperative are the owners and guardians of the Warsaw Housing Cooperative buildings and estates. They are entitled to live in this collective property on the basis of the membership share and housing contribution, which were intended to finance part of the cost of building our houses and to cover the cooperative duties. Membership shares and housing contributions (not to be confused with deposits) are part of the value of the cooperative buildings. The membership share and housing contribution can be reimbursed only when a new member—accepted in place of the leaving member—pays his/her membership share and housing contribution. (…) Immediate tenants' needs (keeping order and cleanliness in the estates, staircases, flat allocation and repairs, maintenance of laundry room and bathhouse, maintenance of water and sewage systems, housing maintenance, etc.) can only be ensured if tenants pay their fees in a timely manner. The WHC has never had any cash reserves. If anything was left from the rent paid by the tenants before the war, it was taken by the Bank Gospodarstwa Krajowego to cover the interest rates on the loan' (Communiqué of 22 December 1939).

Thus, the cooperative members explained to the residents how the proceeds from the rent were used: 40%—administration and residents' benefits, 40%—interest rates on loans for the construction, 20%—depreciation of buildings (repayment of long-term loans, major overhauls of buildings, complementary housing contribution of the Cooperative for the construction of new buildings).

The report also presents the tragic state of Żoliborz and Rakowiec estates at the beginning of the war. The destruction of the first of the estates was estimated at about PLN 150,000, and the destruction of the second one exceeded PLN 1 million. Sixteen apartments in Żoliborz had to be evacuated, and 48 required major refurbishment. One hundred and eighty-six flats and a social building in Rakowiec were burned completely, and out of the remaining 120 flats 37 needed repair (Szymański 1989, p. 207). 'Each time the repairs are postponed, apart from inflicting additional harm on the tenant who is the most affected, the destruction of the joint property is aggravated, causing the loss of the savings of all members, invested in shares and contributions', wrote the Cooperative Board (Communiqués of 22 December 1939).

The war did not deter the activists and representatives of the WHC management from continuing their activities. The houses were still constructed, though, of course, on a smaller scale; after all, people in need of a roof over their head became more and more numerous. Barbara Brukalska was asked to finish construction plans for the Colony IX. The tenants who were in arrears on rent worked on the construction of this Colony. Youngsters aged 14–20 were employed to level the land. At the end of September 1940, the flat allocation rules were laid down. Out of 76 completed flats which were put into use, 50 were allocated to the Rakowiec fire victims whose flats were burnt down during the bombing. The official flat handover was planned, and the mayor of Warsaw was invited to attend the ceremony. We do not know, however, whether this ceremony did take place. The Social Construction Firm [*Społeczne Przedsiębiorstwo Budowlane (SPB)*] also worked in Rakowiec to rebuild ruined houses. Tenants cooped up in overpopulated flats. That is why the team of SPB workers decided to volunteer and work 2–3 h a day, without remuneration, in order to finish 6 flats where fire victims could move in. It should be mentioned here that the SPB workers earned pre-war wages which, in the face of constantly rising food and fuel prices, were insufficient to live on. The hot soup served at the WHC canteen could be considered as extra payments and benefits in addition to these low wages.

In order to avoid raising rental rates, the WHC management decided to suspend the cultural activities of the 'Glass Houses' Association which, by the decision of the German occupiers, could not officially operate anyway. The WHC management referred more and more often to non-institutional forms of social and self-help activities.

The WHC lavished particular care on its employees held in captivity as well as their families. Already in December 1939, the WHC management set the amount of benefits for such families. On 6 June 1941, the WHC Management Board held a briefing for all employees, during which the board members appealed to the entire team of employees and residents for help in organising and running an eatery, based on institutional subsidies, supplies of vegetables from the Horticultural Centre and external assistance. The Board also stated that the team of employees should be treated as a whole and 'the Board demands to share with the team responsibility for the institution. The Board also pledged to avoid redundancies and, if necessary, they would transfer employees to other institutions. The Board members also stated that they would not limit their care solely to the team of employees, but would also organise public works for adolescents and other recipients of social assistance' (Szymański 1989, p. 227).

These assertions seemed to be a formality, a symbolic gesture that confirmed what the Cooperative had already been implementing as part of self-help and self-organisation activities. For example, on 16 April 1940, the WHC management decided to employ the WHC youth in the Horticultural Centre and thus not to report them to the German labour office (*Arbeits Amt*). The wages varied widely and were often irregular, but were supplemented with a one-course meal. Particular attention was also being paid to the upbringing of young people who, under wartime conditions, could lose their morale very easily and give into despair.

Zofia Topińska saw this gardening work as a source of friction and conflict among young people, which somehow had been averted. In addition to the *Ausweis*, young people received a very modest remuneration (PLN 0.10/h), which did not cover the price of rationed goods. Adult cooperative members demonstrated a pedagogical stance: young people should work altruistically for the benefit of the cooperative and the most deprived people. After all, the Horticultural Centre—which provided the community kitchen with food, while the community kitchen, in turn, organised food for the most needy—should be treated as a common good and run collectively and jointly. 'But those teenagers for whom earning was a necessity saw it as pedagogical doctrinairism and were of a different opinion. The conflict was somehow averted when, with the tacit consent of the cooperative, it was allowed to consume crops on the spot, and whoever wanted could get their own patch (8–10 m^2) to cultivate food for themselves' (Topińska 1984, p. 115).

The teenagers' gardening skills, their knowledge, dexterity as well as self-discipline turned out to be extremely helpful. The premises of the Horticultural Centre belonging to the WHC were also enlarged. The WHC extended protection and care over allotment garden estate (new waterworks were installed at the site) which was previously supervised by Edward Osóbka.

There were many party, military and scout organisations operating within the WHC premises. Among them, the extremely active Polish Defenders Command [*Komenda Obrońców Polski (KOP)*], whose headquarters were located in the flat of Henryk Gaudasiński, KOP Commander, at 16 Krasińskiego Street. The KOP published an underground periodical titled 'Polska żyje!' [*Poland is Still Alive!*] (its chief editor was Witold Hulewicz). The KOP started with the print run of 6000 copies and eventually reached 40,000 copies. It attracted a large part of the Żoliborz intelligentsia and youth. Both the propaganda and the military support were provided by the WHC activists Janina Święcicka and the guardian of the 1st Colony, Jakub Bajurski, who became, after the war, the author of many volumes of memoirs (never published) describing the life of the Żoliborz residents. Jakub Bajurski's manuscripts can be found in the archives of the WHC History Chamber.

The KOP carried out organisational, training and propaganda activities, but was also involved in stockpiling arms. To this end, a group of committed Żoliborz residents was formed, composed of, among others, young pupil of socialist youth in Żoliborz—Zbigniew Wróblewski, Zygmunt Gaudasińki (son of the KOP commander, Henryk Gaudasińki), Jakub Bajurski and Stanisław Marchewka (guardians of the WHC colonies). The weapons were stored in the first WHC Colony, and Bajurski was appointed its warehouseman. The Żoliborz residents, if they had any weapon, they could bring it to the central heating station. The weapons could be hidden in case of the searching of premises, to prevent German reprisals. The weapons were collected by the members of the KOP, and preserved and stored in the building at 10, Krasińskiego Street, and explosives were hidden in the Colony VII.

All combat, organisational and underground operations ran smoothly and efficiently in Żoliborz. 'The secret behind this efficiency and skilfulness—writes Zygmunt Gaudasiński—was the fact that this distinguished cooperative was inhabited at that time by a very high percentage of people with extensive revolutionary experience, people having progressive political views, who trusted each other, when the need arose to fight against the occupying forces. Most residents, through common activities, created a common socialist family before the war, thus getting to know each other very well. Hence their ability: to establish contacts easily and undertake underground activities' (Gaudasiński 1970, p. 231).

The political and patriotic imperative, shaped and perpetuated in the Żoliborz housing estate, took on a new meaning during the war. The socialised individualism that developed in Brukalska's architectural discourse during the occupation was filled with a new signified (*signifié*).

10.2 Performance for the Initiated—Underground Education in Poland

It was the Żoliborz branch of the Workers' Friends of Children Association (RTPD) which took care of the children under the age of 14. As of 11 December 1939, the 'Bolesław Limanowski' Primary School began to operate under wartime conditions. In January 1940, the occupier ordered the closure of all middle and high schools. The teacher of middle and high school, Romana Lubodziecka, joined later by Jerzy Kreczmar, organised secret learning[1] at the secondary school level. The entire official activity of the RTPD was suspended on 31 August 1940. However, these establishments continued their activities despite many difficulties. Attempts were made to include kindergartens in a network of municipal institutions, but the condition was to employ a priest and to include religion into the preschool education. The WHC kindergarten operated therefore more as a children's playground, and its official task was to feed the children. Thanks to that, the WHC Social Services could exercise control over it and it operated as a common room. However, the task which the kindergarten staff had to face was highly challenging and difficult. The idea was to make children feel safe, calm and balanced in the kindergarten. And under war conditions, it was not an easy task. The 'Child's Health' primary care outpatient clinic of the Workers' Friends of Children Association (RTPD), banned by the occupier, was transformed into Dr. Landy's private medical practice. Landy's activity was expanded, because in addition to medical care, the unit also organised material assistance and distributed milk and cod liver oil stored in the event of war. Landy also kept a register of breastfeeding mothers and those with surplus of breast milk, so that they could breastfeed other infants. Most importantly, however, Dr. Landy's

[1]Secret learning courses [*tajne komplety*] were illegal gatherings of young people (in students' or teachers' private apartments) in order to learn, under the supervision of educators, the pre-war middle school or high school curricula and pass the maturity exam [*egzamin dojrzałości*].

practice was attended by all those (even living outside the housing estate) who took part in saving Jewish children and sought protection and shelter for them. In order to obtain false baptismal certificates for Jewish children, a parish priest, Father Zygmunt Truszczyński, was contacted. Father Zygmunt Truszczyński issued, in the church chancery, false baptismal certificates to the names of dead Gypsy children and adults.

From the 1939/1940 school year to 1941/42, the secret middle and high school-level courses were organised in private premises of the Żoliborz district, most often in private apartments, and as of the 1942/43 school year mostly in WHC social premises. Underground courses were organised by Ossowskis and Nina Assorodobraj (who was, after the war, professor of sociology and wife of history professor, Witold Kula). The middle and high schools were supported both financially and in terms of housing by the WHC. Over 200 students attended the Żoliborz middle and high schools during the war. Students passed secondary school final exams [*matura*]: 2 people in the 1940/1941 school year; 13 people in the 1941/1942 school year; 21 people in the 1942/1943 school year; and only 10 people in the 1943/1944 school year. In total, 46 students passed the secondary school final examinations in Żoliborz during the German occupation, including 31 girls.

The employees of the 'Glass Houses' Association and the teachers from Żoliborz schools were transferred to the Main Ad Hoc Financial Assistance Committee and to the WHC Social and Educational Fund, as well as to the Warsaw Social Welfare Committee.

Secret university-level courses were extremely popular in Warsaw, and dedicated and courageous lecturers risked their lives teaching under the occupation. However, the secret university-level lectures and seminars were organised to a lesser degree in Żoliborz, even though, of course, a lot of Żoliborz youth attended the underground university, polytechnic and commercial courses.

There were few underground university-level courses organised in Żoliborz, mainly in the field of humanities, and lectures and seminars focused mainly on Polish studies, history and sociology. The reason why the underground courses were not as popular in Żoliborz as in other Warsaw's districts was: transport problems and too large distance from Warsaw's centre.

In total, in German-occupied Warsaw, 587 students received master's degrees and engineering diplomas, 33 students received doctorates, and 29 academics were qualified as assistant professors [*habilitacja*]. In total, approximately 10,000 students attended secret university-level courses. All the certificates issued under the occupation were recognised after the war, despite the fact that there was no official transcript of records. In fact, any document bearing names could be a potential list of people to be arrested by the Gestapo. 'It was a huge success of Polish teachers and professors. The underground education was well organised by the Secret Teaching Organisation [*Tajna Organizacja Nauczycielska, TON*], headed by Czesław Wycech, member of the Polish peasant's party, who was the Speaker of the Sejm after the war. Mass secret education was a phenomenon that had no equivalent in other countries occupied by Germany—and it must be mentioned here that in many western countries, the Nazis did not ban education nor abolished schools. In my opinion—recalls Krzysztof Dunin-Wąsowicz—the greatest achievement of the underground education was not related to providing knowledge but rather to maintaining the educational system, which encouraged students to observe a certain canon of moral norms and rules of conduct. The efforts of teachers who risked their lives teaching under the Nazi occupation, as well as the secret military and political organisations saved a large part of the youth from demoralisation—a great threat in the inhumane time of occupation' (Dunin-Wąsowicz 2012, p. 70).

Secretly conducted education was not only about informal, i.e. illegal transfer of specialist knowledge and expertise. It was about giving hope and enlarging the circle of common struggle and that in turn limited the feeling of dread, a desperate sense of hopelessness and absolute despair. It was about *the humanities of hope, praxis*, which would help the young generation to rebuild the post-war order. Krzysztof Dunin-Wąsowicz, whose memoirs I have mentioned here, was a child who came from a military family and lived in the so-called Żoliborz Oficerski [*Officers' Żoliborz*]. However, he remembers perfectly well the intellectual climate and social atmosphere of the so-called cooperative Żoliborz [*Żoliborz Spółdzielczy*], where secret activities, lectures and meetings were held on a regular basis.

> Already in 1940, the well-known historian and socialist activist, Adam Próchnik (PhD) gave lectures on the history of the French Revolution and the Spring of Nations in the WHC premises in Wilson Square. Maria Ossowska (associate professor at that time), in turn, gave lectures in her apartment (in the WHC Colony III) for a dozen or so students on a wide range of topics which had one common denominator: the science of morality. Maria Ossowska held these secret teaching courses for several years, and after—she joined the systemically organised and structured underground university level courses at Warsaw University' (Dunin-Wąsowicz 1984, p. 75).[2] Already at the beginning of 1940, the action plan was drawn up in Próchnik's apartment to help socialists who had formed battle groups (Wudzki 1948, p. 218). At the same time, Próchnik held secret middle school courses for Żoliborz youth, and in the 1941/1942 school year, he introduced a new subject to the curriculum: the history of social movements. The publishing house of the Secret Teacher Organisation published two brochures by Próchnik: *Kryzys kapitalizmu* [*The Crisis of* Capitalism] and *O demokracji społecznej* [*On Social Democracy*]. At that time, Stanisław Ossowski would hardly ever come to Żoliborz, because he had to go into hiding. The underground authorities of the right-wing nationalist organisations sentenced him to death in absentia for his socialist activity in the past. Besides, there were much conflict and discord within the Polish resistance movement. Dunin-Wąsowicz recalls in his memoirs: 'The scene with Prof. Ludwik Widerszal will be etched on my memory forever. He was of Jewish descent. He gave lectures on ancillary sciences of modern history and the history of diplomacy. We would often go together to attend secret courses held in his flat through Asfaltowa Street in Warsaw's Mokotów district. He was an outstanding educator, a very devoted lecturer concerned about the high standard of his lectures. He was murdered in front of his nine-month pregnant wife in his apartment on 13 June 1944 by Polish underground soldiers, who were told that it was necessary to kill this Jewish-Masonic communist. To this day, it is not known who ordered the murder of Prof. Ludwik Widerszal' (Dunin-Wąsowicz 2012, p. 69). Dunin-Wąsowicz mentions in his memoirs many such heinous crimes perpetrated against his fellow countrymen. And he adds that these events 'were linked to the list of 'the enemies of the Polish nation' drawn up by the National Armed Forces [*Narodowe Siły Zbrojne*]. The list included Tadeusz Manteuffel [a leading and popular Polish historian—*author's note*] and the Ossowskis (Dunin-Wąsowicz 2012, p. 69).

---

[2]The meetings mentioned by Dunin-Wąsowicz took place in the flat no 136, in the Colony IV. It was a social dwelling of the WHC.

Despite the fact that the 'Glass Houses' Association was not functioning officially, the estate residents' energy could be felt and sensed. The WHC residents did everything in their power to make the lives of children and young people, in time of war and havoc, a little bit safer, more cheerful and carefree, reminiscent of their once stable existence. Maria Wieman continued to give eurhythmics classes. She played the piano in unheated rooms, with fingers paralysed from the cold, in winter. She received pay for her work, a bowl of hot soup to be eaten on the spot and another one to carry away in her aluminium military flask. Zofia Topińska also mentioned 'granny Komorowska' who taught girls how to repair carpets, make buttons and slippers made of string, as well as prepare fancy salads. 'Granny Komorowska gave these girls old-fashioned warmth, love and affection. They liked to listen to her instructions that one should always learn something, do something for others, help their mother' (Topińska 1984, p. 122). The library was still operating, and children would willingly go there. Younger kids would want to listen to stories read aloud by adults, while the older ones would read by themselves in silence. Bookbinding courses and lettering classes were organised. And the artistically talented children made aesthetic pictorial catalogues for books for small children who could not read yet. The 'Baj' puppet theatre still operated, despite the fact that adult actors had to either go into hiding or leave Warsaw. The theatre was run by children who helped adult actors at the backstage. Literary soirees, attended at that time mainly by young people, were still organised. It was here where young writers of the war generation—Krzysztof Baczyński and Tadeusz Gajcy—made their literary debuts. The Żoliborz kids used to organise theatrical performances (Zbyszek Cybulski, a famous actor who later appeared in numerous outstanding films by the best Polish directors, such as Andrzej Wajda's *Popiół i diament* [*Ashes and Diamonds*], 1958—played in these performances). Henryk Ładosz organised, in the Colony IV, 'live word soirees', short theatrical forms and recitation of poems. He often invited children to his own flat for music lessons and played them classical music from his records. Ludwik Berger was very active in the cooperative movement—he was a theatre director by profession and a scout, social activist and member of the resistance movement by passion who was actively involved in the theatre activities of the Childcare and Youth Section of the Warsaw Social Self-Help Committee.

Jan Szymański described the social and club activities in detail, which had moved to private apartments, but now took on a different, more political and educational tone. A small house in the 1st Colony, in which the 'Glass Houses' Association's seat had been located in the past, housed a library employing 12 staff. There were double exits and secret wardrobes, thanks to which one could easily leave the premises or hide themselves unnoticed. This allowed the colony residents to organise relatively safe political and educational meetings. There was a youth library on the first floor of this house, and young musical bands used to meet in the building barrack of the Colony IX.

Ossowskis also organised regular meetings with Tadeusz Kotarbiński, which took place every Thursday. Each time, they were held in a different apartment, but always within the premises of the Colony IV. It was a kind of seminar attended by, among others, Nina Assorodobraj (sociologist), Zenon Kanabus (physician),

Joanna Landy-Brzezińska (educator), Stanisław Landy (local government's employee), Janina Ładosz (writer), as well as permanent Żoliborz activists: Helena Syrkus and Szymon Syrkus and Tołwiński. Young members of the Architecture and Urban Planning Studio (PAU) were also invited to these meetings, including Jan Strzelecki.[3] During these seminars, Strzelecki's concept of humanistic socialism which turned out to be particularly important for the Polish post-war humanities was being fleshed out. Strzelecki skilfully and elaborately combined socialism with Christian thought, looking for hope in both traditions in order to save humanist values. His concept of humanistic socialism became the basis for many polemics and important scientific discussions within the framework of post-war sociology in Poland (the dispute between Ossowski and Hochfeld in 1947–1948 concerned the understanding of this concept).

In 1940, the 'Płomienie' [*Flames*] group was founded in Żoliborz, whose cocreator was the twenty-year-old Jan Strzelecki. The group was composed of more than twenty people: students, members of the Union of Independent Socialist Youth [*Związek Niezależnej Młodzieży Socjalistycznej*]. The 'Płomienie' group included: Jan Nowicki, Maciej Weber, Jan Pohoski (studied economics at the secret university-level courses of Free Polish University [*Wolna Wszechnica Polska*], died in the Warsaw Uprising), his sister—Ewa Pohoska (studied ethnography and sociology, writer and painter, editor of 'Droga' [*The Road*] literary magazine, executed in January 1944)—and Karol Lipiński (he saw in his sociological studies a way to understand modern times; executed during the extermination of prisoners in Lublin, in July 1944). Meetings and seminars were organised in private apartments, during which works of Abramowski, Georges Sorel, Marx, Emmanuel Mounier and Jacques Maritain were discussed.

Krzysztof Dunin-Wąsowicz wrote: 'we used to meet at home. Also at my home. I just listened, because they were smarter than me, Jan Strzelecki and Karol Lipiński above all—they were the most eminent leaders of the 'Płomienie' group. (…) We analysed different educational methods—it had an inspiring dimension for the future. I also took part in the discussion 'Catholicism and socialism'. We invited Władysław Bartoszewski and the pre-war columnist, Witold Bieńkowski, who after the war was the editor-in-chief of the Catholic weekly titled 'Dziś i Jutro' [*Today and Tomorrow*] and a member of the Polish Sejm. During the German occupation, Bartoszewski

[3]In addition to the activities described here, it is worth noting that during the war, he participated in the Warsaw Uprising, and he fought in Żoliborz. After 1945, he was the editor of the ZNMS 'Płomienie' magazine and the author of the famous manifesto titled *O socjalizmie humanistycznym* [*On humanistic socialism*] which, in 1946, triggered a discussion in the press (in which Adam Schaff, Julian Hochfeld and Stanisław Ossowski, among others, were involved). The press discussion ended with the end of democracy and freedom of speech and the consolidation of Stalinism in Poland in 1948. In 1949, Strzelecki graduated in sociology, defended his doctoral dissertation in 1964 and became assistant professor in 1987. He worked at the Institute of Philosophy and Sociology of the Polish Academy of Sciences in Warsaw. He was an adviser to the solidarity movement (in August 1980, he supported the work of the Expert Committee at the Presidium of the Inter-Enterprise Strike Committee [Prezydium Międzyzakładowego Komitetu Strajkowego]), a member of the Alain Touraine research team and a co-author of *Solidarité. Analyse d'un mouvement social, Pologne 1980–81*, Paris 1982.

and Bieńkowski were members of a small organisation of lay Catholics—the Polish Rebirth Forum [*Forum Odrodzenia Polski*]. It was headed by Zofia Kossak-Szczucka, a writer of historical novels. It turned out to be a futile debate, because Bieńkowski began to talk about… the mystical body of Christ. And we wanted to talk about the attitude of Catholics to the land, industry and state reforms. On one matter we were in total agreement, both Catholics and socialists: we need to help Jews. Bieńkowski and Bartoszewski were active within Żegota,[4] and headed the Jewish Department of the Government Delegation for Poland [*Delegatura Rządu Rzeczypospolitej Polskiej na Kraj*]. (…) During the occupation, Jan Strzelecki was interested in Christian Personalism. It was a new philosophical trend, born in France during the great economic crisis at the turn of the 1920 and 1930s. For Marxists, this crisis proved the necessity of eliminating capitalist structures. The adherers of Christian Personalism claimed that the economic collapse was a poignant symptom of the human crisis. And in order to overcome this crisis—we need not a social revolution but a moral one. The war has shown what is the lowest point of debasement mankind can reach, what acts of cruelty humans are capable of inflicting upon others. That is why Jan Strzelecki supplemented the noun of socialism with the adjective humanistic. However, for the 'Płomienie' milieu, discussions about Austro-Marxism were much more inspiring' (Dunin-Wąsowicz 2012, p. 59–60). Dunin-Wąsowicz mainly referred to the achievements of the Social Democratic Workers' Party of Austria (*Sozialdemokratische Arbeiterpartei Österreichs*—SDAPÖ).[5] They were fascinated by the Red Vienna (*Rotes Wien*) project implemented there and the Karl Marx-Hof building for 1400 apartments, and its premises included many amenities (shops, laundromats, kindergartens and a library). They were interested in the Spanish Civil War and used to read André Malraux and Joseph Conrad.

The Żoliborz youth, brought up in the spirit of social activism and patriotism, felt unsatisfied and powerless due to the limitations resulting from the German occupation. The actions described herein can be easily called self-study and intellectual work. Such work prepared young people morally not only to fight against the occupying forces, but above all to answer the question of what to do next after a possible victory and how to further plan the future. The Żoliborz youth was looking for the causes of social changes that unfolded before their eyes, programming a new political thought and shaping a new social sensitivity. But apart from that, young people were eager to fight and lead an armed struggle, here and now, which would bring an imme-

[4]'Żegota' is the code name of the Council to Aid Jews, the only state organisation in Europe that helped Jews during World War II. 'Żegota' was operating underground in several regions of Poland, involving people of various political options, denominations and beliefs.

[5]In 1919, *Sozialdemokratische Arbeiterpartei Österreichs* won the elections to the city council of Vienna by the overwhelming victory, appointing for the first time in history the mayor of the capital. Dominated by the Social Democrats, the city council of Vienna carried out, during the interwar period, many projects to improve the workers' existence. Thanks to the Social Democratic government's policy, Vienna managed to create a model of a welfare town that was a model for the workers' movement from around the world. The Social Democrats carried out tax, school and construction policy reforms in Vienna and introduced care for the elderly, mothers and children. In 1929, 55% of SDAPÖ activists (out of 718 thousand in the whole country) lived in Vienna.

diate effect. Young people were involved in various operations of the 'Gray Ranks' [*Szare Szeregi*] organisation which stemmed from the scouts. They participated in various subversive operations of the Home Army [*Armia Krajowa*], carried out sabotage, painting—with red paint—anti-German graffiti: 'Pawiak pomścimy' [*We will avenge the Pawiak prison atrocities*], 'Śmierć okupantom' [*Death to the German occupiers*] or 'Oktober'—which was to remind Germans their defeat during the World War I. On 3 May 1943, 'Gray Ranks' organised a megaphone campaign. The Nazi propaganda megaphone (the so-called szczekaczka, loudspeaker) placed on Wilson Square was intercepted by the 'Wawer' subversive unit of the Home Army. It was an important, symbolic event. The intercepted megaphone system served to broadcast Polish patriotic programme. At first, Polish patriotic songs were played and then words that bolstered the civilian morale. Those who stopped for a moment in the square, those who listened from their balconies and windows—this short speech gave them all hope for survival. Then, the sounds of the Polish national anthem came out of the loudspeakers. Everyone joined in singing, all those random people who gathered in the square. 'For a very long time Żoliborz did not see such a solemn ceremony' (Poterański 1970, p. 128). As a result of the minor-sabotage operations carried out by the Żoliborz youth, the supplies of raw material were destroyed in one of the occupier's factories, German telephone connections were destroyed, and several liquidation operations were carried out against spies and blackmailers [*szmalcowniks*], who benefited from blackmailing and denouncing residents who helped Jews. It is worth emphasising that despite the attempts to idealise the Żoliborz community and create a myth of solidarity with the Jewish population, as if Żoliborz was some 'safe oasis' or a peaceful enclave, blackmailing or denouncing Jews and those who helped them happened also here, as in every other district of Warsaw and as in any other German-occupied city.

Young people identified themselves with various military organisations and organised military training within the WHC premises. The underground press—on all kinds of topics and across the political spectrum—was also printed here. Young people were very much involved in the distribution of the underground press. The underground operations, very well organised and coordinated, revealed their full potential during the outbreak of the Warsaw Uprising, which started in Żoliborz. The Warsaw Uprising turned out to be the most tragic event of World War II for both the district and its residents.

It is worth highlighting, however, that the Polish resistance movement encompassed the entire city of Warsaw. 'It would be an exaggeration to overstate the specificity of the Żoliborz district during the occupation. The underground organisations, both affiliated with the Communist Party and People's Guard [*Gwardia Ludowa (GL)*], operated throughout Warsaw, and were especially active in workers' communities in Wola, Targówek, Bródno, Marymont' (Zarzycki 1970, p. 308)—writes Janusz Zarzycki, focusing on both the militant and the ideological communist and socialist organisations. He can see, however, the specificity of this district in perfectly prepared—both housing and technical—conditions. There were a lot of utility space, social and common rooms within the WHC premises, and that allowed its residents to carry out many operations, without risking the lives of the occupants of private flats. The WHC residents also appreciated the importance of the publishing activity.

That is why Dunin-Wąsowicz could say without exaggeration that 'in every second home in Żoliborz, there was either a printing press or a radio station, or secret training or secret courses were taking place' (Dunin-Wąsowicz 2012, p. 57). When taking a closer look at one of the organisations operating within the WHC premises, one could see how it operates. The so-called Five fellows from Żoliborz (including Antonina Sokolicz-Merklowa and Edward Bonisławski, among others) founded the Association of Friends of the Soviet Union [*Towarzystwo Przyjaciół ZSRR*], and in March 1940, they began to publish the 'Wieści z Świata' [*News From the World*] magazine. Bonisławski would hide, during the day, his radio transmitter in his cellar, and another active member of the Association would listen to the news, at night. Together, they edited the magazine, which was also printed on the duplicating machine by night. At the time of its publication, technical assistance was provided by a polytechnic student under the *nom de guerre* of 'Zosia' and a law student under the *nom de guerre* of 'Janek'. Both of them (the boy and the girl) were arrested in July 1940. At the same time, the 'Five fellows from Żoliborz' got involved in the work for the middle school students of 'Spartacus' group, who published the 'Strzała' [*Arrow*] magazine, distributing at the same time another magazine, titled 'Brygada Wolności' [*Freedom Brigade*]. All these organised collaborative networks and interdependencies between them resulted, firstly, from the belief in the power of propaganda, i.e. words that encourage and mobilise to action, and secondly, they were a manifestation of a great willingness to act, to take action and not to be passive. In addition to publishing the underground press, they also organised a food and clothing collection campaigns.

10.3 Social Construction Firm [*Społeczne Przedsiębiorstwo Budowlane (SPB)*] and Architecture and Urban Planning Studio [*Pracownia Architektoniczno-Urbanistyczna (PAU)*]—Enterprise for War Survival and Creative Workshop

The role of the Social Construction Firm had to change during the occupation. The SPB developed its activity mainly within the WHC premises, but it had to adapt to the new situation under wartime conditions. The SPB's main purpose was to safeguard the existing jobs and to create the new ones. Throughout the entire war, the SPB was headed by Stanisław Tołwiński. The SPB, legally and in accordance with the statute, conducted construction and investment works for social institutions in Warsaw, as well as throughout the entire general government. Soon, orders for repair works, indispensable after wartime destruction, started flooding in from across Warsaw to the SPB. But the most serious client was the mutual insurance company [*Powszechny Zakład Ubezpieczeń Wzajemnych*], as well as the 'Społem' Union of Consumer Cooperatives [*Związek Spółdzielni Spożywców 'Społem'*]. The investments of the above clients required the employment of the largest number of workers, intellectuals and young people involved—during the war—in politics or the Polish underground resistance movement. Under the direction of Jan Strzelecki, members of the 'Płomienie' group worked as a separate brigade. Szymański wrote in his book that there was a very specific atmosphere at the construction site, because workers would engage constantly in political discussions or would read the underground press. There was also community kitchen there, giving away meals not only to the SPB employees, but also to the Jews hiding in Żoliborz.

Among the Social Construction Firm workers, there was, for example, Juliusz Rydygier—a cooperative activist and member of the Polish Socialist Party and, later, member of the Communist Party of Poland. In the interwar period, Juliusz Rydygier was arrested numerous times and imprisoned for his communist activities. On 5 January 1942, he organised, in his Żoliborz flat, a founding meeting of the Polish Workers' Party. On 30 October 1942, he was arrested by the Germans and, after an investigation at the Gestapo detention centre in Aleja Szucha, transported to the Auschwitz camp where he was murdered. Rydygier worked at the SPB as a warehouseman. His biography, including his work for the SPB and his political engagement, seem to be the typical common denominator of the biographies of almost all Żoliborz residents.

The SPB was a shelter for young people, for the intelligentsia of Jewish origin, but also for many scholars, architects and artists. Thus, the Social Construction Firm gave them protection in the form of *Ausweis*—an employment card allowing them to avoid being deported to work in Germany. It gave the possibility of physical survival ('at 12 noon, it was a solemn moment at the SPB office, as all employees were given a bowl of tasty and chunky soup' (Durko 1970, p. 238)) and a shared moment of creative work. It was a safe haven for urban planners, sociologists and hygienists from the Polish Society of Housing Reform, for co-workers of the 'Dom-Osiedle-Mieszkanie' monthly. A joke has been doing the rounds around Warsaw that the SPB required higher education diploma even from janitors.

The stories about well-organised and efficient help for Jews during the occupation, which Stanisław Tołwiński organised through the Social Construction Firm, were described on the Polish Righteous website.[6]

In January 1940, the Architecture and Urban Planning Studio was established, in which both cooperative activists and architects, urban planners, geographers and economists worked hand in hand—in short, all those who could shape the visions of a modern city. Initially, the Studio was located in a private flat of the architect Juliusz Żakowski (18 Krasińskiego St., on the second floor). After expanding and significantly enlarging the circle of collaborators, Tołwiński assigned PAU two additional premises at Krasińskiego St. 'We have gathered, within the Architecture Studio, a wider group of employees, formerly working at the Institute of Social Economy and the Institute of Social Affairs. We have created an institution called the Institute of Social Planning, which managed to get even some money from the Government Delegation for Poland (…). And this money was used for some economic, social, and political studies compiled for the State National Council

[6]https://sprawiedliwi.org.pl/en/stories-of-rescue/social-construction-enterprise-known-its-wide-rescue-activities-story-stanislaw-tolwinski.

[*Krajowa Rada Narodowa*]' (Tołwiński 1970, p. 210).[7] More than 100 theoretical, critical and conceptual studies have been created in the PAU Studio. The architects from the 'Praesens' group continued to work on a city design project for Warsaw's future functional urban area. Helena Syrkus and Szymon Syrkus inaugurated the series of lectures with the report titled 'Social services as a factor shaping the housing estate'. This was the PAU's official task. Unofficially, they worked on various aspects of the functioning of modern cities. The PAU carried out theoretical work and analysed sociological concepts of the city (by Robert Ezra Park, among others); they also discussed the research studies carried out by the Chicago school. On 20 October 1941, Ossowski prepared a work project at PAU: 'the series of lectures would cover selected issues from the following works: Sorokin P., Zimmermann CC., *Principles of Rural—Urban Sociology*. In particular, comparative studies on social attitudes, on the role of institutions and the culture of the city and the countryside would be interesting. Gist N., Halbert L., *Urban Society*. Social ecology of the city: shaping social relations in the city, planning for the urban community. Zorbaugh, Harvey Warren, *A sociological study of Chicago's Near North Side*. Social institutions and social activity. City and society. Reformative activities and city life: McKenzie, *The Metropolitan Community*, Mumford, *Technics and Civilisation* and *The Culture of Cities*. Burgess E., *Can Neighbourhood work have a scientific basis?*' (Udział Stanisława Ossowskiego w pracach PAUw. Wybrane materiały 1976, p. 307 [Participation of Stanisław Ossowski in the works of PAU. Selected materials].

Many of the employed urban planners, architects and scientists in the field of social sciences have explained and popularised global research. They knew the scientific literature perfectly well. They worked on it, making use of their critical thinking skills. Of course, they did not forget about their own local empirical and conceptual research.

The PAU was headed by Szymon Syrkus, his wife Helena was his deputy, and Kazimierz Leon Toeplitz (son of Teodor) was the secretary. The works were divided into three study groups: studies on the entire region, the capital city of Warsaw (this study group was led by Jan Chmielewski) and Warsaw's northern residential districts were led by Zygmunt Skibniewski. Further works were carried out on the concepts of the development of the Żoliborz district (mainly by Barbara Brukalska and Stanisław Ossowski), Rakowiec housing estate (the team was headed by Helena Syrkus, and its members included: Tołwiński, Chmielewski, Piotrowski, Spychalski).[8] In his memoirs, Stanisław Tołwiński pointed out that in subsequent years, the

[7]The State National Council was established on the New Year's Eve of 1944. It was formed by members of the Polish Workers' Party (PPR) and was a kind of self-proclaimed Polish Parliament. The State National Council described itself as 'the actual political representation of the Polish nation, authorised to act on behalf of the nation and direct its fate until Poland has been liberated from occupation'. The State National Council was composed, during its first meeting, of Władysław Gomułka, Stanisław Szwalbe, Stanisław Tołwiński, Ignacy Loga-Sowiński, Franciszek Jóźwiak, Zenon Kliszko.

[8]The PAU work has been described in detail by Helena Syrkusowa who organised particular issues and topics discussed. As it turns out today, the architect took very conscientious notes from lectures and presentations made by her colleagues, and also kept, for example, 'Notatkę w sprawie

PAU went through various stages: 1940/1941 was the year of searching for their own ways, 1941/1942 was the year of theoretical sketches, and 1942/1943 was the year of design projects. From the point of view of Stanisław Tołwiński, who was the first post-war mayor of ruined Warsaw, the activity of the PAU was invaluable. 'We have prepared a general concept of Warsaw's self-government organisation based on the planned economy assumptions, especially in terms of satisfying the basic needs of the population (housing, health and social care, supplies, culture and education), municipal public utility companies (...) The structure of Warsaw's residential districts, based on local government units, was developed in greatest detail' (Tołwiński 1965, pp. 9–10). After all, all the work on the development of the capital city of Warsaw was prepared by the PAU Studio. Tołwiński seemed to be the best candidate for the office of post-war mayor of Warsaw, razed to the ground, but hopeful to be rebuilt.

The Architecture and Urban Planning Studio Group (*Grupa* PAU) presented not only concrete housing solutions and public space design patterns, but more importantly designed the future shape of culture. And without understanding the intellectual atmosphere of the Studio, it would be difficult to grasp the true meaning of housing estates, principles of cooperativism and the humanistic socialism that was developed there. It was in the Studio that the collaboration between Ossowski and Brukalska flourished. They both worked on the design of the post-war Żoliborz estate reconstruction, which resulted in Brukalska's *Social Rules of Housing Estate Design* and Ossowski's *Towards New Forms of Social Life* (Ossowski 1968, pp. 325–364) being published. Ossowski's brochure was released under pseudonym in 1943, and a legal reprint was issued after the war in 1946. Under German occupation, in the interest of preserving the Żoliborz atmosphere and in the hope of a better post-war world, the following works were written: *Organizacja przestrzeni i życie społeczne w przyszłych osiedlach* [Spatial Organisation and Social Life in Future Housing Estates], *Ogólne zagadnienia dotyczące współżycia zbiorowego w dzielnicy pracy* [General Issues Regarding Collective Life Within The Working District], *Odbudowa stolicy w świetle zagadnień społecznych* [Reconstruction of Warsaw in the Light of Social Issues], *Urbanistyka i socjologia* [Urban Planning and Sociology][9] and important work of Maria Ossowska titled *Wzór demokraty* [The Ideal Democrat] which in 1946 was published as a brochure by Biblioteka Oświaty Robotniczej [Workers' Education Library] titled *Wzór obywatela w ustroju demokratycznym* [The Ideal Citizen of a Democratic State].

Stanisław Ossowski had also drawn plans and models of future culture. He considered the German occupation as a time of 'Manichean moods' reflecting the hope that war will topple the old order (ruthless capitalism, poverty, backwardness) and build a new social reality. 'The will to live—Ossowski wrote in *The Manichean Moods*—is the will to live in a new, reborn world' (Ossowski 1967a, b, p. 192), i.e.

zamówienia miejskiego dla Pracowni Architektoniczno-Urbanistycznej (Warszawa, 1 X 1941)' [Note regarding the city order for the Architecture and Urban Planning Studio (Warsaw, 1 October 1941)'. (See Syrkus 1976, pp. 229–328).

[9]All of the essays were published in Ossowski (1967c).

a world of social rebirth. Yet, he was afraid that the 'rebirth' could be governed by the state that the ideal government will be imposed by another 'monstrosity'.

While awaiting the rebirth of a new world and the implementation of a new vision of the future society, Ossowski envisioned a new urban culture, considered the relationship between sociology and urban planning, and authored underground articles on school and university scholarships that would contribute to rebuilding civic culture and attitudes after the war.[10] Such works reflected Ossowski's hope for changing the system and creating conditions in which the socialist activist could reach beyond the boundaries of the estate. They also expressed his strong conviction that critical thought and theoretical reflection are both active forms of social activity, changing the world through designing its desired form.

At the time, Ossowski often wrote about a new society, a society of the future where there would be no proletariat as we know it. Yet, Ossowski was far from adopting science-like laboratory ambitions or building a utopia. Rather, his writings (similar to those of Brukalska and Ossowska) should be approached as critical thought, an opposition to the bourgeois, undemocratic reality, as a struggle to eliminate stark inequalities and ensure equal participation in culture (defined not as creative domains, such as science or art, but rather as the right to freely express moral attitudes and everyday lifestyles). The conflict surrounding Ossowski's social activism continues up to this day and may be summarised by the following question: 'Ossowski was a democrat, socialist close to Abramowski's ideals, supporter of planning, whose humanistic ideals were rooted in socialism. Should he be considered an ivory tower intellectual, a utopian, a Marxist or simply a scholar who treated Marx seriously?' (Madajczyk 1999, p. 437).

Today's postmodern perspective makes us doubt the engineering role of social sciences and the rationality of new social order projects. Those doubts, however, should not bring on an all-encompassing critique of modernity, an external critique of those who are convinced that modernity has ended. Bauman's late twentieth century claims about postmodernity being modernity minus the delusions were mainly a result of him becoming aware of the delusions inherent in various forms of ideocracy. Those, however, are not sufficient grounds for entirely doubting the emancipatory purpose of the humanities. Obviously, it takes on different forms today and sets different goals. Yet, there is no doubt that the laboratory metaphor has been revived. Today's researchers increasingly often cite thinkers such as Charles Sanders Pierce, Aby Warburg and Claude Lévi-Strauss, all of whom equated their scholarly activity with the laboratory. Suffice it to say that Paul Rabino, Andrew Lakoff and Stephen Collier founded the Laboratory of Contemporary Anthropology. Ryszard Nycz and Roma Sendyka adopted the laboratory metaphor for the Polish anthropology litera-

[10] The following works were written during that period: *Kultura robotnicza* [Workers' Culture] (1943), *O drogach upowszechniania kultury na najwyższym poziomie* [On the Means of Popularising High Culture] (1944), *Z zagadnień przyszłej polityki stypendialnej* [On the Issues of Future Scholarship Policy] (1944), *Socjologia w świecie powojennym* [Sociology in the Post-War World] (1946). All of the above essays were published in: Ossowski, Dzieła [Collected Works], vol 6, 1970.

ture studies. Also, the Laboratory of Contemporary Humanities was founded at the Wrocław's Institute of Cultural Studies to explore non-anthropocentric humanities (Kil et al. 2017).

10.4 Żoliborz Socialism—'To Be on the Side of the Suffering'

'Żoliborz socialism is also socialism which, faced with the Nazi threat, was able to defend democracy by all means, even by force' (Krzysztof Dunin-Wąsowicz).

In March 1942, the Warsaw Housing Cooperative, like all other housing and construction cooperatives in Warsaw, lost their autonomy and were placed under the supervision of the German compulsory administration of Franz Koepisch, who sent letters to the cooperative on 9 March 1942.

During the war, the population structure changed in Żoliborz. Firstly, the inhabitants of Poznań and the Pomerania region, expelled from their homes, came to live here, and secondly, the Jewish population was hiding here during the war, the number of whom increased noticeably after the liquidation of the Warsaw ghetto. In 1942–1943, workers who lost their jobs and did not have an employment card, for fear of being deported to perform forced labour in Germany, most often left the city. The domestic servants had disappeared, because the intelligentsia could no longer afford to maintain them. While workers tried to cope with the situation, by taking various physical jobs, the intelligentsia and representatives of liberal professions lost their livelihoods—that is why the SPB gave them so much hope.

Janusz Durko—an employee of the SPB secretariat, son of a socialist who, as a child brought up in the Żoliborz milieu of leftist activists, listened to adult political debates—cannot understand to this day how it was possible to exclude people because of nationality and, above all, because of race. In the years 1934–1938, he studied history at the University of Warsaw in a period which, he says, 'covered the university with shame'. He saw 'ghetto benches', supervised by the right-wing youth using sticks with razors inserted in the top. As a sign of solidarity with the persecuted, he would seat in benches designated for the Jewish students. He would run away, along with the Jews, chased by the National Democracy paramilitary groups.

Durko remembers the war in Żoliborz mainly through experiences related to the survival strategies of the Jewish population. 'No, they did not hide. They walked around openly. It's astonishing, you don't hear anything about that, nobody writes about it… that Warsaw had anything like this enclave, a closed environment, where no one even considered going to the ghetto… all of them survived' (Durko 2009, p. 53).

The WHC management organised false documents for many Jewish families and transferred these families from one colony to another one within the area of the WHC. In the light of other reports, the most astonishing thing is that the population of Jewish descent has become such an integral part of the Żoliborz culture and education, through its development and joint work, that it would not occur to anyone to denounce their co-workers. Many escapees from the ghetto passed through the apartment of Janusz Durko and his wife, Janina. Janina Durko helped them find a job. And when someone was in danger that a hostile neighbour would denounce them, the Durkos sought another shelter: another apartment, SPB seat or an adapted garage at Senatorska Street.

Piotr Topiński, in turn, who was the son of Zofia Topińska, a teacher at 'Szklany Domek' preschool, author of the book *The Promises of Żoliborz*, recalls Żoliborz as an amazing district. 'This way Jewish children were led out of the Warsaw Ghetto, and not only children. In the preschool which was managed by my mother, Jewish children learned Polish, were provided with false documents and directed further; according to my knowledge, an important stopover for those exhausted children was the Monastery and the Institute for the Blind in Laski. The Mother Superior of the Convent, still in this position in the 1970s, was Sister Maria, a Jewish survivor from the Holocaust, and in addition, a former Communist, she adds with laughter. I was amazed at their mutual good relations. My mother was never religious. Already after the death of my parents, I dug up a few documents—Ausweis, Kennkarte bearing their photos and different names. What seemed to Julian Brysz, a Jewish child who told me the story, a manifestation of the actions of 'good people', from the sociologist's point of view is something unheard of. Warsaw is occupied, absolute terror reigns in the country, flats are being searched at night, German soldiers execute random people in the streets, while at Suzina street, there are Jewish children from the ghetto who lost their parents, do not know Polish, and have no place to hide, playing, surrounded by non-Jewish kids from Żoliborz. At that time, Żoliborz was quite a closed district, the inhabitants knew each other and it was unlikely that children from preschool would not later tell their parents about the new, strange children whom they met. These children had to stay somewhere at night, someone took care of them, fed them, dressed them. It was a large preschool for 150–200 children. (…) And yet no information leaked to the Germans and some attempts at denouncing them, probably made by blackmailers, had to be 'neutralised' by 'boys'.[11] Julek Brysz, Krzysztof Pomian, Wiktor Jassem—were the Jewish children who attended the WHC preschool, but there were also non-Jewish children from the WHC estate in the same preschool, famous today, who can be enumerated: Zbigniew Zapasiewicz, Andrzej Krzysztof Wróblewski, Jarosław Abramow-Newerly. None of them mentions these events, because they were just ordinary and natural acts of help and kindness. Nothing to speak of! Numerous people [of Jewish origin—*author's note*], aware of the danger, worked in the preschool. Dr. Aleksander Landy was the preschool doctor, who, I think, was the mastermind behind the underground operations, and Ms. Maria Arnoldowa ran a preschool library (I still wonder today what was the purpose of keeping a library in a preschool?). Ms. Maria Arnoldowa was always accompanied by young Janka Cygańska. Art classes with preschoolers were conducted by Hanka Rembowska. Throughout the German occupation, music

[11]Topiński probably meant by 'boys' members of various underground militias; they were responsible for liquidating informers and blackmailers.

appreciation and eurhythmics classes were given by two exceptional teachers: Maria Wieman and Maria Zukermann, later called Cukierówna. Throughout the German occupation, Maria Wieman hid a young girl, Franciszka Leszczyńska, who later became a well-known composer. (…) Ida Merzan worked here as a preschool teacher; not only did she work under her true name, but she also had a very strong Yiddish accent. The Żoliborz district gave Jews a chance for survival. My favourite teacher was Ms. Miller—I do not remember her name. (…) Dr. Feliks Kanabus, surgeon, was our neighbour. His wife, who was a great professor of paediatrics and always ready to help, was Dr. Aleksander Landy's niece. Both of them were awarded the medal of 'Righteous Among the Nations'. Dr Kanabus was famous for foreskin restoration performed among Jews. Probably there was no other place like the Żoliborz housing estate, where one could become famous in the field of medicine which could lead to the execution at any time. His fame was rather dangerous, but he was imprisoned only under Communist Poland, not before (for alleged espionage, but eventually he was cleared of accusations and rehabilitated)' (Topiński 2012).

This socialist housing estate built in the spirit of social activism turned out to be incredibly integrated and well-organised internally, despite many ideological differences. On the one hand, pre-war social and political activists got used to underground resistance because under partitioned Poland, they had to go underground and fight. And then, after Poland regained independence, they were active in left-wing organisations that were delegalised at the time in Poland, as associated with Bolshevism, the biggest dread of the Polish authorities. On the other hand, Żoliborz residents used perfectly well the learned and tested mechanisms of self-organisation and mutual assistance. They learnt not only close neighbourly ties, but also the principle of trust based on cooperativism; they did not get in their mutual ways, applied to everything admirable social distance and, at the same time, commitment. 'I remember—Piotr Topiński recalls—when in one of our conversations, my father's friend and underground collaborator, Witold Rogala, recipients of the medal of Righteous Among the Nations from Żoliborz told the story about the conflict he had with two janitors who hid from him, the WHC administrator at the time, free rooms in attics, basements, and outhouses of the WHC colonies III and IV. He himself suffered from housing deficit, while free rooms were indispensable in underground operations, and he sensed it was a scam and the invisible hand of the black market. It was not until the end of the war that he learnt that these premises were used all the time for underground operations. He was ashamed of this conflict. Various people, equipment, are printing presses were hidden there. Everyone was affiliated with the Polish Socialist Party (PPS) and it was not safe to know too much. One of the caretakers' name was Marchewka and he managed our backyard until his retirement, the other, Pawłowski, supervised the adjacent backyard. Simple people, nameless heroes. I do not even know their names' (Topiński 2012).[12]

In the memories of those who survived the atrocities of World War II, Żoliborz was a good place. Many called it an 'oasis' because the Germans rarely ventured

[12]The history of the Topiński family was also described by their son Piotr on the website of the Polish Righteous https://sprawiedliwi.org.pl/en/stories-of-rescue/story-rescue-topinski-family.

into the streets of the district. The soldiers of the Home Army and the scouts (ready as always for 'minor-sabotage' operations) found shelter here. Everyone knew each other, and they knew even too much about each other. According to estimates by Krzysztof Dunin-Wąsowicz, as many as 2000 Jews were hidden here during the war, probably more than in any other district of Warsaw. They could even lead a normal life here. It must be remembered, however, that Dunin-Wąsowicz writes about the entire district and therefore also about Żoliborz Dziennikarski [*Journalists' Żoliborz*], Żoliborz Oficerski [*Officers' Żoliborz*], Żoliborz Urzędniczy [*Clerks' Żoliborz*] and Spółdzielczy Żoliborz [*Cooperative Żoliborz*]. In the latter, the criteria of the WHC founders for selecting members of the cooperative were based on the principle of cultural, social, trade union or political involvement. In spite of the often fundamental differences in views and political visions (mainly between communists and socialists), but also in spite of the differences in relation to the Soviet Union, everyone shared the conviction that one must stand on the side of the suffering in these cruel times.

Social and educational activities, undertaken with great care before the war, shaped within the period of 10 years the conditions in which civic attitude and social sensitivity could develop freely, without intrusive pathos, and be tested during the war. Young people took an active part in underground operations, issuing underground press, attending secret courses and being members of socialist groupings, dreaming about attacking the Nazis and helping the suffering.

It is not true that everyone survived and that it was possible to feel safe in Żoliborz. There were also among the WHC tenants people who could be dangerous—either for profit or out of fear. They were, however, so few that others knew about them and the WHC residents tried to keep an eye on them. This, in turn, meant that they were afraid of the neighbours' opinion. During the Nazi terror, no one could feel safe. Certainly not the Jewish population, neither those who hid them. The punishment for aiding Jews was death penalty for the entire family. What really strikes me here, however, is, above all, a natural social attitude, a concern for the survival of those whom the cruel Nazi regime tried to trap. That is why, when mentioning Żoliborz, people do not write about heroism, altruism and sacrifice. The literature gives us a picture of great deeds, and extremely courageous and devoted people who perished. Help others, be a part of underground organisation, fight the enemy, live honestly and in dignity despite terror of war, fear and hunger, without giving into despair—these are certainly not easy strategies, very difficult to implement. They had been, however, symbolically practiced and tested, because people always think about them in terms of responsibility, collective good, the future of some more objective and shared world based on intersubjective experience.

Helping Jews during the occupation is not just an event from the distant past. On the initiative of the Institute of History of the Polish Academy of Sciences and in cooperation with the project titled 'The Polish Righteous—Recalling Forgotten History' carried out in today's Żoliborz, workshops are organised addressed to young people, during which knowledge about these acts of solidarity and social sensitivity becomes part of the historical experience. 'Interactive workshops are aimed at encouraging young people to learn about the past in a creative manner.

Warsaw's Żoliborz district, where workshop participants will be walking, will be discovered on the basis of people's accounts, photos, maps, objects and places. Thus a complex image of the district, where many inhabitants helped Jews, will be created. The workshop participants will also try to answer the question, why was it here, in this district of Warsaw, that so many people were ready to risk their and their family's lives to save others' (Zawadzka 2010).

During the war, Żoliborz residents reached their civic maturity in the atmosphere of being obliged to risk their own safety when it came to saving others.

10.5 Aftermath of World War II: Warsaw Housing Cooperative—The End

On 1 January 1944, the first plenary meeting of the State National Council [*Krajowa Rada* Narodowa] was held, during which its presidium was elected. Bolesław Bierut, who just returned from Moscow, became the KRN's chairman, and Edward Osóbka was co-opted as KRN's vice chairman. The meeting was also attended by Tołwiński, Szwalbe and Święcicka. On 1 August 1944, the Warsaw Uprising broke out in Żoliborz. The combat group of the OW PPS (Combat Organisation of the Polish Socialist Party) clashed with German soldiers. The Social Construction Firm (SPB) played a significant role in the Warsaw Uprising: the SPB materials and equipment were used by Warsaw insurgents. Many residents of the housing estate and employees of its institutions took part in the operation. A deep-water well located within the WHC premises supplied water to the estate throughout the entire period of the paralysis of the city.

At the beginning of September 1944, the buildings of the first WHC Colony were completely destroyed, and the buildings of the 7th Colony were burnt down. The colonies VIII and VI had been partially saved from the havoc of war. The colonies III, IV and V suffered the least. The Rakowiec housing estate was almost completely destroyed. Out of three hundred flats built with great dedication, only eight buildings survived the war.

Filip Springer sums it up: 'Even if all workers' flats, not only in Rakowiec, but also those in Żoliborz, Koło and Praga [workers' districts of Warsaw—*author's note*] were razed to the ground, it would still be a small loss on the city's scale. Social activists failed to achieve much. The achievements of twenty years of their work constitute only five percent of all flats built at that time. From the perspective of the glossy door handles of the townhouses in Warsaw's centre and Mokotów district, from the perspective of the crystal-clear water of the ponds in their courtyards, from the height of the green terraces on the roofs—you can simply fail to notice them' (Springer 2015, p. 119). However, after nearly a hundred years of Warsaw Housing Cooperative's existence, we continue to discuss the achievements of the Żoliborz activists. And this book proves this point. They did not build much—it is true. But they tested city-forming strategies, developed urban self-management, initiated civic and neighbourhood housing practice and also proved that on the periphery of the capitalist system, you can build autonomous zones, which are not escapist survival strategies or remote enclaves, but civic self-organised communities.

Already in January 1945, the WHC residents themselves initiated clearing the debris and reconstructing the housing estate, and in February of the same year, they organised the Tenant Council. However, there was an extremely significant change. After the formation of political power and state institutions, the Cooperative became one of the institutions of the state. It lost its bottom-up, self-governing character. And it lost its appeal to many of its activists. Many cooperative members, such as Bierut, Osóbka, Tołwiński, Hochfeld and Szwalbe, were transferred to the state apparatus or other central government institutions. Bolesław Bierut became the chairman of the State National Council and then, in the years 1944–1947, served as Deputy President of the Republic of Poland. During the most difficult period of Stalinism in post-war Poland, i.e. from 1947 to 1952, Bierut served as President. In Szwalbe's memoirs, Bierut seemed in the 1920s to be an 'avid and ideological cooperative member, but later, in the 1930s, he disappeared mysteriously due to his work for the Comintern in Moscow and then in Bulgaria. During the war, Bierut returned to Poland. Szwalbe recounts that Bierut asked him at that time for help and he received it from the WHC structures, but after the war, as President of the Republic of Poland, and later the chief enforcer of Stalinism in Poland, he did not return the favour. He returned as a completely different man and his cooperative ideals were replaced by doctrinairism' (Topelitz 2004, p. 334). Edward Osóbka-Morawski, in turn, became Prime Minister of the Provisional Government of National Unity [*Tymczasowy Rząd Jedności Narodowej*]. From July to December 1944, he chaired the Polish Committee of National Liberation [*Polski Komitet Wyzwolenia Narodowego, PKWN*], and in the years 1944–1947, he was appointed Prime Minister (he was also given the role of Minister of Foreign Affairs and Agriculture, and Minister of Public Administration). He was also a member of the State National Council and, in 1947–1952, member of the Legislative Sejm. In 1970, after the bloody suppression of workers' protests (the Polish 1970 Strikes), he left the Polish United Workers' Party. Stanisław Tołwiński was also a member of the State National Council in the years 1945–1947 and then a member of the Legislative Sejm and the First Term Sejm (1952–1956). From 1945 to 1950, he was the mayor of the capital city of Warsaw. In April 1997, the Yad Vashem Institute posthumously honoured him by granting him the title of Righteous Among the Nations. Stanisław Szwalbe's biography was very similar; in 1943–1952, he was a member of the State National Council and the Legislative Sejm, serving as the Deputy Marshal of the Sejm. As of 1947, he was also a member of the Polish Council of State (*Rada Państwa*). Disappointed by the state apparatus and the central management of cooperative movement, he later became an activist of the cooperative movement and tried to reform it. Julian Hochfeld spent the first years of war in the Soviet Union, then in the Middle East and finally in London, from where he returned to Poland in 1945. As other WHC activists, he also became a member of the State National Council, the Legislative Sejm, and then the First Term Sejm (1952–1956) and the Second Term Sejm (1957–1961). In 1948, when the communist state apparatus saw him as a not fully dedicated state's officer and fearing political purges, Hochfeld undertook scientific work: first, at the Main School of Planning and Statistics [*Szkoła Główna Planowania i Statystyki*], and then at the School of Foreign Service [*Szkoła Główna Służby Zagranicznej*]. In 1951, he became the head of the Chair of Historical Materialism at the University of Warsaw and became Zygmunt Bauman's mentor. In the years 1962–1966, he was the deputy director of UNESCO's Social Science Department in Paris. Other devoted ideologists fell into the trap of commitment and hope that they could rebuild the country and arrange socialism on their own, not Soviet conditions. And the most doctrinaire among them was of course Bierut.

In the first year after the liberation, the Żoliborz housing estate was rebuilt, and the laundry room and bathhouse were reopened. The expansion of the 9th Colony was planned and the foundation stone was laid for the construction of the 11th Colony (92 flats were handed over to tenants on 1 April 1947, and six weeks later, the next 90). These were the first cooperative houses built after the war.

In January 1947, the WHC Supervisory Board approved the plan for the year 1947 (they wanted to build a House for the Lonely—in the 13th Colony and in a new 12th Colony). In February, the 'Smallest Flat' exhibition was opened by the Deputy Marshal of the Sejm, Stanisław Szwalbe, and the mayor of Warsaw, Stanisław Tołwiński, in the 'Tęcza' cinema in Żoliborz. It is not clear whether all these activities were still the effect of grass roots and independent decisions of activists, or already a top-down plan of politicians. On 26 April 1948, the Workers' Housing Development Administration [*Zakład Osiedli Robotniczych*] was created, a central investment authority in the housing construction industry subordinate to the Ministry of Reconstruction. Since then, the housing construction industry started to be completely centralised. The Workers' Housing Development Administration (ZOR) has become a contractor for state building construction projects and housing policy. As a result, in January 1949, the enrolment of new members to the Warsaw Housing Cooperative was suspended, and the construction activity was suspended until 1956. The subsequent attempts at self-organisation and civic self-management, both formal and informal, require a different, new narrative, more political than a city-forming one, and therefore their description and analysis must be cut short here.

References

Dunin Wąsowicz K (1984) Na Żoliborzu 1939–1945 [In Żoliborz 1939–1945]. Książka i Wiedza, Warszawa

Dunin-Wąsowicz K (2012) Historyk, socjalista, pamiętnikarz [Historian, socialist, diarist]. Książka i Prasa, Warszawa, Instytut Wydawniczy

Durko J (1970) Ludzie i sprawy w latach okupacji [People and matters during the occupation]. In: Anonymous (ed) Żoliborz wczoraj, dziś, jutro [The past, present and future of Żoliborz]. Książka i Wiedza, Warszawa, pp 237–245

Durko J (2009), Poles who rescued Jews during the Holocaust. Recalling forgotten history chancellery of the President of the Republic of Poland, Museum of the History of Polish Jews, Łódź, p 53. https://sprawiedliwi.org.pl/sites/default/files/sprawiedliwi_2009.pdf

Gaudasiński Z (1970) Komenda Obrońców Polski w latach 1939–1941 na Żoliborzu [Polish defenders command in 1939–1941 in Żoliborz] In: Anonymous (ed) Żoliborz wczoraj, dziś, jutro [The past, present and future of Żoliborz]. Książka i Wiedza, Warszawa, pp 230–233
Kil A, Małczyński J, Wolska D (2017) Ku laboratorium humanistycznemu [Towards the humanistic laboratory]. Teksty Drugie [Second Texts] 163:274–285
Madajczyk C (1999) Klerk czy intelektualista zaangażowany [An independent or engaged intellectual]. Wydawnictwo Poznańskie, Poznań
Ossowski S (1967a) Ku nowej kulturze [Towards a new culture]. In: Ossowski S, Dzieła [Collected works], vol 3. Państwowe Wydawnictwo Naukowe, Warszawa, pp 164–185
Ossowski S (1967b) Z nastrojów manichejskich [From the manichean spirits]. In: Ossowski S, Dzieła [Collected works], vol 3. Państwowe Wydawnictwo Naukowe, Warszawa, pp 187–197
Ossowski S (1967c) Dzieła [Collected works], vol 3 [Issues of social psychology]. Państwowe Wydawnictwo Naukowe, Warszawa
Ossowski S (1968) Dzieła [Collected works], vol 5. Państwowe Wydawnictwo Naukowe, Warszawa
Poterański W (1970) Żoliborz w czasie okupacji [Żoliborz under occupation]. In: Anonymous (ed) Żoliborz wczoraj, dziś, jutro [The past, present and future of Żoliborz]. Książka i Wiedza, Warszawa, pp 102–137
Springer F (2015) 13 Pięter. Wydawnictwo Czarne Wołowiec
Syrkus H (1976) Ku idei osiedla społecznego 1925–1975 [Towards the idea of the social housing estate 1925–1975]. Państwowe Wydawnictwo Naukowe, Warszawa
Szymański JA (1989) Warszawska Spółdzielnia Mieszkaniowa 1921–1970. Zarys dziejów [Warsaw housing cooperative 1921–1970. History outline]. Wydawnictwo Spółdzielcze, Warszawa
Toeplitz KT (2004) Rodzina Toeplitzów. Książka mojego ojca [The Toeplitz family. My father's book], Wydawnictwo Iskry, Warszawa
Tołwiński S (1965) Pierwsze pięciolecie odbudowy i gospodarki samorządowej Warszawy (1945–1950) [The first five-year reconstruction period and local government economy of Warsaw (1945–1950)]. In: Biuletyn Informacyjny Prezydium Rady Narodowej m. st. Warszawy [Information bulletin of the presidium of the national council of the capital city of Warsaw], January, no 41 (special issue), pp. 9–10
Tołwiński S (1970) Społeczne Przedsiębiorstwo Budowlane i Pracownia Architektoniczno-Urbanistyczna [Social Construction Firm and Architecture and Urban Planning Studio] In: Anonymous (ed) Żoliborz wczoraj, dziś, jutro [The Past, Present and Future of Żoliborz]. Książka i Wiedza, Warszawa
Topińska Z (1984) Żoliborskie obietnice: o wychowawczym środowisku WSM [The promises of Żoliborz: the formative environment of the WHC]. Wydawnictwo Spółdzielcze, Warszawa
Topiński P (2012) Opowieść żoliborska. Opowieść czwarta: bohaterska [The Żoliborz story. The fourth, heroic story], http://studioopinii.pl/archiwa/18456/4
Wudzki L (1948) Zjednoczenie klasy robotniczej [The unification of the working class], 'Życie Osiedli WSM', December 1948, p. 218
Zarzycki J (1970) O specyfice Żoliborza – trudnościach i perspektywach jego rozwoju [On the specificity of Żoliborz—the difficulties and prospects for its development]. In: Anonymous (ed) Żoliborz wczoraj, dziś, jutro [The past, present and future of Żoliborz]. Książka i Wiedza, Warszawa, pp. 309–312
Zawadzka M (2010) Warsztaty „Żoliborz – Ogród Sprawiedliwych Wśród Narodów Świata". https://sprawiedliwi.org.pl/pl/aktualnosci/warsztaty-zoliborz-ogrod-sprawiedliwych-wsrod-narodow-swiata)

Chapter 11
Conclusions

Abstract Focusing on these historical, yet successful (and to some extent exemplary) urban strategies allows us to think about residents as shareholders of the urban commons and to prove that the interests of these communities cannot be subordinated to economic discourse. After all, urban residents, both in the past and today, are the main players who formulate the right to the city. This does not mean, however, that cities are to become spaces for exclusively 'joyful encounters'. The city must employ a kind of 'empty oppression', it needs an institutionalised framework, something the citizens–residents can rely on. This is why I was reluctant to engage in a hard-hitting critique of modernity. My analysis shows that the city can be a radical space; an area where certain entities can exercise their agency; a political idea; a common good; a contemporary form of living-in-common; a space for urban collectives, urban subversions, and autarkies. This is why I consider the urban strategies developed in Żoliborz as extremely inspiring for both engaged activist researchers and curators, and urban reformers, cultural animators and representatives of urban movements alike; in other words, for all those whom Miessen would describe as critical spatial practitioners. This is why I situate my research in the area of critical urban studies, participatory humanities and utopian studies—an area that Ruth Levitas has described as *The Imaginary Reconstitution of Society*.

Keywords Urban commons management · Common goods · Urban lifestyles · Urban activists · Urban entity · Codwelling

The modernist urban project presented in this book—notwithstanding its emancipatory and educational ambitions, indeed, its modernist spirit—cannot, naturally, serve as a model to be followed in our times. And it was not discussed for this purpose. While I believe in the usefulness of exercises in history, I am also aware that the position of architects and that of the inhabitants (or city users) as well as the very reflection on urban life are today different than in the modern era. Architects' convictions and certainty about their power to shape social relations and to bring about social change, proved pernicious. So did the belief that social housing is the most important objective of urban policies. The case of the Żoliborz designers, however, demonstrates that not all modernist architects shared this hope. The discussion of the myth of the Pruitt–Igoe estate—Charles Jencks saw its demolition as the fall of

M. Matysek-Imielińska, *Warsaw Housing Cooperative*, The Urban Book Series,
https://doi.org/10.1007/978-3-030-23077-7_11

modernism—shows clearly that it was not the fault of architecture, but the failure of immature urban institutions which segregated, marginalised and controlled residents, stripping them of dignity and agency as urban citizens.

Following Hayden White's advice,[1] my suggestion concentrates on leaving this history behind by adopting a perspective, which is not critical of modernism. Although I do not share the enthusiasm of Le Corbusier's adherents, and his style of housing architecture appears to me inhumane and neglecting the inhabitant, I still believe in what Markus Miessen called 'crossbench practice' (Miessen 2017, p. 144), which is both reflexive and critical toward pursued and planned urban projects. This strategy does not aim at abandoning the already organised urban space but enlivens it through practice, codesign and dialogue with users of the city, and it surrounds the organised city with theoretical thought.

Indeed, our awareness of the modern illusions of order-making (let us remember that Bauman defined post-modernity as modernity minus its illusions[2]) should not lead to architecture's social helplessness. In the 1970s, Tafuri (1976) noted that capitalism deprived architecture of its ideological purpose, and that architecture, which returns to pure form, a non-ideological architecture, is merely a cold form that modern cities turn into, filled with office buildings, towers, monumental public edifices, museums and art centres, and with the 'Bilbao effect'. It is the architecture without social illusions. So, why do we need this history lesson?

My interpretation of the historical urban phenomenon presents both its actors and the stage through completely new analytical categories. Inspired by Levitas's (2013) advice to construct archaeological, ontological and architectural modes in studying utopias, I called the founders of the Żoliborz housing estate—its ideologues, social workers and architects—activists and critical spatial practitioners. Contrary to their Western European counterparts of the time, but very much like today's activists, they had not been agents of the welfare state nor could they rely on state's political support. They did not apply detailed and ready-made scenarios, nor did they implement a utopia.

The essence of their practical methodology was agency. They did not create forms (as these proved highly imperfect, indeed, unsatisfactory) but actions, and often only conditions for effective practice. And they were pragmatic, though not lacking ideals. Active and pursuing modern illusions, Żoliborz activists followed their non-dogmatic idealism and social mission. Their budget was so small and the people for whom they designed the project so poor that they had to apply methods adequate to the socio-economic context. They had to think flexibly and react openly to unexpected changes; they had to establish connections and flows, points of communication and modes of social inclusion. They considered the very circumstances, identified major issues and created conditions for sociocultural change. As a way of goal-pursuing, the pragmatic attitude requires great amounts of imagination, engagement, flexibility and readiness

---

[1] Hayden White argues that European historical consciousness is embittered by the unfulfilled hopes of the Enlightenment, especially those of the French Revolution (White 1973, p. 434).

[2] Postmodernity is modernity reconciled to its own impossibility—and determined, for better or worse, to live with it (Bauman 1995, p. 98).

to modify, experiment, test and learn. And even though the project of Ruth Levitas was inspiring, I am still far from seeing Żoliborz as utopia. While this was an essentially modernist urban project, its practitioners cannot be called 'urban surgeons' whose task would have been merely to remove poverty from the city. The proper metaphor here is not 'scalpel' but 'needle'. The 'urban acupuncture[3] serves as a tool as much for the pre-war activists of Żoliborz as for those in our times. The city cannot be transformed entirely. Nevertheless, one can think of creating new places—mini-projects integrated with the wholeness of the city. Therefore, I believe this tool can be useful for local actions that constitute a part of the broader multifaceted network of urban politics. The Żoliborz project, as the city within the city, had an impact way beyond its own locality. It created abilities to practice democracy both within and outside its boundaries. Thus, my analysis is based on workshop, clubroom and laboratory metaphors. I believe they can inspire other, more contemporary urban studies.

When state and municipal authorities today refrain from the active use of space as an instrument of reducing social divisions and initiating urban culture, Lefebvre's thought should be particularly remembered: today, more than ever, struggle for the city is linked with spatial issues. The activists of Żoliborz had known that well. What differentiates them, however, from today's campaigners are, for example, the greater humility and the pragmatism of the latter. The activists of our times are not socialists or communists (and these categories have little meaning these days anyway), they do not seek revolution and neither do they support old-style mobilisation and propaganda strategies.

Today's activists also know that great (surgical) projects of urban planning, social change or education are political pipe dreams. Unlike the modernist designers of social order, who had style, history, technology and ideology on their side, contemporary activists, architects and town planners are forced to rely on effective yet relatively small-scale urban strategies based on trust in tiny local communities. They have to apply 'urban acupuncture'. Those active nowadays must seek the support of residents as allies in the struggle over the city.

Shortly after World War II, when this 'autonomous zone' and self-government had been taken over by state authorities, the Żoliborz initiative was contained. The estate's life, massified and nationalised, became subject to top-down control. In the 1960s, Julian Hochfeld wrote that the causes of the fall of the great cooperative idea lie 'in the great expansion of the WHC, loss of its pioneering role, in an outflow of many of the most prominent activists to the broad social arena, in the inevitable bureaucratisation of the ideas' (Hochfeld 1962, p. 10), and not in the cooperative's internal politics and adopted strategies. Today, of course, it is impossible to tell what an alternative history of the cooperative would have been like, if the totalitarian regime had not been in

[3]We can find such 'medical metaphors' both in urban studies theorists as well as literary authors. Among the latter, a Polish 19th-century writer Bolesław Prus, in his famous novel *The Doll*, wrote about the city with 'its own anatomy and physiology' (see the chapter 'Grey Days and Baneful Hours'). Another popular metaphor is that of 'urban tissue' or 'urban fabric.' The surgical metaphor, therefore, should not come as a surprise. On the other hand, the notion of 'urban acupuncture' is used by McGuirk (2014). In Polish literature see, for example, Krajewska (2012).

place. We cannot tell if it would have survived in an unchanged form. Undoubtedly, the cooperative's democratic and communal qualities resulted from the pioneers' solidarity and necessity as well as from the zeal and satisfaction with making the new order, organising new reality day by day; from collective action and the establishment of institutions. The residents created these qualities from the very beginning: they both witnessed the beginnings of the initiative and made it alive. Such spontaneity and enthusiasm could not be shared by successive generations, which would have entered a ready-made institutionalised reality based on fixed norms and established ways of being not subject to negotiation. Carnivals of resistance against existing reality and rituals cannot last forever. Over time, institutions become objectivised and alienating to city users. At the same time, one must admit that the schools of Żoliborz had effectively produced conditions and experience which enabled the practice of everyday-life democracy, and equipped the subsequent generations with a 'sociological imagination'. They educated the new resident in the new estate.

Yet, like Bakhtin's carnival, the performative democracy, as Elżbieta Matynia argues (Matynia 2009), is a transient phenomenon, unable to last continuously.

In the interwar period, the WHC was not only a joyful and alterative form of the political. It also succeeded in containing human passions and organising them within frameworks of established forms and conventions. There was plenty of civic creativity and myriads of liberation dreams in the cooperative's vital legacy. It was in Żoliborz where the intellectual potential of democratic dissent leading up to the 'Solidarity' movement was forged. And it was no coincidence. Memoirs of Żoliborz residents, written at the beginning of the twenty-first century, i.e. those memoirs that did not go back as far as the pre-war times, were based on the experience of spontaneous periods of 'carnivals of freedom' initiated by people who opposed the communist regime. However, such memoirs contain a certain tactic of emptying the Żoliborz history of its political radicalism and Left-wing orientation. Marek Rapacki, for instance, wrote about 'miserable Wanda Wasilewska' or 'communist believer' Juliusz Rydygier. He also mentions the WHC school: 'the first secular school in the country'. As he says: 'It is not a tradition, which today would be claimed by many, but neither good intentions of the school founders nor their pure engagement, patriotism and democratic sentiment should be questioned' (Rapacki 2009).

Another enthusiast of the Żoliborz project and an estate resident linked her memories with the slogan of 'socially-oriented individualism' and also applied the political neutralisation in her account: 'Connections between the churches of Żoliborz with «Solidarity», as well as martyrdom of father Popiełuszko, changed the image of Żoliborz that shifted from being mainly the symbol of the secular intelligentsia to the symbol of freedom struggles, trans-local patriotism and a religious site. It was here that Mr. Kaczyński with several other people—Nowicki, Mazowiecki, Geremek, Stelmachowski—started their activity. Michnik and Kuroń also joined. There were also Wałęsa, Olszewski and Jan Józef Lipski' (Zaremba 2009, p. 127).

Yet the founders of the Żoliborz cooperative republic had sought a different kind of 'socially oriented individualism'. It is worth noting that the slogan had been strongly promoted by Barbara Brukalska in her architectural discourse which focused on new type of residential project driven by the idea of housing reform. Multiple initiatives in the estate matched this approach.

It is clear now that in Żoliborz, we have seen transformation from the community of the secular intelligentsia and non-Marxist socialism to the site of a resistance movement (flavoured with the spiritual dimension of the Catholic Church). Liberation dreams, which had been created here, did not dwindle but rather exploded anew, providing new strengths and new hopes. The democratic opposition movement did not emerge in the vacuum: the neighbourhood provided an already shaped public sphere with the civil-society ethos and dissent. This proves that, while Żoliborz changed its *signifié* and was somehow politically neutralised, in the final decades of communism, due to the still existing political zeal, the estate became an arena of bottom-up transformations. For the quoted author, the 'mental dismantling' of the once socialist Żoliborz and its 'cooperative republic' may have appeared as a right way, yet she still valued the fact that the estate kept the atmosphere, which allowed one to 'live the old-time intelligentsia ethos, which makes me feel home and allows me to pursue the "socially-oriented" individualism' (Zaremba 2009, p. 131).

Krzysztof Pomian explained the ideological gap between anti-communist oppositionists and the pre-war Left: 'I remember my conversation with Kisiel in Paris in the 1970s, after the press conference when I talked about a meeting of the Flying University crashed by the secret service: a squad with batons entered the apartment and beat up people who participated in the meeting. Then Kisiel, with his characteristic chortle and common sense, said to me: "What's the matter, Krzysztof? Why don't they go to church? We organise our meetings in churches, talk much more openly, yet we've had no police raids." I do not remember how I responded. But I can recall what I thought to myself after that conversation: that is precisely the point not to go to church of the Church—capitalised or not. And it was all about our autonomy not just from the PZPR [Polish United Workers' Party] but from the Church, too! Well, during the martial law the need for such autonomy vanished, which also meant the decline of the whole secular-Left tradition' (Pomian and Walenciak 2009).

For Pomian, 'the WHC is inseparable from the PPS [Polish Socialist Party] and from the Left, of which the communists had been only a tiny faction, present in the history of the WHC from the start but merely as its small part. It was the PPS that formed the political base of the WHC, along with—as it was then called—the secular left, because secularism played a crucial role in the cooperative' self-awareness. These people were not necessarily non-believers, but they always emphasised the separation of Church and state as their vital principle. So, when somebody says "WHC", then one must add the PPS and the RTPD (Workers' Society for the Friends of Children)' (Pomian and Walenciak 2009). And one must also say that the cooperative aspired to autonomy from the bourgeois state and the capitalist system.

Had the cooperative completed its mission? Can we say that its Left-wing and socially-sensible purpose had been reached? Or, perhaps, it ended up as an elitist project for the intelligentsia, which declared a social orientation while remaining separate from the real problems of the working class?

Whereas the WHC part of Żoliborz was immersed in the sunlight, enjoyed the green spaces and the neat white buildings of the cooperative's colonies, on the other side of the district shanty barracks formed colonies of poverty housing. Around 1922, a shanty town was constructed, which served the needs of returning migrants and repatriates. Sometime later, it was transformed into the residence for the homeless and the unemployed. Formally, it was under the supervision of the Polish Red Cross. Jan Dąbrowski described the place as 'fifty or more grim brown coffins' where each room was populated by three persons living in dirt and damp. In these shacks 'rats would drink up children's milk at night' and 'the barrack's supervisor with physical terror forced inhabitants to humility and obedience' (Dąbrowski 1965, cited in Kormanowa 1970, p. 70). These shanty barracks were not the only area that contrasted with the modern Żoliborz of the WHC estate. In the 1930s, large colonies of poverty houses were built in the Marymont neighbourhood. Contemporaries wrote of the neighbourhood that it 'steeps in shanties, the horrible housing poverty and cultural backwardness' (Dembiński 1939). Edward Szymański even wrote a poem titled *Three Cities*, describing the shanty barracks of Żoliborz, the picturesque view of the WHC estate, and the muddy misery of Marymont. Would such a vicinity be conducive to the free development of lofty ideas by the progressive intelligentsia and socially considerate cooperators?

It seems that the answer is 'no'. That is why Barbara Brukalska designed portable barracks and proposed a modest housing programme for the homeless. She wished to ensure the humane aspect of the barrack rooms, and 'in some cases even the educational one: so that one learns how to dwell. That is why the plan—though being based on a limited programme and assuming a very modest standard of living—contains all elements necessary to develop a social life of regular housing estate' (Brukalska 1948, p. 159). Brukalska was fully aware of the fact that unkempt and ill-maintained buildings would become the 'most miserable of shelters', cause diseases, moral damage and humiliation, while perpetuating this condition. With care and sensitivity she decided to make an impact through her plans and architectural visions. She avoided unnecessary exaltation and moralising. She identified problems and found solutions. Her architecture was not that of greatness but one focused on praxis: tackling poverty, stimulating new lifestyles, belonging to urban culture, and one which could educate about how to live urban-style. This educating aspect did not assume dwellers' ignorance, it was not intrusive either. Instead it stressed cooperation and responsiveness to their needs.

The cooperative aimed at eliminating the housing hunger by creating decent urban-life conditions for the working class. The estate had been intended for the workers, that is, as a homogeneous neighbourhood. Over time it happened, however, that the majority of its residents belonged to the liberal professions (artists, journalists, scholars, teachers) (Tables 11.1 and 11.2). One might thus say that the WHC estate in Żoliborz failed to fulfil its mission and it became an elitist place for the working intelligentsia. The WHC leaders tried to keep the neighbourhood's working-class profile by signing mutual contracts with trade unions and even patronage agreements. The influx of the intelligentsia continued up until 1937. Its share grew from 55% in 1929 to 74% in 1937. In 1938, the workers constituted only 31% of the Żoliborz estate but as much as 86.5% in Rakowiec. However, we may look at this phenomenon from another angle. The Rakowiec housing estate, built in 1930s, much more modest and thus much more affordable for the working class, possessed similar modernist qualities yet no model of urban culture characteristic for Żoliborz developed there. While we can describe specific features of the WHC's activist profile, no residential ethos emerged in Rakowiec. It seems that it required a great deal of residents' self-awareness and the sense of bonds, should World War II have not destroyed such possibilities. No doubt, active citizenship, organising and meeting cultural needs as well as participation in the social life of the estate were features present only in Żoliborz. At the same time, one must admit that, as a working-class neighbourhood, Rakowiec did not turn into an isolated enclave, mainly thanks to its democratic, self-governing and social programme (Table 11.3).

The Żoliborz estate, therefore, succeeded in avoiding homogeneous status (i.e. working class only), and developed a multi-class social composition of residents—a multitude (made of people living of their own labour and lacking capital) connected by collective work for democratic conditions, for the institution of the estate's community and for the production of the common in the city. The class diversity of Żoliborz—its dual profile, linking the intelligentsia with the working class—and the social homogeneity of Rakowiec do not, in my opinion, prove the WHC failed. Attempts at social democratisation and egalitarianism occurred parallel to aspirations to the free choice of community and neighbours. Similar efforts to reconcile both such tendencies can be seen in Scandinavia or the Netherlands where small homogeneous neighbourhoods formed larger heterogeneous estates.

According to Janusz Ziółkowski, Polish town-planners before World War II were free of the illusion that the spatial proximity of incomers from diverse social backgrounds would automatically produce social proximity. The latter was rather a result of common democratic experiences, practicing citizenship and participation in collective rituals. Important were the efforts to create adequate conditions for such bonds to develop. Ziółkowski's observation is of crucial significance though not common among urban sociologists in Poland. It should not be ignored, even if we consider the context of the 1960s when Ziółkowski wrote that 'founding the socially diverse housing estate is perhaps the largest "sociological" achievement of Polish urban planners, despite many deficits of residential housing in Poland' (Ziółkowski 1965, p. 248).

To sum up: one of the WHC's estates lost its working-class character and another (Rakowiec) did not meet the minimal standards of the smallest apartment. The latter, instead, provided flats most needed by the workers, and was built as a result of cooperative work in organisation, construction and architectural codesign.

The Żoliborz estate also falsifies a sociological myth based on the common belief that spatial proximity leads to social proximity more easily among the so called lower strata whose social relations form, allegedly, in a more spontaneous manner. 'In turn, the upper strata are more distanced, reluctant to express overt confidence and trust, and more selective in choosing partners of proximate social contacts' (Ziółkowski 1964, p. 204). Undoubtedly, this belief is not unfounded. Nonetheless, it is contradicted by developments that occurred in the community of Żoliborz, that

Table 11.1 Residents of the WHC estate in Żoliborz[a]

Year	Residents	Men		Women		Adults		Children		Subtenants		Domestic servants	
	Total	Number	%	Number	%	Number	%	Number	%	Number	%	Number	%
1929	894	412	46	482	54	538	60	356	40	39	4	47	5
1930	2069	905	44	1664	56	1331	64	738	36	178	9	92	4
1931	2737	1157	42	1580	58	1794	66	943	34	139	5	158	6
1932	3675	1549	42	2126	58	2486	68	1189	32	34	1	200	5
1933	3906	1626	42	2280	58	2639	68	1267	32	63	2	220	6
1934	4154	1712	41	2442	59	2863	69	1291	31	31	1	235	6
1935	4107	1692	41	2415	59	2866	70	1241	30	39	1	226	6
1936	3949	1620	41	2325	59	2813	71	1136	29	62	2	215	5
1937	3931	1610	41	2321	59	2891	74	1040	26	–	–	–	–
1938	4289	1780	42	2509	58	3055	71	1234	29	–	–	–	–

[a]Drawn up on the basis of WHC reports for the years 1929–1938

– No data available

The compilation prepared by Elżbieta Mazur

Table 11.2 Structure of employment of the residents of the Żoliborz housing estate in 1929–1938[a]

Employees	1929		1930		1932		1933		1934		1935		1936		1937		1938	
	Number	%	Number	%	Number	%	Number	%	Number	%	Number	%	Number	%	Number	%	Number	%
Industrial workers	53	25	131	26	265	25	266	24	223	18	239	18	174	14	172	14	236	17
Communal workers	26	12	40	8	101	10	91	8	105	9	121	9	91	7	90	7	109	8
Other workers	20	10	50	10	23	2	22	2	33	3	44	3	55	4	63	5	79	6
Workers in total	99	47	221	44	389	37	379	34	361	30	404	30	320	25	325	26	424	31
Employees of social institutions	58	28	131	26	134	13	144	13	155	12	190	14	152	12	154	12	155	14

(continued)

Table 11.2 (continued)

Employees	1929		1930		1932		1933		1934		1935		1936		1937		1938	
	Number	%	Number	%	Number	%	Number	%	Number	%	Number	%	Number	%	Number	%	Number	%
State officials	45	21	126	25	437	40	489	43	545	44	618	45	596	48	591	47	603	43
Liberal profes-sions	8	4	25	5	107	10	114	10	175	14	157	11	190	15	197	15	204	15
White-collar workers In total	111	53	282	56	678	63	747	66	875	70	965	70	938	75	942	74	962	69
All employ-ees	210	100	503	100	1067	100	1126	100	1236	100	1369	100	1258	100	1267	100	1386	100

[a]Drawn up on the basis of WHC reports for the years 1929–1938
Note No data for the year 1931 available
The compilation prepared by Elżbieta Mazur

Table 11.3 Structure of employment of the residents of the Rakowiec housing estate in 1935–1938[a]

Employees	1929		1930		1932		1933		1934		1935		1936		1937		1938	
	Number	%	Number	%	Number	%	Number	%	Number	%	Number	%	Number	%	Number	%	Number	%
Industrial workers	53	25	131	26	265	25	266	24	223	18	239	18	174	14	172	14	236	17
Communal workers	26	12	40	8	101	10	91	8	105	9	121	9	91	7	90	7	109	8
Other workers	20	10	50	10	23	2	22	2	33	3	44	3	55	4	63	5	79	6
Workers in total	99	47	221	44	389	37	379	34	361	30	404	30	320	25	325	26	424	31
Employees of social institu-tions	58	28	131	26	134	13	144	13	155	12	190	14	152	12	154	12	155	14

(continued)

Table 11.3 (continued)

Employees	1929		1930		1932		1933		1934		1935		1936		1937		1938	
	Number	%	Number	%	Number	%	Number	%	Number	%	Number	%	Number	%	Number	%	Number	%
State officials	45	21	126	25	437	40	489	43	545	44	618	45	596	48	591	47	603	43
Liberal profes-sions	8	4	25	5	107	10	114	10	175	14	157	11	190	15	197	15	204	15
White-collar workers In total	111	53	282	56	678	63	747	66	875	70	965	70	938	75	942	74	962	69
All employ-ees	210	100	503	100	1067	100	1126	100	1236	100	1369	100	1258	100	1267	100	1386	100

[a]Drawn up on the basis of WHC reports for the years 1929–1938
Note: No data for the year 1931 available
The compilation prepared by Elżbieta Mazur

is, the cooperative action, collective caring about social institutions, a bottom-up approach, repeated authoritative speech, socially oriented individualism, internal attractiveness of the estate, solidarity of the founders as well as the community's sense of responsibility for its own achievements.

All this proves that democracy rests not on the personal or social competences of a class, a social stratum or a professional group, but rather on certain structural conditions based above all on the principle of civic equality. Therefore, the only way to fight deficits of democracy is to have more democracy and to learn democracy through practicing it (Biesta 2011).

Interestingly, the sense of the success of the experiment in social housing has persisted up to our times in the new residents of Żoliborz, those who cannot remember the pre-war period, and it manifests itself in contemporary efforts to revive the atmosphere of the past.

I analysed the WHC estate as cultural performance, looking at complexities of everyday life and the moments of the community's self-reflection as presented in commentaries written for the *Życie WSM* bulletin, in film, in memoirs, in plans of social life, in various ritual dramatic expressions of art and community activities. This kind of research perspective allows analysis of the estate's urban cultural models through metaphors of workshop, clubhouse and laboratory.

This approach also made it possible to perceive the founders of the Żoliborz project as urban activists claiming the right to the city (Lefebvre, Harvey) and solving old and new urban questions (Castells, Merrifield) by forming the multitude, which will then create its own living conditions through producing the common (Hardt and Negri). My interpretation draws on these contemporary analytical categories.

The estate took form of a comprehensive, total and efficient social action with civic and emancipatory educational ambitions. Żoliborz can be seen, then, as the transformative potential of performance, a project founded on the intellectual currents of socialism and the cooperative movement: its authors challenge capitalist culture, and the residents undergo radical changes. Decorum and social scenery were meant to complement the new identity of the progressive worker, to shape his modern taste and style. By educating tenants how to dwell, by creating proper housing conditions for them and by including them in public life, a challenge was posed to 'well-informed citizen' and the passive tenants were made into active residents. This could happen due to various ways of shaping the habits and collective rituals that enabled the residents to enter new roles, either on a daily basis or on special occasions. Lectures and meetings, debate clubs and theatre shows, social games and collective work, organisation of space in the neighbourhood and volunteering for the local school—all these situations allowed residents to enter new interactions, dramatised in their forms, which enabled safe (mediated through collective ritual) participation in, and introduction into and sustaining the transformed social relations. In this urban 'autonomous zone' the residents practiced the 'self-development of living labour', exercising their citizenship in intellectual, emotional and physical dimensions; they shaped their habits of openness, dialogue and practical performative democracy.

The process of designing the culture of Żoliborz and forging the citizen through organised forms of neighbourly life somewhat resembles the management of performance, which is about designing, testing and producing 'cultural products' in form of social events, educational campaigns, events and public talks. Organisational cultural performance was essentially regulated by the cooperative's authority that supervised particular projects through its special representatives working in multiple organisations, associations and institutions divided into commissions, sections and departments. Yet the performance of Żoliborz was significantly different from today's performance that is technocratic and 'effects-oriented'. The former was, first and foremost, of 'amateur' nature.

The Żoliborz experiences posed a challenge to architecture, sociology and pedagogy, pushing these discourses towards concrete actions. Thus, I shall argue that they served as performative tools—as action-oriented architecture, sociology and pedagogy.

While such preposterous history reveals new areas of interpretation, it also suggests new directions of action: one can see in today's urban phenomena and movements new opportunities—more institutionalised, less emotional, but equally radical.

This book had no intention to fall into the utopia of interdisciplinarity, that is, 'undertaking studies that should lead to an integrated description of the whole subject area, a contemporary version of the old ideal focused on complete, reliable and objective account' (Nycz 2006, pp. 29–30). For this reason, historians of Polish socialism or the cooperative movement will not find much inspiration here. Historians of architecture will also notice that my analysis concentrates only on some aspects of the Żoliborz project necessary to understand it not merely as an urban-architectural phenomenon but as an attempt in creating urban culture. So, my description is far from integrating multiple perspectives in a balanced manner. Certainly, urban studies scholars, too, might think that my use of their concepts and theories was not sufficient. Placing my study on a map of academic division of labour, they overlap connections between three fields: urban studies (inspired mainly by David Harvey, Henri Lefebvre, Manuel Castells and Andy Merrifield), cultural studies of community (critical social theory of Antonio Negri and Michael Hardt) and activist cultural studies (Skórzyńska 2016). The contemporary utopian turn in social sciences also provided significant inspiration (Immanuel Wallerstein's 'utopistics' and Ruth Levitas's methodology of *The Imaginary Reconstitution of Society*) along with performance studies (most importantly, Richard Sennett and John McKenzie). While these theoretical fields are certainly interlinked, my perspective is not situated in any of them unambiguously and exhaustingly. My study, due to its subject matter, is located between particular areas of academic knowledge; it resists clear classification although it draws inspiration from each of them. The analysed phenomena are of interest for multiple disciplines, but in each field they are explored in a unique way.

References

Bauman Z (1995) Modernity and ambivalence. Polity Press, Cambridge

Biesta G (2011) From teaching citizenship to learning democracy. In: Biesta G (ed) Learning democracy in school and society: education, lifelong learning, and the politics of citizenship. Sense Publishers, Rotterdam, Boston, Taipei, pp 5–16

Brukalska B (1948) Zasady społeczne projektowania osiedli mieszkaniowych [Social rules for the design of housing estates]. Developed at the Institute of Social Economy. Wydawnictwo Ministerstwa Odbudowy, vol 22, Warszawa

Dąbrowski J (1965) W barakach na Żoliborzu [In the shanty barracks of Żoliborz]. In: Dąbrowski J, Kosowski J (eds) Niepiękne dzielnice. Reportaże o międzywojennej Warszawie [Unpretty districts: stories of warsaw between the world wars]. PIW, Warszawa, pp 124–129

Dembiński H (1939) Z zagadnień urbanistycznych i społecznych Żoliborza [Urban and social issues of Żoliborz]. Życie WSM, May

Hochfeld J (1962) Więź społeczna wuesemowców [The social ties of WHC activists]. Życie Osiedli WSM, Jan:9

Kormanowa Ż (1970) Postępowe tradycje inteligencji żoliborskiej 1918–1939 [Progressive traditions of Polish intelligentsia 1918–1939]. In: Anonymous (ed) Żoliborz wczoraj, dziś, jutro [The past, present and future of Żoliborz]. Książka i Wiedza, Warszawa, pp 67–101

Krajewska J (2012) Akupunktura tożsamości krajobrazu miejskiego – Londyn – Paryż – Berlin [Acupuncture of urban landscape's identity: London – Paris – Berlin]. In: Petryszyn H, Sochacka-Sutkowska E (ed) Tożsamość krajobrazu miasta [Identity of the urban landscape]. Zachodniopomorski Uniwersytet Technologiczny, Szczecin, pp 69–76

Levitas R (2013) Utopia as method. The imaginary reconstitution of society. Palgrave, MacMillan, New York

Matynia E (2009) Performative democracy. Paradigm Publishers, London

McGuirk J (2014) Radical cities: across Latin America in search of a new architecture. Verso, London-New York

Miessen M (2017) Crossbenching: towards a proactive mode of participation as a critical spatial practice. Centre for Research Architecture, Goldsmiths College/University of London, London. http://research.gold.ac.uk/20980/1/VIS_thesis_MiessenM_2017.pdf. Accessed 7 Sept 2018

Nycz R (2006) Kulturowa natura, słaby profesjonalizm. Kilka uwag o przedmiocie poznania literackiego i statusie dyskursu literaturoznawczego [Cultural nature and weak professionalism: a few notes on the literary object and discourse in literature studies]. In: Markowski MP, Nycz R (eds) Kulturowa teoria literatury. Główne pojęcia i problemy [A cultural theory of literature: major concepts and issues]. Universitas, Kraków, pp 5–38

Pomian K, Walenciak R (2009) Kto zaprowadził lewicę do Kościoła [Who brought the left to the church?]. Przegląd (49). https://www.tygodnikprzeglad.pl/kto-zaprowadzil-lewice-do-kosciola/ Accessed 2 Dec 2013

Rapacki M (2009) Czwarta ćwiartka [The fourth quarter]. In: Chałasińska E, Gawecka A (eds) O pięknym Brzegu. Żoliborz, ludzie, wydarzenia [About the Beautiful River Bank. Żoliborz, People and Events]. Towarzystwo Przyjaźni Polsko-Francuskiej [French Polish Friendship Society], Fundacja 'Żoliborski Fundusz Lokalny', Warszawa, pp 56–62

Skórzyńska A (2016) Czy możliwe jest kulturoznawstwo aktywistyczne? Partycypacja w perspektywie filozofii praxis [Are activist cultural studies possible? Participation in the praxis philosophy perspective]. In: Matysek-Imielińska M (ed) Uczestnictwo, ruch, wspólnota [Participation, movement, community]. Prace Kulturoznawcze, vol 19. Wydawnictwo Uniwersytetu Wrocławskiego, Wrocław, pp 15–40

Tafuri M (1976) Architecture and Utopia. In: Design and capitalist development (trans: La Penta BL). MIT Press, Cambridge, MA

White H (1973) Metahistory. In: The historical imagination in nineteenth-century Europe. John Hopkins University Press, Baltimore, London

Zaremba IA (2009) Prospołeczny indywidualizm [Socially-oriented individualism]. In: Chałasińska E, Gawecka A (eds) O pięknym Brzegu. Żoliborz, ludzie, wydarzenia [About the Beautiful River Bank. Żoliborz, People and Events]. Towarzystwo Przyjaźni Polsko-Francuskiej, Fundacja "Żoliborski Fundusz Lokalny', Warszawa, pp 124–135

Ziółkowski J (1964) Osiedle jako system społeczny [The housing estate as a social system]. Studia Socjologiczne 4(2):204, 183–209

Ziółkowski J (1965) O socjologicznej wyobraźni urbanisty [On the sociological imagination of the urban planner]. In: Ziółkowski J, Urbanizacja, miasto, osiedle. Studia socjologiczne [Urbanisation, city, estate]. Państwowe Wydawnictwo Naukowe, Warszawa, pp 227–251

GPSR Compliance

The European Union's (EU) General Product Safety Regulation (GPSR) is a set of rules that requires consumer products to be safe and our obligations to ensure this.

If you have any concerns about our products, you can contact us on ProductSafety@springernature.com

In case Publisher is established outside the EU, the EU authorized representative is:

Springer Nature Customer Service Center GmbH
Europaplatz 3
69115 Heidelberg, Germany

Batch number: 10845468

Printed by Printforce, the Netherlands